Ergebnisse der Mathematik und ihrer Grenzgebiete

Band 77

To
Berjis, Marjan, Shirin,
Helen, Alan, Jeffrey, Michael

Preface

In recent years there has been a large amount of work on invariant subspaces, motivated by interest in the structure of non-self-adjoint operators on Hilbert space. Some of the results have been obtained in the context of certain general studies: the theory of the characteristic operator function, initiated by Livsic; the study of triangular models by Brodskii and co-workers; and the unitary dilation theory of Sz.-Nagy and Foiaş. Other theorems have proofs and interest independent of any particular structure theory. Since the leading workers in each of the structure theories have written excellent expositions of their work, (cf. Sz.-Nagy-Foiaş [1], Brodskii [1], and Gohberg-Krein [1], [2]), in this book we have concentrated on results independent of these theories. We hope that we have given a reasonably complete survey of such results and suggest that readers consult the above references for additional information.

The table of contents indicates the material covered. We have restricted ourselves to operators on separable Hilbert space, in spite of the fact that most of the theorems are valid in all Hilbert spaces and many hold in Banach spaces as well. We felt that this restriction was sensible since it eases the exposition and since the separable-Hilbert-space case of each of the theorems is generally the most interesting and potentially the most useful case.

We wanted to make this book readable by mathematicians with a working knowledge of measure theory, complex analysis, and elementary functional analysis, whether or not they have seriously studied operators on Hilbert space. For this reason we have included some well-known material, (e.g., the spectral theorem). There are only two places in the text where the exposition is not self-contained: we require the elements of $\mathscr{H}^p$ theory in Chapter 4 and some facts concerning operators in a Schatten p-class in Chapter 6. In both cases we discuss the theorems which are required and refer to standard texts for the proofs.

Each chapter contains a section entitled "Additional Propositions" which consists of a list of results closely related to the material of the chapter and provable by methods similar to those in the text. These have two purposes: they mention a number of useful facts which could not be covered because of lack of space, and they can be regarded as problems whose solution will help to deepen the reader's understanding.

The theorems presented in this book are the accomplishment of hundreds of mathematicians. We have attempted to indicate the development of the subject in "Notes and Remarks" sections, but we realize that such attempts cannot be entirely adequate. Many people have contributed ideas which have become so much a part of the subject that it is difficult to trace their history. We owe a great debt to the many unnamed mathematicians who contributed, as well as to those cited in the references.

We are grateful to Chandler Davis, George Duff, and Israel Halperin of the University of Toronto for making the administrative arrangements that facilitated our collaboration. We are also very grateful to the students and colleagues who made suggestions and pointed out errors after reading preliminary drafts of this book. Abie Feintuch, Peter Fillmore, Ali A. Jafarian, Eric Nordgren and Helen Rosenthal have been particularly helpful in this regard.

It is difficult to express the depth of our gratitude to Paul Halmos. As an inspirational teacher and mathematician, he has strongly influenced all of our mathematical work; as an editor for Springer-Verlag, he has been an extremely constructive critic; and as a friend, he has been our main source of encouragement during the times when completing this book seemed to be an impossible task. We hope that this book justifies, to some extent, his efforts on our behalf.

Toronto, October, 1972   Heydar Radjavi
                                                                  Peter Rosenthal

Heydar Radjavi · Peter Rosenthal
Department of Mathematics, University of Toronto
Toronto, Canada

AMS Subject Classifications (1970):
Primary   46 L 15, 47-02, 47 A 15   Secondary   46-02

ISBN 0-387-06217-3 Springer-Verlag New York Heidelberg Berlin
ISBN 3-540-06217-3 Springer-Verlag Berlin Heidelberg New York

Heydar Radjavi · Peter Rosenthal

Invariant Subspaces

Springer-Verlag New York Heidelberg Berlin
1973

Contents

Chapter 0. Introduction and Preliminaries

Most of this chapter consists of certain basic definitions and results that will be required later. With the exception of Section 0.4 the results are all very well-known. The reader may wish to use the other sections merely for reference regarding notation.

0.1 Hilbert Space

Throughout this book we shall be concerned with operators on the unique (up to isomorphism) separable, infinite-dimensional complex Hilbert space, generally denoted by $\mathscr{H}$. The hypothesis that this is the space is to be regarded as implicitly contained in the statement of all results except for those explicitly concerning the finite-dimensional case. (Most of the results actually hold for non-separable Hilbert spaces too, and many are valid for Banach spaces, but the exposition is generally simpler in the case of separable Hilbert space.) We shall have occasion to consider certain concrete realizations (e.g. $\mathscr{L}^2(0,1)$ and ℓ^2) of $\mathscr{H}$.

We assume familiarity with the geometry of Hilbert space as treated in Rudin [1, Chapter 4], Halmos [2], and Berberian [1]. The inner product on $\mathscr{H}$, denoted $(\cdot, \cdot)$, is linear in the first variable and conjugate-linear in the second. The direct sum of the Hilbert spaces $\mathscr{H}$ and $\mathscr{K}$ is denoted $\mathscr{H} \oplus \mathscr{K}$, and elements of $\mathscr{H} \oplus \mathscr{K}$ are written in the form $x \oplus y$ with $x \in \mathscr{H}$ and $y \in \mathscr{K}$. The direct sum of the spaces $\{\mathscr{H}_\alpha\}_{\alpha \in \Lambda}$ is denoted $\sum_{\alpha \in \Lambda} \oplus \mathscr{H}_\alpha$, and its elements are written $\sum_{\alpha \in \Lambda} \oplus x_\alpha$ with $x_\alpha \in \mathscr{H}_\alpha$ for all α.

A *linear manifold* in $\mathscr{H}$ is a subset of $\mathscr{H}$ which is closed under vector addition and under multiplication by complex numbers. A *subspace* of $\mathscr{H}$ is a linear manifold which is closed in the norm topology; the *trivial subspaces* are $\{0\}$ and $\mathscr{H}$. If $\mathscr{S} \subset \mathscr{H}$ then the *span* of $\mathscr{S}$, denoted $\bigvee \mathscr{S}$, is the intersection of all subspaces containing $\mathscr{S}$. The subspaces $\mathscr{M}$ and $\mathscr{N}$ are *complementary*, or are *complements* of each other, if $\mathscr{M} \cap \mathscr{N} = \{0\}$

and $\{x+y: x \in \mathcal{M}, y \in \mathcal{N}\} = \mathcal{H}$. The *orthogonal complement* of the subspace $\mathcal{M}$ is $\mathcal{M}^{\perp} = \{x \in \mathcal{H}: (x, y) = 0$ for all $y \in \mathcal{M}\}$, and $\mathcal{M} \ominus \mathcal{N} = \mathcal{M} \cap \mathcal{N}^{\perp}$.

An *operator* on $\mathcal{H}$ is a bounded (i.e., continuous) linear transformation with domain $\mathcal{H}$ and range a subset of $\mathcal{H}$. We use the term *linear transformation* for linear maps that may be unbounded. The Banach algebra of all operators on $\mathcal{H}$ is denoted $\mathcal{B}(\mathcal{H})$. The identity operator is simply denoted by 1, and the adjoint of A is denoted by A^*. If $\mathcal{M}$ and $\mathcal{N}$ are complementary subspaces, then the *projection on $\mathcal{M}$ along $\mathcal{N}$* is the operator P defined by $P(x+y) = x$ for $x \in \mathcal{M}, y \in \mathcal{N}$; (continuity of P follows from the closed graph theorem: cf. Riesz—Sz. Nagy [1]). The projection on $\mathcal{M}$ along $\mathcal{M}^{\perp}$ is called simply *the projection on $\mathcal{M}$*. We assume the basic properties of operators as discussed in Halmos [2], Berberian [1], and Taylor [1]. Convergence of the net $\{A_\alpha\}$ to A in the uniform, strong, and weak operator topologies on $\mathcal{B}(\mathcal{H})$ is denoted $\{A_\alpha\} \Rightarrow A$, $\{A_\alpha\} \to A$, and $\{A_\alpha\} \rightharpoonup A$ respectively. Corresponding to a decomposition $\mathcal{H} = \sum \oplus \mathcal{H}_\alpha$ of $\mathcal{H}$, there is the usual representation of each $A \in \mathcal{B}(\mathcal{H})$ as an operator matrix; (cf. Halmos [3]).

A *unitary* operator U is an automorphism of $\mathcal{H}$; i.e., $U \in \mathcal{B}(\mathcal{H})$ is unitary if and only if U is invertible and $(Ux, Uy) = (x, y)$ for all $x, y \in \mathcal{H}$. The operators A and B are *unitarily equivalent* if there is a unitary operator U such that $A = UBU^{-1}$. We shall generally be interested in unitarily invariant properties of operators, and will often fail to distinguish between unitarily equivalent operators. The operators A and B are *similar* if there is an invertible operator S such that $A = SBS^{-1}$.

0.2 Invariant Subspaces

Definition. The subspace $\mathcal{M}$ is *invariant* under the operator A if $Ax \in \mathcal{M}$ for every $x \in \mathcal{M}$. The collection of all subspaces of $\mathcal{H}$ invariant under A is denoted $\text{Lat } A$; if $\mathscr{S} \subset \mathcal{B}(\mathcal{H})$, then $\text{Lat } \mathscr{S} = \bigcap_{A \in \mathscr{S}} \text{Lat } A$.

For any $A \in \mathcal{B}(\mathcal{H})$, $\text{Lat } A$ is obviously closed under the formation of intersections and spans, and it follows that $\text{Lat } A$ is a complete sublattice of the lattice of all subspaces of $\mathcal{H}$. Note that the trivial subspaces are in $\text{Lat } A$ for every A.

The basic motivations for the study of invariant subspaces come from interest in the structure of operators and from approximation theory. The Jordan-canonical-form theorem for operators on finite-dimensional spaces can be regarded as exhibiting operators (to within similarity) as direct sums of their restrictions to certain invariant subspaces. The fact that every matrix on a finite-dimensional complex vector space is unitarily equivalent to an upper triangular matrix follows immediately from the existence of non-trivial invariant subspaces for operators on finite-dimensional spaces, (cf. Halmos [4]). If $\mathscr{H}$ is any Hilbert space, $A \in \mathscr{B}(\mathscr{H})$, and $\mathscr{M} \in Lat\, A$, then the representation of A with respect to the decomposition $\mathscr{M} \oplus \mathscr{M}^\perp$ of $\mathscr{H}$ is upper triangular:

$$A = \begin{pmatrix} A_1 & A_2 \\ 0 & A_3 \end{pmatrix},$$

where $A_1 = A|\mathscr{M}$ (the *restriction* of A to $\mathscr{M}$) and where A_2 and A_3 are operators mapping $\mathscr{M}^\perp$ into $\mathscr{M}$ and $\mathscr{M}^\perp$ respectively. Thus it is not surprising that there exist various relations between the structure of A and of *Lat A*.

If $A \in \mathscr{B}(\mathscr{H})$ and $x \in \mathscr{H}$, then $\bigvee_{n=0}^{\infty} \{A^n x\}$ is easily seen to be invariant under A. Therefore knowledge of *Lat A* gives information about the vectors which can be approximated by linear combinations of $\{A^n x\}$.

The vector x is *cyclic* for A if $\bigvee \{A^n x\} = \mathscr{H}$, and $\mathscr{M}$ is a *cyclic subspace* for A if $\bigvee \{A^n x\} = \mathscr{M}$.

Definition. The subspace $\mathscr{M}$ is *hyperinvariant* for A if $\mathscr{M} \in Lat\, B$ for every B which commutes with A.

Knowledge of the hyperinvariant subspaces of A can give information about the structure of the *commutant* of A, (the set of all operators B such that $AB = BA$).

Much less is known about invariant subspaces than is unknown; the results presented in this book may eventually be fragments of a much more comprehensive theory. The most fundamental unsolved problem, the *invariant subspace problem*, is: does every operator have a non-trivial invariant subspace? A related question is the *hyperinvariant subspace problem:* does every operator that is not a complex multiple of 1 have a non-trivial hyperinvariant subspace? Some partial results on the invariant subspace and hyperinvariant subspace problems are obtained in Chapters 5, 6 and 8.

We begin with some basic facts.

Theorem 0.1. *If $A \in \mathscr{B}(\mathscr{H})$ and P is any projection onto $\mathscr{M}$, then $\mathscr{M} \in Lat\, A$ if and only if $AP = PAP$.*

Proof. If $\mathscr{M} \in Lat\, A$ and $x \in \mathscr{H}$, then APx is contained in $A\mathscr{M}$, and since $A\mathscr{M} \subset \mathscr{M}$ it follows that $P(APx) = APx$. Conversely, if $AP = PAP$ and $x \in \mathscr{M}$, then $Px = x$ and $Ax = PAx$. Since $P(Ax) = Ax$, $Ax \in \mathscr{M}$, and therefore $\mathscr{M} \in Lat\, A$. $\square$

Theorem 0.2. *If $A \in \mathscr{B}(\mathscr{H})$ and P is the projection on $\mathscr{M}$ along $\mathscr{N}$, then $\mathscr{M}$ and $\mathscr{N}$ are both in $Lat\, A$ if and only if $AP = PA$.*

Proof. By Theorem 0.1 $\{\mathscr{M}, \mathscr{N}\} \subset Lat\, A$ if and only if $AP = PAP$ and $A(1 - P) = (1 - P) A(1 - P)$, (since $1 - P$ is a projection on $\mathscr{N}$). The second equation is equivalent to $A - AP = A - PA - AP + PAP$, or $0 = -PA + PAP$. The first equation gives $0 = -PA + AP$. $\square$

Definition. The subspace $\mathscr{M}$ *reduces* A if $\mathscr{M}$ and $\mathscr{M}^{\perp}$ are both in $Lat\, A$.

It is easy to see that $\mathscr{M}$ reduces A if and only if $\mathscr{M} \in (Lat\, A) \cap (Lat\, A^*)$, (Proposition 0.1). Theorem 0.2 implies that $\mathscr{M}$ reduces A if and only if the projection on $\mathscr{M}$ commutes with A. Reducing subspaces are generally easier to treat than arbitrary invariant subspaces, (cf. Chapter 7).

0.3 Spectra of Operators

We require a number of facts about spectra of operators. It is convenient to consider the more general situation of spectra of elements of Banach algebras.

Definition. If $\mathfrak{A}$ is a complex Banach algebra with identity 1, and if $x \in \mathfrak{A}$, then the *spectrum* $\sigma(x)$ of x is the set of all complex numbers λ such that $x - \lambda$ has no inverse in $\mathfrak{A}$; ($x - \lambda$ is an abbreviation for $x - \lambda \cdot 1$). The *resolvent set* $\rho(x)$ is the complement of $\sigma(x)$ in $\mathbb{C}$, (the set of complex numbers), and the *spectral radius* $r(x)$ is $\sup\{|\lambda| : \lambda \in \sigma(x)\}$. For each x, $\sigma(x)$ is a non-empty compact subset of $\mathbb{C}$ and $r(x) = \lim_{n \to \infty} \|x^n\|^{1/n}$; (cf. Rudin [1], Rickart [1]). The element x is *quasinilpotent* if $\sigma(x) = \{0\}$. In the case where $\mathfrak{A}$ is commutative, the Gelfand theory provides a useful characterization of $\sigma(x)$.

Theorem 0.3. *If $\mathfrak{A}$ is a complex, commutative Banach algebra with identity and $x \in \mathfrak{A}$, then*

$$\sigma(x) = \{\phi(x) : \phi \text{ is a homomorphism of } \mathfrak{A} \text{ onto } \mathbb{C}\}.$$

Proof. The proof can be found in any book that treats Banach algebras, (e.g., Rudin [1], Rickart [1]). $\square$

For $A \in \mathscr{B}(\mathscr{H})$, $\sigma(A)$ denotes the spectrum of A relative to the Banach algebra $\mathscr{B}(\mathscr{H})$. The Gelfand theory can be applied to operators by considering the spectrum relative to commutative subalgebras of $\mathscr{B}(\mathscr{H})$; the following theorem makes this possible.

Theorem 0.4. *If $\mathfrak{B}$ is a maximal commutative subalgebra of the complex Banach algebra $\mathfrak{A}$, if $x \in \mathfrak{B}$ and $\sigma_{\mathfrak{A}}(x)$, $\sigma_{\mathfrak{B}}(x)$ denote the spectra of x relative to $\mathfrak{A}$ and $\mathfrak{B}$ respectively, then $\sigma_{\mathfrak{A}}(x) = \sigma_{\mathfrak{B}}(x)$.*

Proof. Clearly if $x - \lambda$ has an inverse in $\mathfrak{B}$ it has an inverse in $\mathfrak{A}$; thus $\sigma_{\mathfrak{A}}(x) \subset \sigma_{\mathfrak{B}}(x)$. Suppose that $(x - \lambda)y = y(x - \lambda) = 1$ for some $y \in \mathfrak{A}$. Then, for each $z \in \mathfrak{B}$, $yz(x - \lambda)y = y(x - \lambda)zy$, so that $yz = zy$. Hence y commutes with every member of $\mathfrak{B}$, and, by maximality, $y \in \mathfrak{B}$. Thus $\sigma_{\mathfrak{B}}(x) \subset \sigma_{\mathfrak{A}}(x)$. ☐

Theorem 0.5 *(Spectral Mapping Theorem). If* $x \in \mathfrak{A}$ *and* p *is any polynomial, then* $\sigma(p(x)) = \{p(\lambda) : \lambda \in \sigma(x)\}$.

Proof. Let $\mathfrak{B}$ be any maximal commutative subalgebra of $\mathfrak{A}$ containing x. Then, by Theorems 0.3 and 0.4,

$$\sigma(p(x)) = \sigma_{\mathfrak{B}}(p(x)) = \{\phi(p(x)) : \phi \text{ a homomorphism of } \mathfrak{B} \text{ onto } \mathbb{C}\}$$
$$= \{p(\phi(x)) : \phi \text{ a homomorphism of } \mathfrak{B} \text{ onto } \mathbb{C}\}$$
$$= \{p(\lambda) : \lambda \in \sigma(x)\}. \quad ☐$$

If A is an operator, there are several possible reasons that $A - \lambda$ may fail to be invertible.

Definition. The *point spectrum* of A, denoted $\Pi_0(A)$, is the set of all $\lambda \in \mathbb{C}$ such that $A - \lambda$ is not one-to-one. If $\lambda \in \Pi_0(A)$, then λ is an *eigenvalue* of A, and there exist non-zero vectors x (*eigenvectors of A*) such that $Ax = \lambda x$.

Note that an invertible operator A must be *bounded below*, in the sense that there exists a constant $k > 0$ such that $\|Ax\| \geq k\|x\|$ for all vectors x.

Definition. The *approximate point spectrum* of A, denoted $\Pi(A)$, is the set of all $\lambda \in \mathbb{C}$ such that $A - \lambda$ is not bounded below. The *compression spectrum* $\Gamma(A)$ of A is the set of all λ such that the range of $A - \lambda$ is not dense.

Clearly $\Pi_0(A) \subset \Pi(A)$, and $\Pi(A) \cup \Gamma(A) \subset \sigma(A)$.

Theorem 0.6. *If* $A \in \mathcal{B}(\mathcal{H})$, *then* $\sigma(A) = \Pi(A) \cup \Gamma(A)$.

Proof. We need only show that $\lambda \notin \Pi(A)$ and $\lambda \notin \Gamma(A)$ implies $\lambda \notin \sigma(A)$. But $\lambda \notin \Pi(A)$ implies, as is easily verified by considering sequential limit points, that the range of $A - \lambda$ is closed. Since $\lambda \notin \Gamma(A)$, the range of $A - \lambda$ is also dense, and it follows that $A - \lambda$ is one-to-one and onto, and thus invertible. ☐

The next result is useful for the study of invariant subspaces. For subsets S of $\mathbb{C}$ we use ∂S to denote the point-set boundary of S; i.e., $\partial S = \bar{S} \cap \overline{(\mathbb{C} \setminus S)}$.

Theorem 0.7. *If* $A \in \mathcal{B}(\mathcal{H})$, *then* $\partial \sigma(A) \subset \Pi(A)$.

Proof. First, $\partial \sigma(A) = \sigma(A) \cap \overline{\rho(A)}$. Suppose that $\lambda \in \partial \sigma(A)$ and $\lambda \notin \Pi(A)$. Choose a sequence $\{\lambda_n\} \subset \rho(A)$ which converges to λ. We claim that there is a $k>0$ and a positive integer N such that $n\geq N$ implies $\|(A-\lambda_n)x\|\geq k\|x\|$ for all x. If this were not the case, then for all positive integers m and N there would exist an $n\geq N$ and a vector x_m of norm 1 such that $\|(A-\lambda_n)x_m\|<1/m$. But

$$\|(A-\lambda)x_m\| \leq \|(A-\lambda_n)x_m\| + \|(\lambda-\lambda_n)x_m\|$$

$$\leq \frac{1}{m}+|\lambda-\lambda_n|,$$

and this implies $\lambda\in\Pi(A)$. Hence there do exist such k and N.

We can now show that $\lambda\notin\Gamma(A)$. For if $x\in\mathcal{H}$, then for each n there is a $y_n\in\mathcal{H}$ with $(A-\lambda_n)y_n=x$. Since $\|(A-\lambda_n)y_n\|\geq k\|y_n\|$, this gives $\|y_n\|\leq(1/k)\|x\|$ for $n\geq N$. Then

$$\|(A-\lambda)y_n-x\| \leq \|(A-\lambda_n)y_n-x\| + \|(\lambda_n-\lambda)y_n\|$$

$$= 0+|\lambda_n-\lambda|\,\|y_n\|$$

$$\leq \frac{1}{k}\,\|x\|\,|\lambda_n-\lambda|.$$

If n is sufficiently large, $(1/k)\|x\|\,|\lambda_n-\lambda|$ is arbitrarily small, and it follows that $\lambda\notin\Gamma(A)$.

Hence $\lambda\in\partial\sigma(A)$ and $\lambda\notin\Pi(A)$ implies $\lambda\notin\sigma(A)$, (Theorem 0.6), which is a contradiction. $\quad\square$

This theorem gives information about the spectra of restrictions to invariant subspaces.

Definition. If $A\in\mathcal{B}(\mathcal{H})$, then the *full spectrum* of A, denoted $\eta(\sigma(A))$, is the union of $\sigma(A)$ and all the bounded components of $\rho(A)$; i.e., $\eta(\sigma(A))$ is $\sigma(A)$ together with the "holes" in $\sigma(A)$.

Theorem 0.8. *If $\mathcal{M}\in\mathrm{Lat}\,A$, then $\sigma(A|\mathcal{M})\subset\eta(\sigma(A))$.*

Proof. First note that $\lambda\in\Pi(A|\mathcal{M})$ implies that there exists a sequence $\{x_n\}$ of unit vectors such that $\|(A-\lambda)x_n\|$ converges to 0. Thus $\Pi(A|\mathcal{M})\subset\Pi(A)$. Now, by Theorem 0.7,

$$\partial\sigma(A|\mathcal{M})\subset\Pi(A|\mathcal{M})\subset\sigma(A).$$

If $\sigma(A|\mathcal{M})$ contained points of the unbounded component of $\rho(A)$, then $\partial\sigma(A|\mathcal{M})$ would have to meet the unbounded component of $\rho(A)$ too, and the above shows that this is impossible. $\quad\square$

It is easy to construct examples showing that $\sigma(A|\mathcal{M})$ need not be contained in $\sigma(A)$; e.g., let A be the bilateral shift and $\mathcal{M}=\mathcal{H}^2$ (see Chapter 3).

Definition. The operator A has *finite rank* if the dimension of $\{Ax : x \in \mathscr{H}\}$ is finite.

Clearly finite-rank operators can be studied using the techniques of finite-dimensional linear algebra, (cf. Proposition 0.5), and therefore we can obtain a great deal of information about them. A wider class of operators to which certain generalizations of finite-dimensional techniques can be applied is the class of compact operators.

Definition. The operator A is *compact*, (or, *completely continuous*), if the closure of $A\mathscr{S}$ is a compact subset of $\mathscr{H}$ whenever $\mathscr{S}$ is a bounded subset of $\mathscr{H}$; equivalently, A is compact if and only if $\{\|Ax_n - Ax\|\} \to 0$ whenever $\{x_n\} \to x$, ($\{x_n\}$ converges weakly to x).

The class of compact operators is a two-sided ideal in $\mathscr{B}(\mathscr{H})$, (Proposition 0.6). We state without proof the well-known spectral properties of compact operators.

Theorem 0.9. *If A is compact, then*
(i) *(The Fredholm alternative)* $\sigma(A) = \{0\} \cup \Pi_0(A)$.
(ii) $\Pi_0(A)$ *is either finite or consists of a sequence converging to* 0.
(iii) *The eigenspace* $\{x : Ax = \lambda x\}$ *is finite-dimensional if* $\lambda \neq 0$.

Proof. The proof can be found in Dunford-Schwartz [1], Halmos [3], and in many other standard texts. □

We require a result on the relation between the spectrum of an operator and the spectrum of its perturbation by a compact operator.

Theorem 0.10 *(Weyl's Theorem).* *If $A \in \mathscr{B}(\mathscr{H})$ and K is a compact operator, then $\sigma(A + K) \subset \sigma(A) \cup \Pi_0(A + K)$.*

Proof. Suppose that $\lambda \in \sigma(A + K)$ and $\lambda \notin \sigma(A)$. Then $A + K - \lambda = (A - \lambda)(1 + (A - \lambda)^{-1}K)$. Since $A - \lambda$ is invertible, $(1 + (A - \lambda)^{-1}K)$ is not invertible, and it follows from the Fredholm alternative that -1 is an eigenvalue of the compact operator $(A - \lambda)^{-1}K$. Hence $A + K - \lambda$ has non-trivial nullspace and $\lambda \in \Pi_0(A + K)$. □

We shall have occasion to consider a slightly wider class of operators.

Definition. The operator A is *polynomially compact* if there exists a non-zero polynomial p such that $p(A)$ is compact.

0.4 Linear Operator Equations

Properties of the operator equation $AX - XB = Y$, where A and B are given operators on possibly different Hilbert spaces, will be used several times in the sequel. We begin with a simple lemma.

Lemma 0.11. *If A and B are commuting operators on the Banach space $\mathscr{X}$, then $\sigma(p(A, B)) \subset p(\sigma(A), \sigma(B)) = \{p(\alpha, \beta) : \alpha \in \sigma(A), \beta \in \sigma(B)\}$ for every polynomial p in two variables.*

Proof. The proof is similar to the proof of the spectral mapping theorem. Let $\mathfrak{B}$ be a maximal commutative subalgebra of the algebra of operators on $\mathscr{X}$ containing A and B. By Theorems 0.3 and 0.4,

$$\sigma(p(A, B)) = \{\phi(p(A, B)) : \phi \text{ a homomorphism of } \mathfrak{B} \text{ onto } \mathbb{C}\}$$
$$= \{p(\phi(A), \phi(B)) : \phi \text{ a homomorphism of } \mathfrak{B} \text{ onto } \mathbb{C}\}$$
$$\subset p(\sigma(A), \sigma(B)). \quad \square$$

If $\mathscr{H}$ and $\mathscr{K}$ are Hilbert spaces, (of any finite or infinite dimension), let $\mathscr{B}(\mathscr{K}, \mathscr{H})$ denote the Banach space of all bounded linear operators from $\mathscr{K}$ into $\mathscr{H}$.

Theorem 0.12 *(Rosenblum's Theorem).* *If A and B are operators on the Hilbert spaces $\mathscr{H}$ and $\mathscr{K}$ respectively, and if the operator $\mathscr{T}$ on $\mathscr{B}(\mathscr{K}, \mathscr{H})$ is defined by $\mathscr{T}(X) = AX - XB$, then*

$$\sigma(\mathscr{T}) \subset \sigma(A) - \sigma(B), \ (= \{\alpha - \beta : \alpha \in \sigma(A), \beta \in \sigma(B)\}).$$

Proof. Define operators $\mathscr{A}$ and $\mathscr{B}$ on $\mathscr{B}(\mathscr{K}, \mathscr{H})$ by $\mathscr{A}(X) = AX$ and $\mathscr{B}(X) = XB$. If $\lambda \notin \sigma(A)$, then $(A - \lambda)(A - \lambda)^{-1} X = (A - \lambda)^{-1}(A - \lambda)X$ for all $X \in \mathscr{B}(\mathscr{K}, \mathscr{H})$, and thus $\sigma(\mathscr{A}) \subset \sigma(A)$. Similarly, $\sigma(\mathscr{B}) \subset \sigma(B)$. The operators $\mathscr{A}$ and $\mathscr{B}$ commute, and $\mathscr{T} = \mathscr{A} - \mathscr{B}$. By Lemma 0.11, $\sigma(\mathscr{T}) \subset \sigma(\mathscr{A}) - \sigma(\mathscr{B})$; hence $\sigma(\mathscr{T}) \subset \sigma(A) - \sigma(B)$. $\quad \square$

The notation of the above theorem is retained for the next corollary.

Corollary 0.13 *(Rosenblum's Corollary).* *If $\sigma(A) \cap \sigma(B) = \emptyset$, then for each operator Y from $\mathscr{K}$ to $\mathscr{H}$ there is a unique operator X from $\mathscr{K}$ to $\mathscr{H}$ such that $AX - XB = Y$. In particular, $AX = XB$ implies $X = 0$.*

Proof. This follows immediately from Theorem 0.12, for $\sigma(A) \cap \sigma(B) = \emptyset$ implies $\mathscr{T}$ is invertible. $\quad \square$

We present two applications of Rosenblum's theorem below; it will also be used in subsequent chapters.

If $A \in \mathscr{B}(\mathscr{H})$ and $B \in \mathscr{B}(\mathscr{K})$ we use $A \oplus B$ to denote the operator on $\mathscr{H} \oplus \mathscr{K}$ defined by $(A \oplus B)(x \oplus y) = Ax \oplus By$; matricially, $A \oplus B$ is represented by $\begin{pmatrix} A & 0 \\ 0 & B \end{pmatrix}$.

Corollary 0.14. *If $\sigma(A) \cap \sigma(B) = \emptyset$ and S is an operator on $\mathscr{H} \oplus \mathscr{K}$ which commutes with $A \oplus B$, then $S = C \oplus D$ where C commutes with A and D commutes with B.*

Proof. Let $S = \begin{pmatrix} C & E \\ F & D \end{pmatrix}$. Then $\begin{pmatrix} A & 0 \\ 0 & B \end{pmatrix} \begin{pmatrix} C & E \\ F & D \end{pmatrix} = \begin{pmatrix} C & E \\ F & D \end{pmatrix} \begin{pmatrix} A & 0 \\ 0 & B \end{pmatrix}$ yields, by matrix multiplication, $AE = EB$ and $BF = FA$. Rosenblum's corollary implies $E = 0$ and $F = 0$, and the result follows. □

Corollary 0.15. *If* $\sigma(A) \cap \sigma(B) = \emptyset$, *then the operator* $\begin{pmatrix} A & C \\ 0 & B \end{pmatrix}$ *is similar to* $A \oplus B$.

Proof. We show that there is an operator of the form $\begin{pmatrix} 1 & X \\ 0 & 1 \end{pmatrix}$ that implements the similarity. Every operator of the form $\begin{pmatrix} 1 & X \\ 0 & 1 \end{pmatrix}$ is invertible, (the inverse of $\begin{pmatrix} 1 & X \\ 0 & 1 \end{pmatrix}$ is $\begin{pmatrix} 1 & -X \\ 0 & 1 \end{pmatrix}$), so it suffices to show that there is an X such that

$$\begin{pmatrix} 1 & X \\ 0 & 1 \end{pmatrix} \begin{pmatrix} A & C \\ 0 & B \end{pmatrix} = \begin{pmatrix} A & 0 \\ 0 & B \end{pmatrix} \begin{pmatrix} 1 & X \\ 0 & 1 \end{pmatrix}.$$

This is equivalent to finding an X satisfying the equation $C + XB = AX$. Such an X exists by Rosenblum's corollary. □

0.5 Additional Propositions

Proposition 0.1. For any operator A, $Lat\, A^* = \{\mathcal{M} : \mathcal{M}^\perp \in Lat\, A\}$.

Proposition 0.2. The unit ball of $\mathcal{B}(\mathcal{H})$, ($\mathcal{H}$ separable), is a separable metric space in the weak operator topology.

Proposition 0.3. For any A and B, $\sigma(A \oplus B) = \sigma(A) \cup \sigma(B)$; also, if $\mathcal{M}$ and $\mathcal{N}$ are complementary invariant subspaces for A, then $\sigma(A) = \sigma(A|\mathcal{M}) \cup \sigma(A|\mathcal{N})$.

Proposition 0.4. If A is compact, then there is a sequence of finite-rank operators which converges uniformly to A.

Proposition 0.5. If A is a finite-rank operator, then A is unitarily equivalent to an operator of the form $B \oplus 0$, where B is an operator on a finite-dimensional space.

Proposition 0.6. The class of compact operators is a two-sided ideal in $\mathcal{B}(\mathcal{H})$.

Proposition 0.7 (The Polarization Identity). If $A \in \mathcal{B}(\mathcal{H})$, then

$$(Ax, y) = \tfrac{1}{4} [(A(x+y), x+y) - (A(x-y), x-y) + i(A(x+iy), x+iy)$$
$$- i(A(x-iy), x-iy)].$$

Proposition 0.8. If A and B commute and B is quasinilpotent, then $\sigma(A+B) = \sigma(A)$.

Proposition 0.9. If K is compact and $0 \neq \lambda \in \sigma(K)$, let $\mathcal{M}_n$ denote the nullspace of $(K-\lambda)^n$. Then $\bigvee\limits_{n=1}^{\infty} \mathcal{M}_n$ is finite-dimensional.

Proposition 0.10. The operator A is compact if and only if A^* is compact.

Proposition 0.11. If there exists an orthonormal basis $\{e_n\}_{n=0}^{\infty}$ such that $\sum\limits_{n,m=0}^{\infty} |(Ae_n, e_m)|^2$ is finite, then A is compact and $\sum\limits_{n,m=0}^{\infty} |(Af_n, f_m)|^2 = \sum\limits_{n,m=0}^{\infty} |(Ae_n, e_m)|^2$ for every orthonormal basis $\{f_n\}$.

Proposition 0.12. If $\mathfrak{A}$ is a Banach algebra with identity, and $\mathcal{I}$ is a proper (one- or two-sided) ideal, then the closure of $\mathcal{I}$ is a proper ideal.

0.6 Notes and Remarks

The evolution of the basic ideas concerning Hilbert space and linear operators is discussed in Dunford-Schwartz ([1], [2]).

The elementary but powerful results on spectra in commutative Banach algebras are due to Gelfand [1]. More of the elementary theory can be found in Dunford-Schwartz [2], and a deeper account of the general theory of Banach algebras is presented in Rickart [1].

Theorem 0.8, which generalizes the result of Bram [1] in the case where A is normal, is due to Scroggs [1]; it was independently found by S. Parrott (see Crimmins-Rosenthal [1]). Theorem 0.10 was discovered by Weyl [1]. Theorem 0.12 is due to Rosenblum [1] in the case where A and B are elements of the same Banach algebra; his proof involves computing an integral formula for the resolvent of $\mathcal{T}$. The simple proof presented above was found by Lumer and Rosenblum [1], and the fact that it applied in the case where A and B are operators on different spaces was observed in Radjavi-Rosenthal [5]. Corollaries 0.14 and 0.15 are taken from Rosenthal [2] and Rosenthal [3] respectively; each has undoubtedly been observed by others.

Chapter 1. Normal Operators

The spectral theorem exhibits normal operators as integrals with respect to some of their reducing subspaces. This is the most satisfying structure theorem known for operators on infinite-dimensional spaces: the class of normal operators is a large and useful class of operators and the spectral theorem is an appropriate tool for answering many questions about them. It should be noted, however, that the spectral theorem does not answer all questions about normal operators; in particular, there are still a number of unsolved problems about invariant subspaces of normal operators, (see Section 10.1).

One result which follows from the spectral theorem is the existence of hyperinvariant subspaces for normal operators—this is Theorem 1.16 (Fuglede's theorem), which is also useful in other contexts.

1.1 Preliminaries

The operator A is *normal* if it commutes with A^*, *Hermitian* if $A = A^*$, *positive* if $(Ax, x) \geq 0$ for all x, and *unitary* if $A^* = A^{-1}$. One can study normal operators by first proving the spectral theorem for Hermitian operators and then deriving the general case (cf. Halmos [2]) but we shall present a more direct approach.

Theorem 1.1. *If A is normal, then $\|A\| = r(A)$.*

Proof. By the spectral radius formula (page 4) it is sufficient to show that $\|A^n\| = \|A\|^n$ for infinitely many n. For this it suffices to show that $\|A^2\| = \|A\|^2$; induction then gives $\|A^{2^k}\| = \|A\|^{2^k}$ for all k. This follows from the computation

$$\|A^2\|^2 = \|(A^2)^* A^2\| = \|(AA^*)^*(AA^*)\| = \|AA^*\|^2 = (\|A\|^2)^2 ;$$

(we have used the easily verified fact (cf. Halmos [2], p. 40) that $\|B^*B\| = \|B\|^2$ for every operator B). $\quad\Box$

Theorem 1.2. *If A is normal and $Ax = \lambda x$, then $A^*x = \bar{\lambda} x$.*

Proof. Squaring and expanding inner products shows that $\|(A-\lambda)x\|$ $= \|(A^*-\bar{\lambda})x\|$ for all x, and the result follows. ☐

Theorem 1.3 *(Spectral Mapping Theorem for Normal Operators).* *If A is normal and $p(\cdot,\cdot)$ is any polynomial in two variables, then*

$$\sigma(p(A,A^*)) = \{p(z,\bar{z}) : z \in \sigma(A)\}.$$

Proof. Let $\mathfrak{A}$ be a maximal commutative subalgebra of $\mathcal{B}(\mathcal{H})$ which contains A and A^*; by Theorems 0.4 and 0.3,

$$\sigma(p(A,A^*)) = \{\phi(p(A,A^*)) : \phi \text{ a complex homomorphism of } \mathfrak{A}\}$$

$$= \{p(\phi(A),\phi(A^*)) : \phi \text{ a complex homomorphism of } \mathfrak{A}\}.$$

Since $\{\phi(A):\phi \text{ a complex homomorphism of } \mathfrak{A}\} = \sigma(A)$ we will be done if we show that $\phi(A^*) = \overline{\phi(A)}$ for every ϕ.

Suppose that $\phi(A)=a+bi$, $\phi(A^*)=c+di$. We first show that $b+d=0$. To see this, consider the Hermitian operator $B=A+A^*$. Then $\phi(B)=(a+c)+i(b+d)$. On the other hand, $\sigma(B)$ is real; (if H is Hermitian, then (Hx,x) is real for all x, hence $\Pi(H)$ is real, and the inclusion $\partial\sigma(H) \subset \Pi(H)$ implies that $\sigma(H)$ is real). It follows that $b+d=0$.

To complete the proof we need only show that $a=c$. For this we simply apply the above argument to the operator iA. ☐

1.2 Compact Normal Operators

It follows immediately from Theorem 1.2 that a normal operator on a finite-dimensional space is unitarily equivalent to a diagonal operator, or, in other words, that given a normal operator on a finite-dimensional space there exists an orthonormal basis for the space consisting of eigenvectors of the operator. The same statement is easily proven for compact normal operators on infinite-dimensional spaces.

Theorem 1.4 *(Spectral Theorem for Compact Normal Operators).* *If A is a compact normal operator, then there exists an orthonormal basis $\{e_\alpha\}$ such that each e_α is an eigenvector of A.*

Proof. By Zorn's lemma there is a maximal orthonormal set $\{e_\alpha\}_{\alpha \in I}$ consisting of eigenvectors of A. We need only show that $\bigvee_{\alpha \in I} \{e_\alpha\} = \mathcal{H}$.

Let $\mathcal{M} = \bigvee_{\alpha \in I} \{e_\alpha\}$. By Theorem 1.2 the one-dimensional space spanned by e_α reduces A for each α, and therefore $\mathcal{M}$ reduces A. If $\mathcal{M}^\perp \neq \{0\}$ consider $A|\mathcal{M}^\perp$. The operator $A|\mathcal{M}^\perp$ is compact and normal. If $A|\mathcal{M}^\perp$ were 0, the set $\{e_\alpha\}$ would not be maximal. Thus $\sigma(A|\mathcal{M}^\perp)$

contains a non-zero complex number, by Theorem 1.1, and hence, by the Fredholm Alternative (Theorem 0.9), $\Pi_0(A|\mathcal{M}^\perp)$ is not empty. Thus A has an eigenvector in $\mathcal{M}^\perp$, contradicting the maximality of $\{e_\alpha\}$. □
  This theorem can be reformulated as follows.

Corollary 1.5 *(Spectral Theorem for Compact Normal Operators).* *The operator A is compact normal if and only if there exists an ortho-normal basis $\{e_n\}_{n=0}^\infty$ and a sequence of complex numbers $\{\lambda_n\}_{n=0}^\infty \to 0$ such that $Ax = \sum_{n=0}^\infty \lambda_n(x, e_n)e_n$ for all x.*

  Proof. It is obvious that every operator of the given form is compact and normal. The converse follows immediately from Theorem 1.4; ($\{e_n\} \to 0$ implies $\{\|Ae_n\|\} \to 0$ for compact operators A). □

1.3 Spectral Theorem — First Form

One way of making examples of normal operators is the following. Let (X, μ) be a finite measure space and let $\phi \in \mathscr{L}^\infty(X, \mu)$. The *multi-plication operator* corresponding to ϕ is the operator M_ϕ on $\mathscr{L}^2(X, \mu)$ defined by $(M_\phi f)(x) = \phi(x)f(x)$ for $f \in \mathscr{L}^2(X, \mu)$. It is easily veri-fied that $\|M_\phi\| = \|\phi\|_\infty$ and that $\sigma(M_\phi) = $ essential range of ϕ $= \{\lambda : \mu\{x : |\phi(x) - \lambda| < \varepsilon\} > 0$ for all $\varepsilon > 0\}$. Also $(M_\phi)^* = M_{\bar\phi}$, where $\bar\phi(x) = \overline{\phi(x)}$ for all $x \in X$. Thus $M_\phi(M_\phi)^* = (M_\phi)^* M_\phi = M_{|\phi|^2}$, and M_ϕ is normal.
  One version of the spectral theorem states that every normal operator is essentially a multiplication operator.

Theorem 1.6 *(Spectral Theorem—First Form).* *If A is a normal operator on a separable space, then there exists a finite measure space (X, μ) and a $\phi \in \mathscr{L}^\infty(X, \mu)$ such that A is unitarily equivalent to the oper-ator M_ϕ on $\mathscr{L}^2(X, \mu)$.*

  Proof. We first consider the special case where there exists a cyclic vector f in the sense that $\bigvee_{n,m=0}^\infty \{A^n(A^*)^m f\} = \mathscr{H}$. The general case will then be shown to follow easily from this special case.
  Suppose, then, that $\bigvee_{n,m=0}^\infty \{A^n(A^*)^m f\} = \mathscr{H}$, or, in other words, that the closure of $\{p(A, A^*)f : p$ is a polynomial in two variables$\}$ is $\mathscr{H}$. In this case we shall show that X can be taken as a subset of the complex plane and ϕ the identity function, (i.e., $\phi(z) = z$ for all $z \in X$). The idea behind the proof is the observation that $Ap(A, A^*)f = q(A, A^*)f$, where $q(z_1, z_2) = z_1 p(z_1, z_2)$.

To construct μ we shall use the Riesz representation theorem, which states that given a bounded positive linear functional F on the space $\mathscr{C}(X)$ of continuous functions on a compact Hausdorff space X, there exists a non-negative regular finite Borel measure μ such that $F(g) = \int g\, d\mu$ for all g in $\mathscr{C}(X)$, (cf. Halmos [1], Rudin [1]).

Let $X = \sigma(A)$. The Stone-Weierstrass theorem implies that $\{p(z, \bar{z}) : p$ is a polynomial in two variables$\}$ is uniformly dense in the space $\mathscr{C}(X)$ (cf. Dieudonné [1], Dunford-Schwartz [1]). Clearly we can assume that $\|f\| = 1$ without loss of generality. Define a linear functional on the space of polynomials in z and $\bar{z}$ by $F(p) = (p(A, A^*)f, f)$. Then

$$|F(p)| \leq \|p(A, A^*)\| \cdot \|f\|^2$$
$$= r(p(A, A^*)), \quad \text{by Theorem 1.1,}$$
$$= \sup_{z \in \sigma(A)} |p(z, \bar{z})|, \quad \text{by Theorem 1.3.}$$

Thus F is a bounded linear functional with $\|F\| \leq 1$ on the space of polynomials in z and $\bar{z}$ regarded as a subset of $\mathscr{C}(X)$. Since this set of polynomials is dense in $\mathscr{C}(X)$, F has a unique extension to a bounded linear functional on $\mathscr{C}(X)$.

We must show that F is positive; i.e., that if $g \in \mathscr{C}(X)$ and $g(z) \geq 0$, then $F(g) \geq 0$. For this let $h(z) = \sqrt{g(z)}$ for $z \in X$. The Stone-Weierstrass theorem for real functions implies that for all $\varepsilon > 0$ there exists a polynomial $p(x, y)$ in the real variables x and y, with real coefficients, such that $|[p(x, y)]^2 - [h(x + iy)]^2| < \varepsilon$ for $x + iy \in X$. Let $H = \frac{1}{2}[A + A^*]$ and $K = (1/2i)[A - A^*]$. Then $p(H, K)$ is Hermitian, and

$$\left|([p(H, K)]^2 f, f) - F(g)\right| \leq \sup_{x + iy \in X} \left|[p(x, y)]^2 - [h(x + iy)]^2\right| \leq \varepsilon.$$

But $\quad ([p(H, K)]^2 f, f) = (p(H, K)f, p(H, K)f) = \|p(H, K)f\|^2 \geq 0 \quad$ and therefore $F(g) \geq 0$.

Now the Riesz representation theorem implies that there exists a non-negative regular finite Borel measure μ on X such that $F(g) = \int_X g\, d\mu$ for all $g \in \mathscr{C}(X)$. We show that A is unitarily equivalent to M_z on $\mathscr{L}^2(X, \mu)$.

Define $Up(z, \bar{z}) = p(A, A^*)f$ for each polynomial p, and regard U as a mapping from the set of polynomials $p(z, \bar{z})$ (as a linear manifold in $\mathscr{L}^2(X, \mu)$) into $\mathscr{H}$. Then U is an isometry, since

$$\int_X |p(z, \bar{z})|^2\, d\mu = \int_X p(z, \bar{z})\, \overline{p(z, \bar{z})}\, d\mu$$
$$= F(p(z, \bar{z})\, \overline{p(z, \bar{z})})$$
$$= (p(A, A^*)[p(A, A^*)]^* f, f) = \|p(A, A^*)f\|^2.$$

Since the linear manifold $\{p(A, A^*)f : p$ is a polynomial$\}$ is dense in $\mathcal{H}$, U has a unique extension to a unitary operator from $\mathcal{L}^2(X, \mu)$ onto $\mathcal{H}$. Also, for any polynomial p,

$$U^{-1} A U p = U^{-1} A p(A, A^*) f = U^{-1}\left(U(M_z p(z, \bar{z}))\right) = M_z p(z, \bar{z}).$$

Thus $U^{-1} A U = M_z$.

This completes the proof in the case that A has a cyclic vector. The general case can be reduced to this special case. Since, for any $f \in \mathcal{H}$, $\bigvee_{n,m=0}^{\infty} \{A^n (A^*)^m f\}$ reduces A, Zorn's lemma implies that there exists a collection $\{\mathcal{M}_n\}_{n=1}^{\infty}$ of pairwise-orthogonal subspaces of $\mathcal{H}$ such that $\mathcal{H} = \sum_{n=1}^{\infty} \oplus \mathcal{M}_n$, each $\mathcal{M}_n$ reduces A, and $A|\mathcal{M}_n$ has a cyclic vector for each n; (the fact that the collection is countable follows from the assumption that $\mathcal{H}$ is separable).

By the cyclic case, for each n there is a finite measure μ_n on $\sigma(A|\mathcal{M}_n)$ such that $A|\mathcal{M}_n$ is unitarily equivalent to M_z on $\mathcal{L}^2(X_n, \mu_n)$, where $X_n = \sigma(A|\mathcal{M}_n)$. Dividing by an appropriate positive number we can assume that $\mu_n(X_n) \leq 2^{-n}$. Let $\phi_n(z) = z$ for $z \in X_n$. Then A is unitarily equivalent to

$$\sum_{n=1}^{\infty} \oplus M_{\phi_n} \quad \text{on the space} \quad \sum_{n=0}^{\infty} \oplus \mathcal{L}^2(X_n, \mu_n).$$

We must show how to regard $\sum_{n=1}^{\infty} \oplus \mathcal{L}^2(X_n, \mu_n)$ as $\mathcal{L}^2(X, \mu)$ in such a way that $\sum_{n=1}^{\infty} \oplus M_{\phi_n}$ is unitarily equivalent to M_ϕ on $\mathcal{L}^2(X, \mu)$ for some ϕ.

Re-label the elements of the sets $\{X_n\}$ to make the sets pairwise-disjoint, and then let $X = \bigcup_{n=1}^{\infty} X_n$. Define the measurable subsets of X as the subsets which are countable unions of measurable subsets of the X_n, and if $S = \bigcup_{n=1}^{\infty} S_n$ is a measurable set with $S_n \subset X_n$ for every n, define $\mu(S) = \sum_{n=1}^{\infty} \mu(S_n)$. Then (X, μ) is a finite measure space. If we define the function ϕ on X by $\phi(x) = \phi_n(x)$ for $x \in X_n$, then $\phi \in \mathcal{L}^\infty(X, \mu)$, $(\|\phi\|_\infty = \|A\|)$.

It is easily verified that M_ϕ on $\mathcal{L}^2(X, \mu)$ is unitarily equivalent to $\sum_{n=1}^{\infty} \oplus M_{\phi_n}$ and hence to A. $\quad\square$

We should note that the above theorem remains true (with essentially the same proof) in the non-separable case, although the measure will not be finite in general.

This version of the spectral theorem is very useful; for example the following is a trivial corollary.

Corollary 1.7. *A normal operator is* (i) *Hermitian,* (ii) *unitary,* (iii) *positive,* (iv) *a projection if and only if its spectrum is* (i') *real,* (ii') *on the unit circle,* (iii') *on the non-negative real axis,* (iv') *in the set* $\{0,1\}$.

The following corollary is another useful standard fact.

Corollary 1.8. *A positive operator has a unique positive square root.*

Proof. Let A be a positive operator; we must show that there exists a unique positive operator B such that $B^2 = A$. The existence of such a B follows immediately from Theorem 1.6: if M_ϕ is unitarily equivalent to A, then ϕ is a non-negative function and $(M_{\sqrt{\phi}})^2 = M_\phi$.

To show uniqueness suppose that $A = B^2 = C^2$ with B and C positive and B the square root of A constructed above. Obviously B and C each commute with A. Moreover, since B is a uniform limit of a sequence of polynomials in A, (simply choose a sequence $\{p_n\}$ of polynomials uniformly approximating the square root function on $\sigma(A)$; then $\{p_n(A)\}$ converges uniformly to B), B and C also commute. Now the nullspace $\mathcal{N}$ of A obviously reduces A, B, and C. It is also easy to see that B, C, and $B+C$ each have nullspace $\mathcal{N}$. Thus we need only prove that $B|\mathcal{N}^\perp = C|\mathcal{N}^\perp$. Note that $(B+C)\mathcal{N}^\perp$ is dense in $\mathcal{N}^\perp$, and

$$(B-C)(B+C) = B^2 - C^2 + BC - CB = 0.$$

Thus $B-C=0$ on $\mathcal{N}^\perp$. □

One of the applications of the above corollary is the existence of the polar decomposition—see Proposition 1.1.

1.4 Spectral Theorem — Second Form

A more common statement of the spectral theorem than Theorem 1.6 is the assertion that every normal operator can be represented as an integral with respect to a spectral measure.

Definition. A (complex) *spectral measure* E in $\mathcal{H}$ is a mapping from the σ-algebra of Borel sets in the complex plane into the set of projections in $\mathcal{B}(\mathcal{H})$ such that: $E(\mathbb{C})=1$, E is countably additive in the sense that whenever $\{S_n\}_{n=1}^\infty$ is a countable collection of pairwise-disjoint Borel sets, then $E\left(\bigcup_{n=1}^\infty S_n\right) = \sum_{n=1}^\infty E(S_n)$ (convergence in the strong topology), and E has compact support (i.e., there exists a compact set S such that $E(S')=0$, where S' is the complement of S).

Note that if E is a spectral measure and x and y are fixed vectors in $\mathscr{H}$, then the map from Borel sets to the complex numbers given by $\mu(S) = (E(S)x, y)$ defines an ordinary complex-valued measure. If f is a bounded Borel measurable function on $\mathbb{C}$, we use the notation $\int f(\lambda) d(E_\lambda x, y)$ to denote the integral of f with respect to this measure.

Theorem 1.9. *If f is a bounded, Borel measurable function on $\mathbb{C}$ and E is a spectral measure in $\mathscr{H}$, then there exists a unique A in $\mathscr{B}(\mathscr{H})$ such that $(Ax, y) = \int f(\lambda) d(E_\lambda x, y)$ for all $x, y \in \mathscr{H}$.*

Proof. Define Φ by $\Phi(x, y) = \int f(\lambda) d(E_\lambda x, y)$ for each fixed x and y. Then clearly Φ is linear in x and conjugate-linear in y. Moreover

$$|\Phi(x, x)| \leq \int |f(\lambda)| d(E_\lambda x, x) \leq \|f\|_\infty (E(\mathbb{C})x, x) = \|f\|_\infty \|x\|^2.$$

Thus $\Phi(x, y)$ is a bounded bilinear functional, and therefore there is a unique A such that $(Ax, y) = \Phi(x, y)$ for all x and y. $\quad\square$

We shall use the notation $A = \int f(\lambda) dE_\lambda$ as an abbreviation for the relation $(Ax, y) = \int f(\lambda) d(E_\lambda x, y)$ for all x and y.

We shall need some of the basic properties of spectral measures.

Theorem 1.10. *If $A = \int f(\lambda) dE_\lambda$, then $A^* = \int \overline{f(\lambda)} dE_\lambda$.*

Proof. Fix x and y. Then

$$(A^* x, y) = (x, Ay)$$
$$= \overline{(Ay, x)} = \overline{\int f(\lambda) d(E_\lambda y, x)}$$
$$= \int \overline{f(\lambda)} d(E_\lambda x, y). \quad\square$$

Theorem 1.11. *If $A = \int f(\lambda) dE_\lambda$ and $B = \int g(\lambda) dE_\lambda$, then*

$$AB = \int f(\lambda) g(\lambda) dE_\lambda.$$

Proof. For each fixed x and y,

$$(ABx, y) = \int f(\lambda) d(E_\lambda Bx, y) = \int f(\lambda) dv,$$

where $v(S) = \int g(\mu) d(E_\mu x, E(S)y)$ for Borel sets S. Thus

$$(ABx, y) = \int f(\lambda) g(\lambda) d(E_\lambda x, y). \quad\square$$

If $\phi \in \mathscr{L}^\infty(X, \mu)$ and if, for each Borel set S, we define $E(S)$ to be multiplication by $\chi_S \circ \phi$ on $\mathscr{L}^2(X, \mu)$ (where χ_S is the characteristic function of S), then E is a spectral measure. The standard version of the spectral theorem can be derived from Theorem 1.6 by using this fact.

Theorem 1.12 *(Spectral Theorem—Second Form).* *If A is a normal operator, then there exists a unique spectral measure E such that $A = \int \lambda \, dE_\lambda$.*

Proof. By Theorem 1.6 we can assume that $A = M_\phi$ on $\mathscr{L}^2(X, \mu)$. For each Borel set S define $E(S)$ as multiplication by $\chi_S \circ \phi$. Clearly E is a spectral measure, with support the essential range of ϕ, (i.e., the spectrum of M_ϕ). We must show that $M_\phi = \int \lambda \, dE_\lambda$.

Fix a Borel set S. Then, for any $f \in \mathscr{L}^2(X, \mu)$,

$$\int \chi_S \, d(E_\lambda f, f) = (E(S) f, f) = \int (\chi_S \circ \phi) f \bar{f} \, d\mu.$$

Thus $\int_{\mathbb{C}} F \, d(E_\lambda f, f) = \int_X (F \circ \phi) |f|^2 \, d\mu$ for all measurable simple functions F, and therefore this relation holds for the function $F(\lambda) = \lambda$. Thus

$$\int_{\mathbb{C}} \lambda \, d(E_\lambda f, f) = \int_X \phi |f|^2 \, d\mu = (M_\phi f, f).$$

The polarization identity (Proposition 0.7) implies that $\int_{\mathbb{C}} \lambda \, d(E_\lambda f, g) = (M_\phi f, g)$ for all f and g in $\mathscr{L}^2(X, \mu)$.

The uniqueness is easily proved too. Suppose that $A = \int \lambda \, dE_\lambda = \int \lambda \, dF_\lambda$. Then $A^* = \int \bar{\lambda} \, dE_\lambda = \int \bar{\lambda} \, dF_\lambda$, by Theorem 1.10. Fix a vector x. Then, by Theorem 1.11,

$$\int p(\lambda, \bar{\lambda}) \, d(E_\lambda x, x) = \int p(\lambda, \bar{\lambda}) \, d(F_\lambda x, x)$$

for all polynomials p. The Stone-Weierstrass theorem implies that $\int f(\lambda) \, d(E_\lambda x, x) = \int f(\lambda) \, d(F_\lambda x, x)$ for all functions f continuous on the union of the supports of E and F. Since the measures $(E_\lambda x, x)$ and $(F_\lambda x, x)$ are finite Borel measures and thus are regular (Rudin [1], p. 48), it follows that

$$(E_\lambda(S) x, x) = (F_\lambda(S) x, x) \quad \text{for all Borel sets } S. \quad \square$$

Theorem 1.13. *If* $A = \int \lambda \, dE_\lambda$ *and* S *is a Borel set, then* $E(S)\mathscr{H}$ *reduces* A *and the spectrum of the restriction of* A *to* $E(S)\mathscr{H}$ *is contained in* $\bar{S}$.

Proof. Since $E(S) = \int \chi_S \, dE_\lambda$, Theorem 1.11 implies that $E(S)$ commutes with A, and therefore $E(S)\mathscr{H}$ reduces A. Also, in the notation used at the beginning of the proof of Theorem 1.12, $\sigma(A | E(S)\mathscr{H})$ is the essential range of the product of the functions $\chi_S \circ \phi$ and ϕ restricted to $\phi^{-1}(S)$, and thus is certainly contained in $\bar{S}$. $\quad \square$

Theorem 1.14. *If* $A = \sum_{j=1}^{\infty} \oplus A_j$, *and if* $A_j = \int \lambda \, dE_\lambda^j$ *on* $\mathscr{H}_j$ *for each* j, *then* $A = \int \lambda \, dE_\lambda$, *where* $E(S) = \sum_{j=1}^{\infty} \oplus E^j(S)$ *for each Borel set* S.

Proof. It is obvious that A is normal. Define $E(S)$ as above for each Borel set S; then E is easily seen to be a spectral measure; (for countable additivity note that E is countably additive when restricted to the dense subset of $\sum_{j=1}^{\infty} \oplus \mathscr{H}_j$ consisting of all vectors with only finitely many non-zero coordinates), and $A = \int \lambda \, dE_\lambda$, (since this is true on the dense subset just mentioned). □

The next theorem is useful in multiplicity theory (cf. Halmos [2]) and in other contexts (see Theorem 9.21).

Theorem 1.15. *If E is the spectral measure of A, x is a vector, and v_1 is a finite positive measure that is absolutely continuous with respect to the measure v_2 defined by $v_2(S) = (E(S)x, x)$ for Borel sets S, then there exists a vector y in the reducing subspace of A generated by x such that $v_1(S) = (E(S)y, y)$ for all Borel sets S.*

Proof. Let $\mathcal{M}$ be the reducing subspace of A generated by x. Then, as shown by the first part of the proof of Theorem 1.6, $A | \mathcal{M}$ is unitarily equivalent to M_ϕ on $\mathcal{L}^2(X, \mu)$, where X is a compact subset of the complex plane and $\phi(z) = z$ for $z \in X$. Let f be the element of $\mathcal{L}^2(X, \mu)$ corresponding to x under this unitary equivalence; then $E(S)$ is multiplication by χ_S and $v_2(S) = \int \chi_S |f|^2 d\mu$ for each Borel set S. By the Radon-Nikodym Theorem (see Rudin [1]) there exists a non-negative Borel-measurable function g such that $v_1(S) = \int \chi_S g \, dv_2$ for all S. Thus

$$v_1(S) = \int \chi_S g |f|^2 d\mu = \int \chi_S |\sqrt{g} f|^2 d\mu.$$

Therefore $\sqrt{g} f \in \mathcal{L}^2(X, \mu)$, and $v_1(S) = (\chi_S \sqrt{g} f, \sqrt{g} f)$. If y is the vector in $\mathcal{M}$ corresponding to $\sqrt{g} f$ under the unitary equivalence, then

$$v_1(S) = (E(S)y, y). \quad □$$

1.5 Fuglede's Theorem

We have seen that every spectral projection $E(S)$ of a normal operator A commutes with A. Fuglede's theorem is an important generalization of this fact.

Theorem 1.16 *(Fuglede's Theorem).* *If $A = \int \lambda \, dE_\lambda$ and B is any operator which commutes with A, then B commutes with $E(S)$ for every Borel set S.*

Proof. Fix a Borel set S. We must show that $E(S)\mathcal{H}$ reduces B. Let S' denote the complement of S. Then it suffices to show that both $E(S)\mathcal{H}$ and $E(S')\mathcal{H}$ are invariant under B. We need only do this for $E(S)\mathcal{H}$; the same proof would apply to $E(S')\mathcal{H}$. This is equivalent to showing that $E(S') B E(S) = 0$. By regularity (see Proposition 1.4) it suffices to show that $E(S_1) B E(S_2) = 0$ whenever S_1 and S_2 are closed subsets of S' and S respectively.

Fix S_1 and S_2 as above. Then, since $AB = BA$, it follows that $E(S_1)AB E(S_2) = E(S_1)BA E(S_2)$.

Thus

$$[E(S_1)A E(S_1)] [E(S_1)B E(S_2)] = [E(S_1)B E(S_2)] [E(S_2)A E(S_2)].$$

Now $E(S_1)A E(S_1)$ and $E(S_2)A E(S_2)$ have disjoint spectra by Theorem 1.13, (as operators on $E(S_1)\mathscr{H}$ and $E(S_2)\mathscr{H}$ respectively), and therefore Corollary 0.13 implies that $E(S_1)B E(S_2) = 0$. ☐

Corollary 1.17. *Every non-scalar normal operator has a non-trivial hyperinvariant subspace.*

Proof. Let A be any non-scalar normal operator. Then $\sigma(A)$ contains at least two distinct points z_1 and z_2; (the easiest way to see this is via Theorem 1.6—if the essential range of ϕ contains only one point, then ϕ is obviously a scalar.) Let E denote the spectral measure of A, and let S be the disc of radius $\frac{1}{2}|z_1 - z_2|$ centred at z_1. The projection $E(S)$ is non-trivial, because $\sigma(A|E(S)\mathscr{H})$ and $\sigma(A|(1 - E(S))\mathscr{H})$ are both proper subsets of $\sigma(A)$ by Theorem 1.13. Since $E(S)$ commutes with every operator commuting with A (by Theorem 1.16), $E(S)\mathscr{H}$ is hyperinvariant for A. ☐

The following corollary is often called Fuglede's theorem; Theorem 1.16 can easily be derived from this corollary if the corollary is independently proved.

Corollary 1.18. *If A is normal and $AB = BA$, then $A^*B = BA^*$.*

Proof. Let $A = \int \lambda dE_\lambda$. Then B commutes with the spectral measure $E(S)$, by Fuglede's theorem. Since $A^* = \int \bar{\lambda} dE_\lambda$ (by Theorem 1.10), for each x and y

$$(A^*Bx, y) = \int \bar{\lambda} d(E_\lambda Bx, y) = \int \bar{\lambda} d(BE_\lambda x, y)$$

$$= \int \bar{\lambda} d(E_\lambda x, B^*y) = (A^*x, B^*y) = (BA^*x, y). \quad ☐$$

Fuglede's Theorem has an interesting generalization.

Corollary 1.19 *(Putnam's Corollary).* *If A and C are normal operators, and if B is an operator such that $AB = BC$, then $A^*B = BC^*$.*

Proof. Let

$$S = \begin{pmatrix} A & 0 \\ 0 & C \end{pmatrix} \quad \text{and} \quad T = \begin{pmatrix} 0 & B \\ 0 & 0 \end{pmatrix}.$$

Then S is normal, and multiplying matrices shows that $ST = TS$. Corollary 1.18 implies that $S^*T = TS^*$. Writing this matricially gives the result. ☐

One nice application of Corollary 1.19 is Proposition 1.5.

1.6 The Algebra $\mathscr{L}^\infty$

Theorem 1.6 states that every normal operator is unitarily equivalent to a multiplication operator. Consider a finite measure space (X, μ), and let $\mathscr{L}^\infty$ denote the algebra of all multiplication operators on $\mathscr{L}^2(X, \mu)$; i.e., $\mathscr{L}^\infty = \{M_\phi : \phi \in \mathscr{L}^\infty(X, \mu)\}$. Then $\mathscr{L}^\infty$ is a commutative subalgebra of the algebra of all operators on $\mathscr{L}^2(X, \mu)$.

Theorem 1.20. *The algebra $\mathscr{L}^\infty$ is maximal abelian; i.e., if A is an operator such that $AM_\phi = M_\phi A$ for all $\phi \in \mathscr{L}^\infty(X, \mu)$, then there exists a $\psi \in \mathscr{L}^\infty(X, \mu)$ such that $A = M_\psi$.*

Proof. Suppose that $AM_\phi = M_\phi A$ for all $\phi \in \mathscr{L}^\infty(X, \mu)$. Let f denote the function on X that is identically 1, and let $\psi = Af$. Then, whenever $\phi \in \mathscr{L}^\infty(X, \mu)$, $A\phi = A(M_\phi f) = M_\phi Af = \phi\psi = M_\psi \phi$. Thus A agrees with M_ψ on $\mathscr{L}^\infty(X, \mu)$, which is a dense subset of $\mathscr{L}^2(X, \mu)$. We would be done if we knew that $\psi \in \mathscr{L}^\infty(X, \mu)$, for then M_ψ would be a bounded operator.

Suppose that ψ is not essentially bounded. Let $S_n = \{x : |\psi(x)| \geq n\}$; we are assuming that $\mu(S_n) > 0$ for all natural numbers n. Then $A\chi_{S_n} = \psi\chi_{S_n}$, and therefore

$$\|A\chi_{S_n}\|^2 = \int_{S_n} |\psi|^2 \, d\mu \geq n^2 \mu(S_n).$$

Thus

$$\frac{\|A\chi_{S_n}\|^2}{\|\chi_{S_n}\|^2} \geq \frac{n^2 \mu(S_n)}{\mu(S_n)} = n^2.$$

This contradicts the fact that A is a bounded operator; therefore $\psi \in \mathscr{L}^\infty(X, \mu)$. □

Corollary 1.21. *The algebra $\mathscr{L}^\infty$ is weakly closed.*

Proof. If A is any operator in the weak closure of $\mathscr{L}^\infty$, then obviously A commutes with all operators in $\mathscr{L}^\infty$; hence Theorem 1.20 gives the result. □

We shall see in Chapter 7 that, up to unitary equivalence, every maximal abelian algebra of operators which contains the adjoint of each of its members is an $\mathscr{L}^\infty$.

1.7 The Functional Calculus

If p is a complex polynomial, then, for any operator A, $p(A)$ is defined in the obvious way. If A is a normal operator and f is a bounded Borel-measurable function on $\sigma(A)$, then Theorems 1.9 and 1.12 give a unique definition of $f(A)$.

Theorem 1.22. *If* $A = \int \lambda \, dE_\lambda$, *then*

(i) *the map* $f \to f(A)$ *from the algebra of bounded Borel measurable functions on* $\sigma(A)$ *into* $\mathcal{B}(\mathcal{H})$ *defined by* $f(A) = \int f(\lambda) \, dE_\lambda$ *is an algebra homomorphism, and*

(ii) *if* p *is any polynomial in two variables, then*

$$p(A, A^*) = \int p(\lambda, \bar{\lambda}) \, dE_\lambda.$$

Proof. Part (i) follows from Theorem 1.11 and the obvious linearity of spectral integrals; (ii) follows from Theorems 1.10 and 1.11. ☐

This functional calculus reflects, in many respects, the power of the spectral theorem. In particular, when f is a characteristic function, $f(A)$ is a spectral projection. Corollary 1.17 states that $f(A)$ gives a projection onto a hyperinvariant subspace for A for such functions f.

In Chapter 2 we shall develop a functional calculus for all operators. This more general functional calculus will not be nearly so powerful as the functional calculus for normal operators. However, in certain cases, functions of A will give projections onto hyperinvariant subspaces.

1.8 Completely Normal Operators

Definition. An operator is *completely normal* if it is normal and all its invariant subspaces are reducing; (equivalently, by Proposition 1.7, if its restriction to every invariant subspace is normal).

Clearly every Hermitian operator is completely normal. It is easily shown that every normal operator on a finite-dimensional space is completely normal, and it is easy to construct examples of normal operators which are not completely normal on infinite-dimensional spaces. In Chapter 9 we shall give a necessary and sufficient condition that an operator be completely normal.

Theorem 1.23. *If* A *is a normal operator, and if* $\sigma(A) = \eta(\sigma(A))$ *and* $\sigma(A)$ *has no interior, then* A *is completely normal.*

Proof. Since $\sigma(A)$ is compact, has no interior, and does not separate the plane, Mergelyan's Theorem (Rudin [1], p. 386) implies that for all $\varepsilon > 0$ there exists a polynomial p_ε such that $|p_\varepsilon(z) - \bar{z}| < \varepsilon$ for $z \in \sigma(A)$. Now $\|p_\varepsilon(A) - A^*\| = r(p_\varepsilon(A) - A^*)$, by Theorem 1.1. Then $r(p_\varepsilon(A) - A^*) = \sup_{z \in \sigma(A)} |p_\varepsilon(z) - \bar{z}|$, by Theorem 1.3. Therefore, $\|p_\varepsilon(A) - A^*\| \leq \varepsilon$. Thus there exists a sequence $\{p_n\}$ of polynomials such that $\{p_n(A)\}$ converges uniformly to A^*. If $\mathcal{M} \in Lat A$ it follows trivially that $\mathcal{M} \in Lat A^*$. ☐

Corollary 1.24. *A polynomially compact normal operator is completely normal.*

Proof. Suppose that $p(A)$ is compact. Then $\sigma(p(A))$ is countable (Theorem 0.9), and the spectral mapping theorem (Theorem 0.5) implies that $\sigma(A)$ is countable. Therefore $\sigma(A)$ satisfies the hypotheses of Theorem 1.23. $\square$

Definition. The normal operator A is *diagonable* if the set of eigenvectors of A spans $\mathscr{H}$.

Note that A is diagonable if and only if there is an orthonormal basis consisting of eigenvectors of A. In other words, A has a diagonal matrix with respect to some orthonormal basis. Every compact normal operator is diagonable (by Theorem 1.4); this is the case for polynomially compact normal operators too, (Proposition 1.10).

Theorem 1.25. *Let A be a diagonable normal operator. Then the following are equivalent:*

(i) *Every non-trivial invariant subspace of A contains an eigenvector of A.*

(ii) *Every non-trivial invariant subspace $\mathscr{M}$ of A is spanned by the eigenvectors of $A|\mathscr{M}$.*

(iii) *A is completely normal.*

Proof. (i) implies (ii): Let $\mathscr{M} \in Lat\, A$, and let $\mathscr{N}$ be the smallest subspace of $\mathscr{M}$ which contains all the eigenvectors of $A|\mathscr{M}$. Then clearly $\mathscr{N} \in Lat\, A$, and it follows from Theorem 1.2 that $\mathscr{N} \in Lat\, A^*$. Thus $\mathscr{M} \cap \mathscr{N}^\perp \in Lat\, A$. It follows from (i) that $\mathscr{M} \cap \mathscr{N}^\perp = \{0\}$.

(ii) implies (iii): Theorem 1.2 and (ii) imply that every invariant subspace of A is reducing.

(iii) implies (i): Let $\mathscr{M}$ be a non-trivial invariant subspace of A. Then, since the eigenvectors of A span $\mathscr{H}$, there exists an eigenvector x of A that is not orthogonal to $\mathscr{M}$. Let $Ax = \lambda x$ and write $x = x_1 + x_2$, with $x_1 \in \mathscr{M}$ and $x_2 \in \mathscr{M}^\perp$. Then $x_1 \neq 0$ and $\lambda(x_1 + x_2) = \lambda x = Ax = A(x_1 + x_2) = Ax_1 + Ax_2$. Therefore $Ax_1 - \lambda x_1 = Ax_2 - \lambda x_2$, and, since $Ax_1 - \lambda x_1 \in \mathscr{M}$ and $Ax_2 - \lambda x_2 \in \mathscr{M}^\perp$, it follows that $Ax_1 = \lambda x_1$. Hence $\mathscr{M}$ contains the eigenvector x_1. $\square$

There exist diagonable normal operators that are not completely normal—(see Wermer [1], which also contains some other information about completely normal operators).

1.9 Additional Propositions

Proposition 1.1 (Polar Decomposition). (A *partial isometry* is an operator that is isometric on the orthocomplement of its nullspace.)

If A is any operator, then there exist a unique partial isometry V and a unique positive operator P such that $A = VP$ and V and P have the same nullspace. If A is invertible, then V is unitary.

Proposition 1.2. If λ is an isolated point of the spectrum of a normal operator, then λ is an eigenvalue.

Proposition 1.3. If A is normal and $\mathcal{M}$ is an invariant subspace with an invariant complement, then $\mathcal{M}$ reduces A.

Proposition 1.4. Every spectral measure is *regular*, in the sense that for each Borel set S, $E(S)$ is the projection onto the smallest subspace containing the union of the ranges of the projections $E(T)$ for T a closed subset of S.

Proposition 1.5. Similar normal operators are unitarily equivalent.

Proposition 1.6. The set of invertible operators is arcwise connected.

Proposition 1.7. The restriction of a normal operator to an invariant subspace is normal if and only if the subspace is reducing.

Proposition 1.8. (An operator A is *hyponormal* if $\|A^*x\| \le \|Ax\|$ for all x.)

(i) The norm of a hyponormal operator is equal to its spectral radius.

(ii) A compact hyponormal operator is normal.

Proposition 1.9. If A is normal, then $\sigma(A) = \Pi(A)$.

Proposition 1.10. A polynomially compact normal operator is diagonable.

Proposition 1.11. A unitary operator U is completely normal if and only if there does not exist a reducing subspace $\mathcal{M}$ of U such that $U|\mathcal{M}$ is a bilateral shift; (shifts are defined in Chapter 3).

Proposition 1.12. A normal operator is unitarily equivalent to a multiplication operator on $\mathcal{L}^2(0,1)$ if and only if it has no finite-dimensional eigenspaces. In particular, the only compact multiplication operator on $\mathcal{L}^2(0,1)$ is 0.

Proposition 1.13. If n is an integer greater than 1 and A is a normal operator other than 0, then A has uncountably many normal n^{th} roots; if A is positive, it has a unique positive n^{th} root.

Proposition 1.14. If (X, μ) is a σ-finite measure space, and if $\mathcal{L}^\infty$ is the subalgebra of $\mathcal{B}(\mathcal{L}^2(X, \mu))$ consisting of all operators of the form M_ϕ for $\phi \in \mathcal{L}^\infty(X, \mu)$, then $\mathcal{L}^\infty$ is a maximal abelian subalgebra of $\mathcal{B}(\mathcal{L}^2(X, \mu))$.

Proposition 1.15. If A is a normal operator, then there exist a compact subset X of $\mathbb{C}$ and a Borel measure μ on X such that A is unitarily equivalent to M_ϕ acting on $\mathscr{L}^2(X,\mu)$ for some $\phi \in \mathscr{L}^\infty(X,\mu)$.

Proposition 1.16. If E is a spectral measure, then $E(S_1 \cap S_2) = E(S_1)E(S_2)$ for each pair $\{S_1, S_2\}$ of Borel subsets of $\mathbb{C}$.

1.10 Notes and Remarks

A discussion of the early development of the spectral theorem by Hilbert, F. Riesz, von Neumann, Stone and others can be found in Dunford-Schwartz ([2], pp. 926—927). The version of the spectral theorem given in Theorem 1.6 is due to Segal [1]; the exposition in the text of the proofs of Theorems 1.6 and 1.12 is strongly influenced by the treatment of the Hermitian case in Halmos [5]. Theorem 1.15 is taken from Halmos [2]. Theorem 1.16 (or the equivalent Corollary 1.18) was proven by Fuglede [1], answering a question of von Neumann [2]. Halmos [6] contains a proof very different from Fuglede's, (also to be found in Halmos [2]), and a very elegant proof was discovered by Rosenblum [2]. Dunford [1] generalized Theorem 1.16 to the case of spectral operators on Banach spaces, and the proof presented in the text (which applies to Dunford's version too) is from Radjavi-Rosenthal [5]. Corollary 1.19 was found by Putnam [1], who modified Fuglede's proof of Theorem 1.16; the proof presented in the text is due to Berberian [3].

An interesting discussion of spectral measures and a derivation of the spectral theorem for normal operators from the (more transparent) case of Hermitian operators is given in Berberian [2]. Halmos [2] contains a thorough discussion of the theory of spectral multiplicity, obtaining the unitary equivalence classes of normal operators in terms of multiplicity functions. A proof of the spectral theorem by Banach algebra techniques (via the Gelfand-Naimark theorem) can be found in Dunford-Schwartz [2]. The proof of Theorem 1.20 given above is due to M. Schreiber, and appears in Brown-Halmos [1].

The results about completely normal operators presented in § 1.8 and Proposition 1.11 are from Wermer [1]; Scroggs [1] contains some additional information. Proposition 1.3 is an unpublished result of R. G. Douglas and Carl Pearcy. Proposition 1.8 was independently discovered by Stampfli [1], Ando [1], and Berberian [4], and Proposition 1.12 is from Brown-Halmos [1]. An elementary version of one of Wermer's examples is in Dowson [1].

Chapter 2. Analytic Functions of Operators

We have observed in Section 1.7 that there is a very rich functional calculus for normal operators. We now return to the study of arbitrary bounded operators on Hilbert space and investigate the possibility of defining functions of such an operator. Let A be a bounded operator on $\mathscr{H}$. We already have a definition of $p(A)$ for any complex polynomial p. In this chapter we shall show how to define $f(A)$ whenever f is a function analytic on an open set containing $\sigma(A)$. The functional calculus which we shall develop is not as rich as the functional calculus for normal operators; nonetheless it has important applications to the study of invariant subspaces. It also has many other important applications which we shall not discuss; (e.g., see Proposition 2.3).

2.1 The Functional Calculus

The motivation for the definitions below is the classical Cauchy integral formula of complex analysis. We therefore first discuss contour integrals of functions with values in a Banach space; (in our subsequent work the functions will have values in the Banach algebra $\mathscr{B}(\mathscr{H})$).

In the following $\mathscr{Y}$ will denote a complex Banach space.

Definition. Let F and α be functions mapping the closed interval $[a,b]$ into $\mathscr{Y}$ and $\mathbb{C}$ respectively. For any partition $P = \{t_0, t_1, \ldots, t_n\}$ of $[a,b]$ (with $t_0 = a$ and $t_n = b$), a *Riemann-Stieltjes sum* associated with P is a sum of the form $S(F, \alpha, P) = \sum_{i=1}^{n} F(t_i')(\alpha(t_i) - \alpha(t_{i-1}))$, where $t_i' \in [t_{i-1}, t_i]$ for each i. Each Riemann-Stieltjes sum is an element of $\mathscr{Y}$. The element R of $\mathscr{Y}$ is the *Riemann-Stieltjes integral* of F with respect to α if for every $\varepsilon > 0$ there exists a partition P_ε such that $\|S(F, \alpha, P) - R\| < \varepsilon$ whenever $S(F, \alpha, P)$ is a Riemann-Stieltjes sum associated with a partition P which refines P_ε. If R is the Riemann-Stieltjes integral of F with respect to α, we write

$$R = \int_a^b F(t) \, d\alpha(t).$$

Theorem 2.1. *If $F:[a,b]\to\mathcal{Y}$ is continuous and $\alpha:[a,b]\to\mathbb{C}$ is a function of bounded variation, then $\int_a^b F(t)\,d\alpha(t)$ exists.*

Proof. Any of the standard proofs of this theorem for the case where $\mathcal{Y}$ is the set of real or complex numbers (cf. T. Apostol [1]) apply to the present more general situation. ☐

We shall be primarily concerned with contour integrals.

Definition. If $F:U\to\mathcal{Y}$ is a continuous function defined on the open subset U of $\mathbb{C}$, and if $C=\{\alpha(t):t\in[a,b]\}$ is a rectifiable arc contained in U , then the *line integral of F along C* is

$$\int_C F\,dz = \int_a^b F(\alpha(t))\,d\alpha(t).$$

(Note that $\alpha(t)$ is a continuous function of bounded variation since C is a rectifiable arc.)

Definition. If U is an open subset of $\mathbb{C}$, then $F:U\to\mathcal{Y}$ is *analytic on U* if, whenever $z_0\in U$, the limit of $(z-z_0)^{-1}[F(z)-F(z_0)]$ as z approaches z_0 exists; i.e., there is a vector $y\in\mathcal{Y}$ with the property that for each $\varepsilon>0$ there exists a $\delta>0$ such that if $|z-z_0|<\delta$ and $z\neq z_0$, then

$$\left\|\left(\frac{1}{z-z_0}[F(z)-F(z_0)]\right)-y\right\|<\varepsilon.$$

If K is any subset of $\mathbb{C}$, then $F:K\to\mathcal{Y}$ is *analytic on K* if there exists an open set U containing K such that f is defined on U and is analytic.

Theorem 2.2. *If C is a simple, closed, rectifiable curve, and if F is a $\mathcal{Y}$ -valued function analytic on an open set U containing C and the interior of C , then $\int_C F\,dz=0$.*

Proof. If ϕ is any continuous linear functional on $\mathcal{Y}$, then $\phi\circ F$ is an analytic complex-valued function on U . Also, the continuity of ϕ implies that $\phi\left(\int_C F\,dz\right)=\int_C(\phi\circ F)\,dz$. But Cauchy's theorem from complex analysis implies that $\int_C(\phi\circ F)\,dz=0$. Therefore $\phi\left(\int_C F\,dz\right)=0$ for every continuous linear functional ϕ on $\mathcal{Y}$. The Hahn-Banach theorem implies that $\int_C F\,dz=0$. ☐

Definition. If $C=C_1\cup\cdots\cup C_n$, where each C_i is a simple, closed, rectifiable curve and where C_i and C_j are disjoint if $i\neq j$, then C is a *path.* If C is a path and if F is a continuous $\mathscr{Y}$-valued function whose domain contains C, then we define $\int_C F\,dz$ by $\int_C F\,dz = \int_{C_1} F\,dz + \cdots + \int_{C_n} F\,dz$.

We now consider the special case where $\mathscr{Y} = \mathscr{B}(\mathscr{H})$. If $A \in \mathscr{B}(\mathscr{H})$, then the resolvent of A, $R_z = (z - A)^{-1}$, is easily shown to be continuous on $\rho(A)$; (Proposition 2.1). The Cauchy integral theorem suggests the definition

$$f(A) = \frac{1}{2\pi i} \int_C f(z) R_z \, dz$$

for analytic functions f, where C is a suitably chosen path. It will be essential to allow the possibility that f be analytic on a disconnected open set, and this makes the description of the choice of C somewhat involved.

Definition. An open subset S of $\mathbb{C}$ is a *Cauchy domain* if it has finitely many components, the closures of its components are pairwise disjoint, and its boundary, ∂S, is a path. We shall assume that ∂S is *positively oriented with respect to* S; i.e., if $\partial S = C_1 \cup \cdots \cup C_n$, then C_i is oriented in a counterclockwise or clockwise direction depending upon whether the component of S whose closure meets C_i is inside or outside C_i.

Lemma 2.3. *If K is a compact subset of $\mathbb{C}$ and U is an open set containing K, then there exists a Cauchy domain S such that $K \subset S$ and $\bar{S} \subset U$.*

Proof. Since K is compact, it is at a positive distance, k, from the closed set $\mathbb{C} \setminus U$. Cover K by open disks with centres in K and radii $k/2$. Some finite collection $\{D_1, \ldots, D_n\}$ of these disks covers K; let $S = \bigcup_{i=1}^{n} D_i$. Then S is obviously a Cauchy domain, (after ∂S, a finite union of arcs of the circles ∂D_i, is suitably oriented), $K \subset S$ and $\bar{S} \subset U$. $\quad\square$

Lemma 2.4 *(The Resolvent Equation).* *If z and z' are in $\rho(A)$, then $R_z - R_{z'} = (z' - z) R_z R_{z'}$.*

Proof. Clearly $(z - A)[R_z - R_{z'}](z' - A) = (z' - A) - (z - A) = z' - z$. Multiplying both sides of this equation by R_z on the left and $R_{z'}$ on the right gives the result. $\quad\square$

Corollary 2.5. *The function R_z is analytic on $\rho(A)$.*

Proof. Fix $z_0 \in \rho(A)$. Then, by Lemma 2.4, for any $z \in \rho(A) \setminus \{z_0\}$

$$\frac{1}{z - z_0}(R_z - R_{z_0}) = -R_z R_{z_0}.$$

The continuity of R_z gives

$$\lim_{z \to z_0} \left(\frac{1}{z - z_0} (R_z - R_{z_0}) \right) = -R_{z_0}^2. \quad \square$$

Definition. If $A \in \mathscr{B}(\mathscr{H})$ and f is a complex-valued analytic function on $\sigma(A)$, then we define $f(A)$ by $f(A) = (2\pi i)^{-1} \int_{\partial S} f(z) R_z dz$, where S is any Cauchy domain containing $\sigma(A)$ whose closure is contained in the domain of f, (such an S exists by Lemma 2.3), and where $R_z = (z - A)^{-1}$ for $z \in \rho(A)$.

We must show that $f(A)$ does not depend upon the choice of S.

Lemma 2.6. *If S_1 and S_2 are Cauchy domains containing $\sigma(A)$, and if f is analytic on $\overline{S}_1 \cup \overline{S}_2$, then*

$$\frac{1}{2\pi i} \int_{\partial S_1} f(z) R_z dz = \frac{1}{2\pi i} \int_{\partial S_2} f(z) R_z dz .$$

Proof. Lemma 2.3 implies that there is a Cauchy domain S_0 containing $\sigma(A)$ with $\overline{S}_0 \subset S_1 \cap S_2$. We shall show that

$$\int_{\partial S_1} f(z) R_z dz = \int_{\partial S_0} f(z) R_z dz = \int_{\partial S_2} f(z) R_z dz .$$

We prove the first equation only; the proof of the second equation is exactly the same.

It follows from Corollary 2.5 that $f(z) R_z$ is analytic on an open set containing $\overline{S}_1 - \sigma(A)$. Suppose that C_1 and C_2 are simple, closed, rectifiable, positively oriented curves such that C_1 is contained inside C_2 and $f(z) R_z$ is analytic on an open set containing the closure of the region between C_1 and C_2. Then $\int_{C_1} f(z) R_z dz = \int_{C_2} g(z) R_z dz$; (introduce arcs in the usual way to reduce to Theorem 2.2).

These considerations, applied to the components of S_0 and S_1, imply that $\int_{\partial S_1} f(z) R_z dz = \int_{\partial S_0} f(z) R_z dz$. $\square$

Theorem 2.7. *If f and g are analytic on $\sigma(A)$, then $f(A)g(A) = (2\pi i)^{-1} \int_{\partial S} f(z)g(z) R_z dz$, where S is any Cauchy domain containing $\sigma(A)$ whose closure is contained in the domains of f and g.*

Proof. Choose a Cauchy domain S_0 containing $\sigma(A)$ such that $\overline{S}_0 \subset S$, by Lemma 2.3. Then

$$f(A)g(A) = \left(\frac{1}{2\pi i} \int_{\partial S_0} f(z)R_z\,dz\right)\left(\frac{1}{2\pi i} \int_{\partial S} g(w)R_w\,dw\right)$$

$$= \left(\frac{1}{2\pi i}\right)^2 \int_{\partial S_0}\int_{\partial S} f(z)g(w)R_z\,R_w\,dw\,dz$$

$$= \left(\frac{1}{2\pi i}\right)^2 \int_{\partial S_0}\int_{\partial S} f(z)g(w)\frac{1}{z-w}[R_w - R_z]\,dw\,dz,$$

by Lemma 2.4.

Since $w \in \partial S$ and $\bar{S}_0 \subset S$, the function $g(w)(z-w)^{-1}R_z$ is an analytic function of w for each fixed z, and thus, by Theorem 2.2, we have

$$f(A)g(A) = \left(\frac{1}{2\pi i}\right)^2 \int_{\partial S}\left(\int_{\partial S_0} \frac{f(z)}{z-w}\,dz\right)g(w)R_w\,dw.$$

The ordinary Cauchy integral formula gives

$$\int_{\partial S_0} \frac{f(z)}{z-w}\,dz = (2\pi i)\,f(w).$$

Hence $f(A)g(A) = \dfrac{1}{2\pi i}\displaystyle\int_{\partial S} f(w)g(w)R_w\,dw.$ $\square$

Corollary 2.8. *For each fixed operator A the mapping from the algebra of functions analytic on $\sigma(A)$ into $\mathcal{B}(\mathcal{H})$ defined by $f \to f(A)$ is an algebra homomorphism. The constant function 1 corresponds to the identity operator under this mapping.*

Proof. The first assertion follows immediately from Theorem 2.7. To prove the second statement let C be the circle $\alpha(t) = r\,e^{it}$, where $r > \|A\|$. Then the operator corresponding to the constant function 1 is

$$\frac{1}{2\pi i}\int_C R_z\,dz = \frac{1}{2\pi i}\int_C \frac{1}{z}\left(1 + \frac{A}{z} + \frac{A^2}{z^2} + \cdots\right)dz =$$

$$\frac{1}{2\pi i}\int_C \frac{1}{z}\,dz + \sum_{n=1}^{\infty} \frac{1}{2\pi i}A^n \int_C \frac{dz}{z^{n+1}} = 1 + 0 = 1.$$ $\square$

Theorem 2.9. *If f is analytic on $\sigma(A)$, then $f(A)$ commutes with every operator which commutes with A.*

Proof. Suppose that $AB = BA$. Then, for each z, $(z-A)B = B(z-A)$. If $z \in \rho(A)$, then multiplying both sides of this equation by R_z on the left and on the right gives $BR_z = R_z B$. Thus, since $f(A)$ is a uniform limit of finite linear combinations of terms of the form R_λ, we conclude that $Bf(A) = f(A)B$. ☐

2.2 The Riesz Decomposition Theorem

Theorem 2.10. *If $\sigma(A) = \sigma_1 \cup \sigma_2$, where σ_1 and σ_2 are disjoint non-empty closed sets, then A has a complementary pair $\{\mathcal{M}_1, \mathcal{M}_2\}$ of non-trivial invariant subspaces such that $\sigma(A|\mathcal{M}_1) = \sigma_1$ and $\sigma(A|\mathcal{M}_2) = \sigma_2$.*

Proof. Since σ_1 and σ_2 are disjoint compact sets, there exist disjoint open sets U_1 and U_2 such that $\sigma_1 \subset U_1$ and $\sigma_2 \subset U_2$. Let f_1 denote the function which is identically 1 on U_1 and identically 0 on U_2, and let f_2 denote the function which is 0 on U_1 and 1 on U_2. Then, by Corollary 2.8,

$$f_1(A) + f_2(A) = 1 \quad \text{and} \quad [f_1(A)]^2 = f_1(A), \quad [f_2(A)]^2 = f_2(A).$$

It follows that $f_1(A)$ and $f_2(A)$ are complementary (not necessarily Hermitian) projections; denote their ranges by $\mathcal{M}_1$ and $\mathcal{M}_2$ respectively. By Theorems 2.9 and 0.2, $\mathcal{M}_1$ and $\mathcal{M}_2$ are invariant under A.

Let $A_1 = A|\mathcal{M}_1$ and $A_2 = A|\mathcal{M}_2$. We must show that $\sigma(A_1) = \sigma_1$ and $\sigma(A_2) = \sigma_2$. We first show that $\sigma(A_1) \subset \sigma_1$. For this suppose that λ is not in σ_1. Choose a Cauchy domain S such that $\sigma_1 \subset S$, $\bar{S} \subset U_1$, and λ is outside all the closed curves bounding S. Note that $\sigma(A_1) \subset \sigma(A)$ by Proposition 0.3, and thus $\sigma(A_1) \cap \partial S = \emptyset$.

Let R'_z denote $(z - A_1)^{-1}$, (as an operator on $\mathcal{M}_1$). We claim that the operator $(2\pi i)^{-1} \int_{\partial S} R'_z(\lambda - z)^{-1} dz$ is the inverse of $\lambda - A_1$. For this first note that

$$(\lambda - A_1)(z - A_1)^{-1} = (z + \lambda - z - A_1)(z - A_1)^{-1}$$
$$= (z - A_1)(z - A_1)^{-1} + (\lambda - z)(z - A_1)^{-1}$$
$$= 1 + (\lambda - z) R'_z.$$

Thus

$$(\lambda - A_1)\left(\frac{1}{2\pi i} \int_{\partial S} R'_z(\lambda - z)^{-1} dz\right) = \frac{1}{2\pi i} \int_{\partial S} (\lambda - A_1)(z - A_1)^{-1}(\lambda - z)^{-1} dz$$

$$= \frac{1}{2\pi i} \int_{\partial S} (\lambda - z)^{-1} [1 + (\lambda - z) R'_z] dz$$

$$= \frac{1}{2\pi i} \int_{\partial S} \frac{1}{\lambda - z} dz + \frac{1}{2\pi i} \int_{\partial S} R'_z dz.$$

Now $(2\pi i)^{-1}\int_{\partial S}(\lambda-z)^{-1}dz=0$, since λ is outside all the closed curves in ∂S. Also, for $z\in\rho(A)=\rho(A)\cap\rho(A_1)$, $\mathcal{M}_1$ is in $Lat(z-A)^{-1}$ and $(z-A_1)^{-1}$ $=(z-A)^{-1}|\mathcal{M}_1$; (since, if $x\in\mathcal{M}_1$, $(z-A_1)^{-1}x=u$ and $(z-A)^{-1}x=v$, then $u\in\mathcal{M}_1$ a priori, and multiplying both sides of both equations by $z-A$ on the left gives $x=(z-A)u$ and $x=(z-A)v$, or $u=v$).

Therefore

$$(\lambda-A_1)\left(\frac{1}{2\pi i}\int_{\partial S}R'_z(\lambda-z)^{-1}dz\right)=\frac{1}{2\pi i}\int_{\partial S}(z-A_1)^{-1}dz$$

$$=\frac{1}{2\pi i}\int_{\partial S}(z-A)^{-1}|\mathcal{M}_1\,dz$$

$$=f_1(A)|\mathcal{M}_1=1|\mathcal{M}_1.$$

We have shown that if λ is not in σ_1, then λ is not in $\sigma(A_1)$; thus $\sigma(A_1)\subset\sigma_1$. In exactly the same way it can be shown that $\sigma(A_2)\subset\sigma_2$. But $\sigma(A_1)\cup\sigma(A_2)=\sigma(A)=\sigma_1\cup\sigma_2$, by Proposition 0.3. Thus $\sigma(A_1)=\sigma_1$ and $\sigma(A_2)=\sigma_2$. This also implies that $\mathcal{M}_1$ and $\mathcal{M}_2$ are non-trivial. □

Corollary 2.11. *If $\sigma(A)$ is disconnected, then A has a complementary pair of non-trivial hyperinvariant subspaces.*

Proof. If $\sigma(A)$ is disconnected, then, since $\sigma(A)$ is closed, there exist disjoint non-empty closed sets σ_1 and σ_2 such that $\sigma(A)=\sigma_1\cup\sigma_2$. Let $f_1(A)$ and $f_2(A)$ be the projections constructed in the proof of Theorem 2.10. Then, by Theorem 2.9, $f_1(A)$ and $f_2(A)$ commute with every operator that commutes with A, and thus their ranges are invariant under every operator which commutes with A. □

2.3 Invariant Subspaces of Analytic Functions of Operators

If f is a polynomial, then clearly $Lat\,A\subset Lat\,f(A)$. This need not be the case for arbitrary analytic functions f. For example, let A denote the bilateral shift of multiplicity 1, (defined in Chapter 3), and let $f(z)=z^{-1}$. Then f is analytic on $\sigma(A)$ and there exist many invariant subspaces of A that are not invariant under $f(A)=A^{-1}=A^*$. Under certain additional hypotheses, however, $\mathcal{M}\in Lat\,A$ implies $\mathcal{M}\in Lat\,f(A)$.

Theorem 2.12. *If $\mathcal{M}\in Lat\,A$, and if f is analytic on $\sigma(A)\cup\sigma(A|\mathcal{M})$, then $\mathcal{M}\in Lat\,f(A)$ and $f(A|\mathcal{M})=f(A)|\mathcal{M}$.*

Proof. Choose, by Lemma 2.3, a Cauchy domain S containing $\sigma(A) \cup \sigma(A|\mathcal{M})$ such that f is analytic on $\overline{S}$. Then

$$f(A) = \frac{1}{2\pi i} \int_{\partial S} (z - A)^{-1} f(z) dz.$$

It follows from the fact that $\partial S \subset \rho(A) \cap \rho(A|\mathcal{M})$, (as in the proof of Theorem 2.10), that $\mathcal{M} \in Lat(z - A)^{-1}$ for $z \in \partial S$. Thus $\mathcal{M} \in Lat f(A)$, since $f(A)$ is a uniform limit of linear combinations of the form $\sum_{j=1}^{n} \alpha_j (z_j - A)^{-1}$ with $z_j \in \partial S$. □

Corollary 2.13. *If f is analytic on $\eta(\sigma(A))$, then $Lat A \subset Lat f(A)$.*

Proof. Let $\mathcal{M} \in Lat A$. Then, by Theorem 0.8, $\sigma(A|\mathcal{M}) \subset \eta(\sigma(A))$. Thus f is analytic on $\sigma(A) \cup \sigma(A|\mathcal{M})$ and $\mathcal{M} \in Lat f(A)$ by Theorem 2.12. □

Theorem 2.14. *If f is analytic and one-to-one on an open set containing $\eta(\sigma(A))$, then $Lat A = Lat f(A)$.*

Proof. First, $Lat A \subset Lat f(A)$ by Corollary 2.13. Let U be a bounded open set containing $\eta(\sigma(A))$ and such that $\overline{U}$ is contained in the given open set. Then $f(U)$ is an open set, and $f(U)$ contains $\sigma(f(A))$, (Proposition 2.5). Also, since $f|\overline{U}$ is a homeomorphism, $f(\overline{U}) \supset \eta(\sigma(f(A)))$. Let g be the inverse of f; i.e., g is an analytic function taking $f(U)$ into U such that $g(f(z)) = z$ for $z \in U$. Then $g(f(A)) = A$ (Proposition 2.6), and thus, by Corollary 2.13, $Lat f(A) \subset Lat A$. □

Corollary 2.15. *If λ is in the unbounded component of $\rho(A)$, then $Lat A = Lat(\lambda - A)^{-1}$.*

Proof. Under these hypotheses the function $f(z) = (\lambda - z)^{-1}$ is analytic on $\eta(\sigma(A))$. This function is one-to-one, and $f(A) = (\lambda - A)^{-1}$, (Proposition 2.6). Hence Theorem 2.14 gives the result. □

Corollary 2.15 is used in proving a result about hyperinvariant subspaces—see Section 6.4.

2.4 Additional Propositions

Proposition 2.1. *If $A \in \mathcal{B}(\mathcal{H})$, then $R_z = (z - A)^{-1}$ is a continuous function from $\rho(A)$ into $\mathcal{B}(\mathcal{H})$.*

Proposition 2.2. *If $f(z) = \sum_{n=0}^{\infty} a_n z^n$ has radius of convergence larger than $\|A\|$, then $f(A) = \sum_{n=0}^{\infty} a_n A^n$.*

Proposition 2.3. If $\eta(\sigma(A))$ does not contain 0, then A has a logarithm, (i.e., there exists an operator B such that $\exp B = A$). Thus A has roots of all orders, (i.e., for each positive integer n there is an operator B such that $B^n = A$). In particular, every invertible operator on a finite-dimensional space has a logarithm and roots of all orders.

Proposition 2.4. There exist invertible operators without square roots.

Proposition 2.5. (Spectral Mapping Theorem). If f is analytic on $\sigma(A)$, then $\sigma(f(A)) = \{f(z) : z \in \sigma(A)\}$.

Proposition 2.6. If f is analytic on $\sigma(A)$ and g is analytic on $\sigma(f(A))$, then $g(f(A)) = (g \circ f)(A)$.

Proposition 2.7. If A is an operator on a finite-dimensional space, and if f is any function analytic on $\sigma(A)$, then $f(A)$ is a polynomial in A.

Proposition 2.8. Corollary 2.15 has a simple direct proof using power series and avoiding the functional calculus.

Proposition 2.9. If A is Hermitian, then there exists a unitary operator U such that $Lat\, A = Lat\, U$.

Proposition 2.10. If $A = \int \lambda\, dE_\lambda$ and f is analytic on $\sigma(A)$, then $f(A) = \int f(\lambda)\, dE_\lambda$.

Proposition 2.11. If f is a function defined on an open set U in the complex plane with values in a Banach space, and if for each continuous linear functional ϕ the function $z \to \phi(f(z))$ is analytic on U, then f is analytic on U.

Proposition 2.12. If f is an analytic function mapping an open connected subset $\mathcal{D}$ of $\mathbb{C}$ into a Banach space, then $\|f(z)\|$ does not attain a maximum on $\mathcal{D}$ unless $\|f(z)\|$ is a constant function.

Proposition 2.13. If there exists a non-constant analytic function f such that $f(A)$ is compact, then A is polynomially compact.

2.5 Notes and Remarks

The functional calculus presented in this chapter was initially developed by F. Riesz [1], who used it to obtain Theorem 2.10. It was extended and applied by a number of subsequent authors, including Lorch [1], Gelfand [1], Dunford ([2], [3]) and Taylor ([2], [3]). In particular, Propositions 2.5 and 2.6 are due to Dunford. Our exposition of this material was influenced by Dunford-Schwartz [1], Riesz-Nagy [1] and

Taylor [1]. Taylor [1] considers the functional calculus for possibly unbounded operators, and Hille-Phillips [1] applies the functional calculus to the study of semigroups. The results of §2.3 are from Crimmins-Rosenthal [1], except for Corollary 2.15 which was discovered earlier by Sarason [1], (cf. Proposition 2.8). Herrero-Salinas [1] contains some further results. Proposition 2.4 is due to Halmos-Lumer-Schäffer [1]; additional information about rootless operators is contained in Halmos-Lumer [1], Deckard-Pearcy ([1], [2]) and Schäffer [1]. Proposition 2.9 was observed in Rosenthal [7]. Proposition 2.11 can be found in Taylor [1] and Proposition 2.12 in Dieudonné [1].

Chapter 3. Shift Operators

This chapter is devoted to the study of two different kinds of operators: bilateral shifts, which are unitary operators, and unilateral shifts, which are restrictions of bilateral shifts to certain invariant subspaces. The results of this chapter have much wider applicability than this would suggest however, for, as we shall see, every operator is unitarily equivalent to a multiple of a "part" of the adjoint of a unilateral shift.

Most of the chapter is concerned with describing the invariant and reducing subspaces of shifts. These results lead to a reformulation of the invariant subspace problem in terms of a problem about factoring operator-valued analytic functions.

3.1 Shifts of Multiplicity 1

Definition. Let $\mathscr{H}$ have an orthonormal basis $\{e_n\}_{n=0}^{\infty}$. Then the unique operator S such that $Se_n = e_{n+1}$ for $n = 0, 1, 2, \ldots$ is called the *unilateral shift of multiplicity* 1, (or simply the *unilateral shift*).

If $x = \sum_{n=0}^{\infty} c_n e_n$ is any vector in $\mathscr{H}$, then $Sx = \sum_{n=0}^{\infty} c_n e_{n+1}$. Thus S is an isometry, with range $\bigvee_{n=1}^{\infty} \{e_n\}$. A trivial computation shows that S^* is characterized by the equations $S^* e_0 = 0$ and $S^* e_n = e_{n-1}$ for $n \geq 1$.

Definition. Let $\mathscr{H}$ have an orthonormal basis $\{f_n\}_{n=-\infty}^{\infty}$. Then the unique operator U such that $Uf_n = f_{n+1}$ for $n = 0, \pm 1, \pm 2, \ldots$ is called the *bilateral shift of multiplicity* 1, (or the *bilateral shift*).

It is easily verified that $U^* f_n = f_{n-1}$ for $n = 0, \pm 1, \pm 2, \ldots$. Thus U is a unitary operator, and $U | \bigvee_{n=0}^{\infty} \{f_n\}$ is (unitarily equivalent to) S.

For the rest of this chapter let C denote the unit circle $\{z : |z| = 1\}$ and D the open unit disc $\{z : |z| < 1\}$.

Theorem 3.1. *Let S and U denote the unilateral and bilateral shifts respectively. Then* $\sigma(S) = \sigma(S^*) = \Pi(S^*) = \bar{D}$, $\sigma(U) = \Pi(S) = C$, $\Pi_0(U^*) = \Pi_0(S) = \Pi_0(U) = \emptyset$, *and* $\Pi_0(S^*) = D$.

Proof. First $\sigma(U) \subset C$ and $\sigma(S) \subset \overline{D}$, since U is unitary and $\|S\| = 1$. Now if $\lambda \in D$, then $\sum_{n=0}^{\infty} \lambda^n e_n \in \mathcal{H}$, and $S^*\left(\sum_{n=0}^{\infty} \lambda^n e_n\right) = \lambda \sum_{n=0}^{\infty} \lambda^n e_n$. Thus $D \subset \Pi_0(S^*)$. Since $\{\|S^{*n}x\|\} \to 0$ for all x, $\Pi_0(S^*) \subset D$, and therefore $\Pi_0(S^*) = D$. Hence $\Pi(S^*) = \overline{D} = \sigma(S^*)$. It follows that $\sigma(S) = \overline{D}$ too. Since S is a restriction of U, $\Pi(U) \supset \Pi(S)$, and since $\Pi(S) \supset C$ and $\Pi(U) \subset C$, it follows that $\Pi(U) = \Pi(S) = C$. The fact that $\Pi_0(U^*)$, $\Pi_0(S)$, and $\Pi_0(U)$ are all empty is easily verified. $\quad\square$

We shall find it very useful to consider particular concrete representations of U and S. Let $\mathcal{L}^2$ denote $\mathcal{L}^2(C, \mu)$, where μ is normalized Lebesgue measure on the circle; (i.e., $\mu(C) = 1$). For each integer n let e_n denote the function $e_n(z) = z^n$. Then $\{e_n\}_{n=-\infty}^{\infty}$ is an orthonormal basis for $\mathcal{L}^2$; (making the change of variable $z = e^{i\theta}$ shows that this is the familiar orthonormal basis associated with classical Fourier series). The space $\mathcal{H}^2$ is, by definition, the subspace $\bigvee_{n=0}^{\infty} \{e_n\}$ of $\mathcal{L}^2$. Multiplication by $e_1(z)$ on $\mathcal{L}^2$ is (unitarily equivalent to) U, and $S = U | \mathcal{H}^2$. The invariant subspaces of S and U will be described in terms of these representations.

We first compute the commutants of U and S. Let $\mathcal{L}^{\infty}$ denote the set of all multiplication operators on $\mathcal{L}^2$, as in Section 1.6. Then $U \in \mathcal{L}^{\infty}$.

Theorem 3.2. *The commutant of U is $\mathcal{L}^{\infty}$.*

Proof. Since $\mathcal{L}^{\infty}$ is abelian, it is contained in the commutant of U. The operator U is multiplication by the independent variable on $\mathcal{L}^2(C, \mu)$; thus $M_\phi = \phi(U)$ for each $\phi \in \mathcal{L}^{\infty}(C, \mu)$, (as defined in Theorem 1.22), and every operator that commutes with U therefore commutes with $\mathcal{L}^{\infty}$. The result now follows from the fact that $\mathcal{L}^{\infty}$ is maximal abelian (Theorem 1.20). $\quad\square$

Definition. The space $\mathcal{H}^{\infty}$ is the subspace of $\mathcal{L}^{\infty}(C, \mu)$ consisting of all bounded measurable functions ϕ such that $\int \phi(z) z^{-n} d\mu = 0$ for $n < 0$; i.e., $\mathcal{H}^{\infty} = \mathcal{L}^{\infty}(C, \mu) \cap \mathcal{H}^2$.

If $\phi \in \mathcal{H}^{\infty}$, then $\mathcal{H}^2 \in \operatorname{Lat} M_\phi$, since obviously $\phi e_n \in \mathcal{H}^2$ for $n > 0$.

Theorem 3.3. *If $\phi \in \mathcal{H}^{\infty}$, then M_ϕ is the strong limit of a sequence of polynomials in U. Hence $M_\phi | \mathcal{H}^2$ is a strong limit of a sequence of polynomials in S.*

Proof. Let $\phi \in \mathcal{H}^{\infty}$, with Fourier series $\sum_{n=0}^{\infty} \alpha_n z^n$. By Fejer's Theorem (Hoffman [1]), the Cesaro means $\{\sigma_k\}$ of the Fourier series of ϕ converge to ϕ in $\mathcal{L}^2(C, \mu)$. Hence a subsequence $\{\sigma_{k_j}\}$ converges to ϕ a.e. (Halmos [1]). Moreover, each σ_k is a polynomial in (non-negative powers of) z, since the negative Fourier coefficients of ϕ vanish. Thus $M_{\sigma_k} = \sigma_k(U)$ for each k. Recall that $|\sigma_k(z)| \le |\phi(z)|$ a.e.

We claim that $\sigma_{k_j}(U)$ converges strongly to M_ϕ. For, if $f \in \mathscr{L}^2$, then

$$\|(M_\phi - \sigma_{k_j}(U))f\|^2 = \int |(\phi - \sigma_{k_j})f|^2 \, d\mu \, ,$$

and, by the Lebesgue dominated convergence theorem (Halmos [1]),
$\lim_{k_j \to \infty} \int |(\phi - \sigma_{k_j})f|^2 \, d\mu = 0$.

Thus M_ϕ is the strong limit of a sequence of polynomials in U; the second assertion of the theorem follows immediately from this fact. $\square$

Definition. An operator T on $\mathscr{H}^2$ is an *analytic Toeplitz operator* if there exists a $\phi \in \mathscr{H}^\infty$ such that $T = M_\phi | \mathscr{H}^2$.

Theorem 3.4. *The commutant of S is the algebra of analytic Toeplitz operators.*

Proof. It is obvious that every analytic Toeplitz operator commutes with S. For the converse, suppose that A is an operator on $\mathscr{H}^2$ such that $AS = SA$. We define an operator $\hat{A}$ on the subset of $\mathscr{L}^2$ consisting of all functions with at most finitely many non-zero negative Fourier coefficients by the formula $\hat{A} U^n f = U^n A f$ for $f \in \mathscr{H}^2$ and n any integer. The fact that A commutes with S implies that $\hat{A}$ is well-defined on this dense subset of $\mathscr{L}^2$. Also

$$\|\hat{A} U^n f\| = \|U^n A f\| = \|A f\| \leq \|A\| \, \|f\| = \|A\| \, \|U^n f\|$$

for all $f \in \mathscr{H}^2$. Thus $\hat{A}$ is bounded on this set and extends to a bounded linear operator commuting with U on all of $\mathscr{L}^2$. By Theorem 3.2 this extension has the form M_ϕ, and thus $A = M_\phi | \mathscr{H}^2$. Since $A e_0 = \phi$ is in $\mathscr{H}^2$, ϕ is in $\mathscr{H}^\infty$. $\square$

3.2 Invariant Subspaces of Shifts of Multiplicity 1

We first consider reducing subspaces, since they are easier to compute than invariant subspaces.

Theorem 3.5. *The unilateral shift is irreducible;* (i.e., *it has no non-trivial reducing subspaces*).

Proof. Suppose that S had a non-trivial reducing subspace; then S would be unitarily equivalent to an operator of the form $S_1 \oplus S_2$. Since S is an isometry, so are S_1 and S_2. If neither S_1 and S_2 were invertible the range of S could not have co-dimension 1. Thus one of S_1 and S_2, say S_1, is unitary. But then, if x is any non-zero vector in the domain of S_1, $\|S^{*n} x\| = \|S_1^{*n} x\| = \|x\|$ for all positive integers n, contradicting the fact that $\{S^{*n} x\} \to 0$ for all x. $\square$

The bilateral shift is a unitary operator, and therefore its spectral subspaces are reducing. In fact, the reducing subspaces of the bilateral shift are precisely its spectral subspaces.

Theorem 3.6. *A subspace $\mathcal{M}$ of $\mathscr{L}^2$ reduces U if and only if there exists a measurable subset N of C such that*
$$\mathcal{M} = \{f \in \mathscr{L}^2 : f = 0 \text{ a.e. on } N\}.$$

Proof. By Theorem 3.2 a projection P commutes with U if and only if $P \in \mathscr{L}^\infty$. The projections in $\mathscr{L}^\infty$ are multiplications by functions with range contained in $\{0, 1\}$, i.e., by characteristic functions. If P is multiplication by χ_M, then the range of P is
$$\{f \in \mathscr{L}^2 : f = 0 \text{ a.e. on the complement of } M\}. \quad \square$$

The subspaces $\bigvee\limits_{n=k}^{\infty} \{e_n\}$, with k an integer, are obviously invariant under U, and such subspaces with $k \geq 0$ are invariant under S. We shall see that *Lat U* and *Lat S* each have many other elements. It is instructive to begin the study of invariant subspaces with those that can be obtained by purely algebraic considerations. Note that U, S, and U^* do not have any finite-dimensional invariant subspaces; (since if $\mathcal{M}$ is a finite-dimensional subspace in *Lat A*, then $A|\mathcal{M}$ has an eigenvector, hence A has an eigenvector, and Theorem 3.1 states that U, S, and U^* have no eigenvectors). On the other hand, *Lat S** does have finite-dimensional elements. We proceed to determine them.

Lemma 3.7. *If $\lambda \in \Pi_0(S^*)$, then the nullspace of $(S^* - \lambda)^n$ is n-dimensional for each positive integer n.*

Proof. It is easily seen that every vector in the nullspace of $S^* - \lambda$ is a multiple of $\sum\limits_{n=0}^{\infty} \lambda^n e_n$, and thus the result is true for $n = 1$. It follows that the nullspace of $(S^* - \lambda)^n$ has dimension at most n.

If $\lambda \in \Pi_0(S^*)$ then, since $|\lambda| < 1$, the vectors
$$x_k^\lambda = \sum\limits_{m=k}^{\infty} \frac{m!}{(m-k)!} \lambda^{m-k} e_m$$

are in $\mathscr{H}^2$, (i.e., $\sum\limits_{m=k}^{\infty} |m(m-1)\dots(m-k+1)\lambda^{m-k}|^2 < \infty$). A computation shows that $\{x_k^\lambda\}_{k=0}^{n-1}$ is a linearly independent set contained in the nullspace of $(S^* - \lambda)^n$. (Note that if $\lambda = 0$ we must equate λ^0 to 1 in the above. This gives the obvious result that the nullspace of S^{*n} is $\bigvee\limits_{k=0}^{n-1} \{e_k\}$.) $\quad \square$

We now characterize the finite-dimensional members of *Lat S**, and thus also the members of *Lat S* which have finite co-dimension.

Theorem 3.8. *Let* $\lambda_1, \ldots, \lambda_m$ *be distinct complex numbers of modulus less than 1, and let* $n(1), \ldots, n(m)$ *be any positive integers whose sum is* n. *Then* $\bigvee_{j=1}^{m} \text{null}(S^* - \lambda_j)^{n(j)}$ *is an n-dimensional invariant subspace of* S^*. *Conversely, every finite-dimensional member of* $\text{Lat } S^*$ *has this form for a suitable choice of* $\lambda_1, \ldots, \lambda_m$ *and* $n(1), \ldots, n(m)$.

Proof. For each j the subspace $\text{null}(S^* - \lambda)^{n(j)}$ is in $\text{Lat } S^*$ and has dimension $n(j)$, by Lemma 3.7. It can easily be verified that $\bigvee_{j=1}^{m} \text{null}(S^* - \lambda_j)^{n(j)}$ is actually an algebraic direct sum, so that its dimension is n.

Now let $\mathcal{M}$ be any finite-dimensional invariant subspace of S^*. Let $\{\lambda_1, \ldots, \lambda_m\}$ be the distinct eigenvalues of $S^*|\mathcal{M}$. We use the Jordan canonical form and Lemma 3.7 (with $n = 1$) to obtain positive integers $n(1), \ldots, n(m)$ such that the dimension of the nullspace of $((S^*|\mathcal{M}) - \lambda_j)^{n(j)}$ is $n(j)$ for each j and

$$\mathcal{M} = \bigvee_{j=1}^{m} \text{null}((S^*|\mathcal{M}) - \lambda_j)^{n(j)} .$$

The nullspace of $(S^*|\mathcal{M} - \lambda_j)^{n(j)}$ is obviously contained in the nullspace of $(S^* - \lambda_j)^{n(j)}$; since, by Lemma 3.7, the dimension of $\text{null}(S^* - \lambda_j)^{n(j)}$ is also $n(j)$, these two nullspaces are equal.   □

We now give a description of the invariant subspaces of U; the reducing subspaces have already been described in Theorem 3.6.

Theorem 3.9. *A subspace* $\mathcal{M}$ *of* $\mathscr{L}^2$ *is a non-reducing invariant subspace for* U *if and only if there is a measurable function* ϕ *on* C, *with* $|\phi(z)| = 1$ *a.e., such that*

$$\mathcal{M} = \phi \mathscr{H}^2 = \{\phi f : f \in \mathscr{H}^2\} .$$

Furthermore, $\phi_1 \mathscr{H}^2 = \phi_2 \mathscr{H}^2$ *with* $|\phi_1| = |\phi_2| = 1$ *a.e. if and only if* ϕ_1/ϕ_2 *is equal a.e. to a constant function.*

Proof. First note that each such $\phi \mathscr{H}^2$ is a closed subspace: if $\{\phi f_n\} \to g$ then, since $|\phi(z)| = 1$ a.e., $\{f_n\} \to g/\phi$. Thus $g/\phi = h$ for some $h \in \mathscr{H}^2$, and $g = \phi h$. Also if f is in $\mathscr{H}^2$, then $U(\phi f) = \phi(U f)$ and therefore $\phi \mathscr{H}^2 \in \text{Lat } U$. Theorem 3.6 implies that $\phi \mathscr{H}^2$ is not reducing.

To prove the converse suppose that $\mathcal{M} \in \text{Lat } U$ and that $\mathcal{M} \notin \text{Lat } U^*$. Then $U\mathcal{M}$ is a proper subspace of $\mathcal{M}$, for $U\mathcal{M} = \mathcal{M}$ would give $U^* U \mathcal{M} = U^* \mathcal{M}$ and therefore $\mathcal{M} \in \text{Lat } U^*$; $(U^* U = 1)$. Choose a vector ϕ in $\mathcal{M} \ominus U\mathcal{M}$ such that $\|\phi\| = 1$. Then, since $U^n \mathcal{M} \subset U\mathcal{M}$ for all positive integers n, $\phi \perp U^n \mathcal{M}$ for $n > 0$. In particular $\phi \perp U^n \phi$, giving $\int |\phi|^2 z^n d\mu = 0$ for $n > 0$. Taking complex conjugates and using the fact that $\bar{z}^n = z^{-n}$ for $z \in C$, shows that $|\phi|^2$ is a function in $\mathcal{L}^1(C, \mu)$ with all its Fourier coefficients 0 except for the 0th. Hence $|\phi|^2$ is a constant a.e., and thus so is $|\phi|$. Since $\|\phi\| = 1$, it follows that $|\phi(z)| = 1$ a.e.

Consider $\bigvee\limits_{n=-\infty}^{\infty} \{\phi e_n\}$, where $e_n(z) = z^n$ for each integer n. This subspace obviously reduces U, and, since $|\phi e_1| > 0$ a.e., Theorem 3.6 implies that $\bigvee\limits_{n=-\infty}^{\infty} \{\phi e_n\} = \mathcal{L}^2$. Now $\bigvee\limits_{n=0}^{\infty} \{\phi e_n\} = \phi \mathcal{H}^2$ is contained in $\mathcal{M}$; thus to prove that $\mathcal{M} = \phi \mathcal{H}^2$ we need only show that $\phi e_n \perp \mathcal{M}$ for $n < 0$. If $f \in \mathcal{M}$ and $n < 0$, then

$$(f, \phi e_n) = \int f \, \bar{\phi} \bar{e}_n \, d\mu = \int f \, e_{-n} \bar{\phi} \, d\mu,$$

and this is 0 since $f e_{-n}$ is $U^{-n} f$, which is orthogonal to ϕ.

Suppose that $\phi_1 \mathcal{H}^2 = \phi_2 \mathcal{H}^2$ with $|\phi_1| = |\phi_2| = 1$ a.e. Then $\phi_1 e_0 = \phi_2 f$ for some $f \in \mathcal{H}^2$, and $\phi_1 g = \phi_2 e_0$ for some $g \in \mathcal{H}^2$. Thus $\phi_1/\phi_2 = f$ and $\phi_2/\phi_1 = g$. Therefore both ϕ_1/ϕ_2 and its complex conjugate ϕ_2/ϕ_1 are in $\mathcal{H}^2$, and we conclude that ϕ_1/ϕ_2 is a constant a.e. $\square$

Corollary 3.10. *If $f \in \mathcal{H}^2$ and f is 0 on a set of positive measure, then $f = 0$ a.e.*

Proof. Suppose that f is not the zero function and that $f = 0$ on M with $\mu(M) > 0$. Let $\mathcal{M} = \bigvee\limits_{n=0}^{\infty} \{f e_n\}$; then $\mathcal{M} \in \text{Lat } U$. If $g \in \mathcal{M}$ and $\{f_k\}$ is a sequence of linear combinations of $\{f e_n\}$ with $\int |g - f_k|^2 \, d\mu \to 0$, then $\int\limits_M |g|^2 \, d\mu = \int\limits_M |g - f_k|^2 \, d\mu \to 0$; hence $g = 0$ a.e. on M. If $f = \sum\limits_{n=0}^{\infty} \alpha_n e_n$ with $\alpha_k \neq 0$, then $(U^*)^{k+1} f$ is not in $\mathcal{M}$, (since it is not even in $\mathcal{H}^2$), and therefore $\mathcal{M}$ is not reducing for U. By Theorem 3.9 $\mathcal{M}$ has the form $\phi \mathcal{H}^2$ for some ϕ with $|\phi| = 1$ a.e. Then $\phi \in \mathcal{M}$ and $\phi = 0$ a.e. on M, contradicting the fact that $|\phi| = 1$ a.e. $\square$

Definition. A measurable function ϕ on C is *inner* if $\phi \in \mathcal{H}^2$ and $|\phi| = 1$ a.e.

The basis elements e_n, $n \geq 0$, of $\mathcal{H}^2$ are clearly inner functions. If $\lambda \in D$, then the function $f(z) = (\lambda - z)(1 - \bar{\lambda} z)^{-1}$ is easily seen to be inner, as is the function

$$f(z) = \exp\left(-\frac{\lambda + z}{\lambda - z}\right)$$

for $\lambda \in C$. Obviously the product of any finite number of inner functions is inner; all inner functions are essentially products of functions of these two types. A *Blaschke product* is an inner function of the form

$$B(z) = c z^k \prod_{j=1}^{\infty} \frac{\overline{\lambda_j}}{|\lambda_j|} \cdot \frac{\lambda_j - z}{1 - \overline{\lambda_j} z}$$

with k a non-negative integer, $c \in C$, and $\{\lambda_j\}$ a sequence of non-zero complex numbers of modulus less than 1 such that $\sum_{j=1}^{\infty} (1 - |\lambda_j|) < \infty$; (this last condition insures the convergence of the infinite product). A *singular inner function* is an inner function of the form

$$S(z) = \exp \left(- \int \frac{w+z}{w-z} d\mu(w) \right),$$

where μ is a finite positive Borel measure on C which is singular with respect to Lebesgue measure. The structure of inner functions is most easily studied by using the fact that for each $f \in \mathcal{H}^{\infty}$ there exists a unique function $\hat{f}$ which is analytic and bounded on D such that $\lim_{r \nearrow 1} \hat{f}(r e^{i\theta}) = f(e^{i\theta})$ for almost all $\theta \in [0, 2\pi)$. The results of elementary complex analysis can then be used to show that every inner function is the product of a singular inner function and a Blaschke product. A very readable discussion of this and related results is in Rudin [1, Chapter 17]; a more detailed account can be found in Hoffman [1] and in Duren [1].

We shall give proofs of those fragments of the theory of inner functions which we require in this chapter. The basic structure theorem mentioned above is needed for an adequate description of *Lat S*, however, and will be used without proof in Chapter 4, (see Example 4.3).

One trivial fact that we shall need is the following: if $f = \sum_{n=0}^{\infty} \alpha_n e_n \in \mathcal{H}^2$, then the function $\hat{f}(z) = \sum_{n=0}^{\infty} \alpha_n z^n$ is analytic on D, (this follows from the boundedness of the sequence $\{\alpha_n\}$). Note also that if f, g, and h are in $\mathcal{H}^{\infty}$ and $f = g h$, then $\hat{f}(z) = \hat{g}(z) \hat{h}(z)$ for all $z \in D$, (since the Taylor coefficients of $\hat{f}$ and $\hat{g}\hat{h}$ are equal to the corresponding Fourier coefficients of f and $g h$, which are equal to each other).

Corollary 3.11. *A non-zero subspace $\mathcal{M}$ of $\mathcal{H}^2$ is invariant under S if and only if $\mathcal{M} = \phi \mathcal{H}^2$ for some inner function ϕ. The function ϕ is determined by $\mathcal{M}$ to within a constant multiplier.*

Proof. If ϕ is inner, then $\phi \mathcal{H}^2 \in Lat\, U$, as in Theorem 3.9. Moreover $\phi \mathcal{H}^2 \subset \mathcal{H}^2$ and thus $\phi \mathcal{H}^2 \in Lat\, S$.

If $\mathcal{M} \in Lat\, S$, then, since $\mathcal{M} \subset \mathcal{H}^2$, it follows from Corollary 3.10 and Theorem 3.6 that $\mathcal{M}$ does not reduce U. Hence, by Theorem 3.9, there exists a $\phi \in \mathscr{L}^\infty(C, \mu)$ with $|\phi| = 1$ a.e. such that $\mathscr{M} = \phi \mathscr{H}^2$. Since $\phi e_0 = \phi$ is in $\mathscr{H}^2$, ϕ is an inner function. The essential uniqueness of ϕ follows as in Theorem 3.9.   ▢

Corollary 3.12. *If ϕ_1 and ϕ_2 are inner functions, then $\phi_1 \mathscr{H}^2 \subset \phi_2 \mathscr{H}^2$ if and only if ϕ_1/ϕ_2 is an inner function.*

Proof. If $\phi_1 \mathscr{H}^2 \subset \phi_2 \mathscr{H}^2$, then $\phi_1 \in \phi_2 \mathscr{H}^2$, and thus $\phi_1 = \phi_2 f$ for some $f \in \mathscr{H}^2$. Since $|\phi_1| = |\phi_2| = 1$ a.e., $|f| = 1$ a.e., and f is inner. Conversely if $\phi_1 = \phi_2 \phi_3$ where ϕ_3 is inner, then $\phi_1 f = \phi_2(\phi_3 f)$ for $f \in \mathscr{H}^2$, and therefore $\phi_1 \mathscr{H}^2 \subset \phi_2 \mathscr{H}^2$.   ▢

The following corollary is stronger than the result (Theorem 3.5) that S has no reducing subspaces.

Corollary 3.13. *The intersection of any two non-trivial members of $Lat\,S$ is non-trivial.*

Proof. If $\mathscr{M}_1 = \phi_1 \mathscr{H}^2$ and $\mathscr{M}_2 = \phi_2 \mathscr{H}^2$, then $\phi_1 \phi_2 \mathscr{H}^2$ is a non-trivial subspace of $\mathscr{M}_1 \cap \mathscr{M}_2$.   ▢

We have described the members of $Lat\,S$ with finite co-dimension in Theorem 3.8. We next show that the inner functions ϕ such that $\phi \mathscr{H}^2$ has finite co-dimension are precisely the finite Blaschke products, i.e., functions of the form

$$\phi(z) = c \prod_{j=1}^{n} \frac{\lambda_j - z}{1 - \overline{\lambda}_j z}$$

with $c \in C$ and $|\lambda_j| < 1$. (Note that we permit $\lambda_j = 0$, allowing for a factor z^k, and that the constants $\overline{\lambda}_j/|\lambda_j|$ for $\lambda_j \neq 0$ have been absorbed into c. The reason that the constants $\overline{\lambda}_j/|\lambda_j|$ are needed in the case of infinite Blaschke products is to insure convergence.)

Theorem 3.14. *If $\lambda_1, \ldots, \lambda_n$ are in D and if*

$$\phi(z) = \prod_{j=1}^{n} \frac{\lambda_j - z}{1 - \overline{\lambda}_j z},$$

then $\mathscr{H}^2 \ominus \phi \mathscr{H}^2$ has dimension n. Conversely, every invariant subspace of S of co-dimension n has this form.

Proof. We shall show that whenever $\lambda_1, \ldots, \lambda_m$ are distinct numbers in D and $n(1), \ldots, n(m)$ are positive integers, then

$$\bigvee_{j=1}^{m} null(S^* - \overline{\lambda}_j)^{n(j)} = (\phi \mathscr{H}^2)^\perp,$$

where

$$\phi(z) = \prod_{j=1}^{m} \left(\frac{\lambda_j - z}{1 - \overline{\lambda}_j z} \right)^{n(j)}.$$

By Theorem 3.8 this is sufficient to prove the result.

Since $\bigvee\limits_{j=1}^{m} \text{null}(S^*-\bar{\lambda}_j)^{n(j)}$ is in $Lat\,S^*$, its orthocomplement has the form $\psi\mathcal{H}^2$, for some inner function ψ, by Corollary 3.11. Note that the vectors $\{x_k^\lambda\}$ introduced in the proof of Lemma 3.7 are multiples of the functions $\{f_k^\lambda\}$, where $f_k^\lambda(z)=(1-\lambda z)^{-(k+1)}$. Thus a function f is in $\psi\mathcal{H}^2$ if and only if f is orthogonal to $f_k^{\lambda_j}$ for all $k\leq n(j)-1$, $j=1,2,\ldots,m$.

We first show that $\phi\in\psi\mathcal{H}^2$; i.e., that, for each p, $\phi\perp f_k^{\lambda_p}$ for $k\leq n(p)-1$. The inner product of ϕ and $f_k^{\lambda_p}$ is

$$\int\limits_C \phi(z)\overline{(1-\bar{\lambda}_p z)}^{-(k+1)}\,d\mu = \int\limits_C \prod_{j=1}^{m}\left(\frac{\lambda_j-z}{1-\bar{\lambda}_j z}\right)^{n(j)}\frac{1}{(1-\lambda_p\bar{z})^{k+1}}\,d\mu.$$

Using the fact that $\bar{z}=1/z$ for $z\in C$ and rearranging shows that this integral is

$$\int\limits_C \prod_{j\neq p}\left(\frac{\lambda_j-z}{1-\bar{\lambda}_j z}\right)^{n(j)}\frac{(\lambda_p-z)^{n(p)}z^{k+1}}{(1-\bar{\lambda}_p z)^{n(p)}(z-\lambda_p)^{k+1}}\,d\mu.$$

The integrand is analytic on $\bar{D}$ (since $k+1\leq n(p)$) and hence this integral is 0 by Cauchy's theorem.

Thus $\phi\in\psi\mathcal{H}^2$, and $\phi=\psi f$ for some $f\in\mathcal{H}^2$. Then $|f|=1$ a.e., since this is the case for ϕ and ψ, and f is also inner. We consider the corresponding functions $\hat{\phi}$, $\hat{\psi}$, and $\hat{f}$ analytic in D, (see the remarks preceding Corollary 3.11). Then

$$\hat{\phi}(z)=\prod_{j=1}^{m}\left(\frac{\lambda_j-z}{1-\bar{\lambda}_j z}\right)^{n(j)}$$

and $\hat{\phi}(z)=\hat{\psi}(z)\hat{f}(z)$ for all $z\in D$. Now λ_j is a zero of order $n(j)$ of $\hat{\phi}$; thus for each j there is a non-negative integer $s(j)\leq n(j)$ such that λ_j is a zero of $\hat{\psi}$ of order $s(j)$ and a zero of $\hat{f}$ of order $n(j)-s(j)$.

Write

$$\hat{\psi}(z)=\prod_{j=1}^{m}\left(\frac{\lambda_j-z}{1-\bar{\lambda}_j z}\right)^{s(j)}\hat{g}(z).$$

Then the function g on C corresponding to $\hat{g}$ is inner. Now we show that $s(j)=n(j)$ for all j. If $s(p)<n(p)$ for any p, then

$$\int\limits_C \psi(z)\overline{(1-\bar{\lambda}_p z)}^{-(s(p)+1)}\,d\mu\neq 0,$$

(by the Cauchy integral formula), contradicting the fact that $(1-\bar{\lambda}_p z)^{-(s(p)+1)}$ is orthogonal to $\psi(z)$. Hence

$$\prod_{j=1}^{m}\left(\frac{\lambda_j-z}{1-\bar{\lambda}_j z}\right)^{n(j)}=\prod_{j=1}^{m}\left(\frac{\lambda_j-z}{1-\bar{\lambda}_j z}\right)^{n(j)}\hat{g}(z)\hat{f}(z)$$

or $1 = \hat{g}(z)\hat{f}(z)$. Thus $\bar{g} = f$, and $\bar{g} \in \mathscr{H}^2$, from which it follows that g is a constant. Therefore f is a constant and $\phi \mathscr{H}^2 = \psi \mathscr{H}^2$. ☐

The following corollary is not at all obvious from the original representation of S as an operator on an arbitrary Hilbert space.

Corollary 3.15. *The unilateral shift has non-trivial invariant subspaces whose orthogonal complements are infinite-dimensional.*

Proof. Let $\phi(z) = \exp((\lambda + z)/(z - \lambda))$ with $|\lambda| = 1$. Then $(\phi \mathscr{H}^2)^{\perp}$ is infinite-dimensional by Theorem 3.14.   ☐

The next result follows immediately from the factorization theory for inner functions; we present an independent proof.

Corollary 3.16. *Suppose that $\mathscr{M}_1$ and $\mathscr{M}_2$ are in $Lat\,S$ with $\mathscr{M}_1 \subset \mathscr{M}_2$ and the dimension of $\mathscr{M}_2 \ominus \mathscr{M}_1$ is greater than 1. Then there exists an $\mathscr{M} \in Lat\,S$ such that $\mathscr{M}_1 \subsetneqq \mathscr{M} \subsetneqq \mathscr{M}_2$.*

Proof. Let $\mathscr{M}_1 = \phi_1 \mathscr{H}^2$ and $\mathscr{M}_2 = \phi_2 \mathscr{H}^2$ with ϕ_1 and ϕ_2 inner. Then ϕ_1/ϕ_2 is inner, by Corollary 3.12. Let $\phi_1/\phi_2 = \phi$. Then $\phi_1 \mathscr{H}^2 \subset \phi \mathscr{H}^2$. Also if $\phi_2 f_0 \in \mathscr{M}_2 \ominus \mathscr{M}_1$, then

$$\int \phi f \bar{f}_0 \, d\mu = \int \phi_2 \phi f \overline{(\phi_2 f_0)} \, d\mu = \int \phi_1 f \overline{\phi_2 f_0} \, d\mu = 0$$

for all $f \in \mathscr{H}^2$, and thus $f_0 \perp \phi \mathscr{H}^2$. Therefore the dimension of $\mathscr{H}^2 \ominus \phi \mathscr{H}^2$ is greater than 1.

We now consider two cases; the first case is where $\hat{\phi}$ has a zero $\lambda \in D$. Then we write $\hat{\phi}(z) = (\lambda - z)(1 - \bar{\lambda} z)^{-1} \hat{f}(z)$, where f is inner, (as in the proof of Theorem 3.14). Then $\mathscr{M} = \phi_2 f \mathscr{H}^2$ is properly between $\mathscr{M}_1$ and $\mathscr{M}_2$.

Now suppose that $\hat{\phi}$ has no zeros in D. Then $\hat{\phi}$ has an analytic square root $\hat{\psi}$ in D. Let $\hat{\psi}(z) = \sum_{n=0}^{\infty} \alpha_n z^n$. We must show that $\hat{\psi}$ corresponds to a function ψ in $\mathscr{H}^2$; i.e., that $\sum_{n=0}^{\infty} |\alpha_n|^2 < \infty$. To prove this note that, for $0 < r < 1$, the functions $\hat{\psi}_r$ and $\hat{\phi}_r$ defined by $\hat{\psi}_r(z) = \hat{\psi}(rz)$ and $\hat{\phi}_r(z) = \hat{\phi}(rz)$ are analytic on $\bar{D}$. Thus ψ_r and ϕ_r are in $\mathscr{H}^2$. Then

$$\sum r^{2n} |\alpha_n|^2 = \int |\psi(rz)|^2 \, d\mu = \int |\hat{\phi}(rz)| \, d\mu = \int |\hat{\phi}(rz)| \cdot 1 \, d\mu$$
$$\leq (\int |\hat{\phi}(rz)|^2 \, d\mu)^{\frac{1}{2}} = \|\hat{\phi}_r\| \, .$$

Let the Taylor coefficients of $\hat{\phi}$ be $\{\beta_n\}$. Then

$$\|\hat{\phi}_r\|^2 = \sum r^{2n} |\beta_n|^2 \leq \sum |\beta_n|^2 = \|\phi\| = 1.$$

Hence $\sum r^{2n} |\alpha_n|^2$ is a bounded function of r for $0 < r < 1$, and it follows that $\sum |\alpha_n|^2 < \infty$. Thus $\hat{\psi}$ corresponds to the $\mathscr{H}^2$ function $\psi = \sum_{n=0}^{\infty} \alpha_n e_n$.

Since $(\hat{\psi})^2 = \hat{\phi}$, it follows that $\psi^2 = \phi$, and ψ is inner. If $\mathcal{M} = \phi_2 \psi \mathcal{H}^2$, then $\mathcal{M}_1 \subsetneqq \mathcal{M} \subsetneqq \mathcal{M}_2$.

(If the full relationship between $\mathcal{H}^2$ and the corresponding space of analytic functions is assumed, then the fact that $\psi \in \mathcal{H}^2$ follows immediately from the boundedness and analyticity of $\hat{\psi}$.) □

3.3 Shifts of Arbitrary Multiplicity

Definition. If α is any cardinal number, then the *bilateral shift of multiplicity* α is the direct sum of α copies of the bilateral shift of multiplicity 1, and the *unilateral shift of multiplicity* α is the direct sum of α copies of the unilateral shift.

An alternate, and obviously equivalent, definition of the bilateral shifts is the following. Let $\mathcal{H} = \sum_{n=-\infty}^{\infty} \oplus \mathcal{K}_n$ where each $\mathcal{K}_n$ is an α-dimensional Hilbert space. Then U is the bilateral shift of multiplicity α if U is a unitary operator which maps $\mathcal{K}_n$ onto $\mathcal{K}_{n+1}$ for all n. If $\mathcal{M}$ is the subspace $\sum_{n=0}^{\infty} \oplus \mathcal{K}_n$ of $\mathcal{H}$, then $U|\mathcal{M}$ is the unilateral shift of multiplicity α.

There are analytic representations of shifts of arbitrary multiplicity similar to those for shifts of multiplicity 1 but a little more difficult to define. These representations will be described in terms of vector-valued $\mathscr{L}^2$ spaces.

Fix a separable (finite- or infinite-dimensional) Hilbert space $\mathscr{K}$. A function f from the unit circle C into $\mathscr{K}$ will be said to be *measurable* if, for each fixed $x \in \mathscr{K}$, the complex-valued function $z \to (f(z), x)$ is Lebesgue measurable (where $(\cdot, \cdot)$ denotes the inner product in $\mathscr{K}$). We note that the function $z \to \|f(z)\|$ is measurable whenever f is; (choose a dense subset $\{x_n\}_{n=0}^{\infty}$ of the unit sphere of $\mathscr{K}$; then, for each $z \in C$, $\|f(z)\| = \sup_n |(f(z), x_n)|$, and the supremum of a countable family of measurable functions is measurable). We define the set $\mathscr{L}^2(\mathscr{K})$ as the collection of all measurable functions f from C to $\mathscr{K}$ such that $\int \|f(z)\|^2 \, d\mu < \infty$, where μ is normalized Lebesgue measure on C. As in the scalar-valued case we make the usual identification of functions which are equal a.e. The polarization identity implies that the function $z \to (f(z), g(z))$ is measurable whenever f and g are in $\mathscr{L}^2(\mathscr{K})$. We define the inner product in $\mathscr{L}^2(\mathscr{K})$ by $(f, g) = \int (f(z), g(z)) \, d\mu$, where the inner product under the integral sign is the one defined on $\mathscr{K}$. (We use the same notation for inner products in $\mathscr{L}^2(\mathscr{K})$ and in $\mathscr{K}$; in each case it should be clear from the context which inner product is meant). Thus

$\mathscr{L}^2(\mathscr{K})$ is an inner-product space; it will be apparent from the discussion below that $\mathscr{L}^2(\mathscr{K})$ is complete.

There is an expansion of functions in $\mathscr{L}^2(\mathscr{K})$ which is analogous to the ordinary Fourier expansion in $\mathscr{L}^2$, and which will prove to be very useful. For each integer n let e_n denote the complex-valued function defined by $e_n(z)=z^n$, as in the representation of shifts of multiplicity 1.

Suppose that $\{x_n\}_{n=-\infty}^{\infty}$ is a sequence in $\mathscr{K}$. We write $f\sim\sum\limits_{n=-\infty}^{\infty}x_n e_n$, and call x_n the nth *Fourier coefficient of* f, if $(f(z),x)=\sum\limits_{n=-\infty}^{\infty}(x_n,x)z^n$ a.e. for every $x\in\mathscr{K}$. If $f\in\mathscr{L}^2(\mathscr{K})$, then the functional ϕ_n defined on $\mathscr{K}$ by $\phi_n(x)=\int(x,f(z))z^n d\mu$ is linear, and

$$|\phi_n(x)|\leq\int|(x,f(z))|d\mu\leq\int\|x\|\,\|f(z)\|\,d\mu\leq\|x\|\,\|f\|\ .$$

Thus there is a unique $x_n\in\mathscr{K}$ such that $\phi_n(x)=(x,x_n)$ for all x. For each x, then,

$$(f(z),x)=\sum_{n=-\infty}^{\infty}\left[\int(f(z),x)z^{-n}d\mu\right]z^n$$

$$=\sum_{n=-\infty}^{\infty}\overline{\phi_n(x)}z^n=\sum_{n=-\infty}^{\infty}(x_n,x)z^n\ .$$

Therefore every $f\in\mathscr{L}^2(\mathscr{K})$ has a uniquely determined Fourier expansion $\sum\limits_{n=-\infty}^{\infty}x_n e_n$. If x and y are any vectors in $\mathscr{K}$, then the functions $x e_n$ and $y e_m$ are orthogonal elements of $\mathscr{L}^2(\mathscr{K})$ for $m\neq n$, since

$$(x e_n,y e_m)=\int(z^n x,z^m y)d\mu=\int z^n z^{-m}(x,y)d\mu=0\ .$$

Thus, if $f\sim\sum\limits_{n=-\infty}^{\infty}x_n e_n$,

$$\|f\|^2=\int(f(z),f(z))d\mu=\int\sum_{n=-\infty}^{\infty}(x_n,f(z))z^n d\mu$$

$$=\int\sum_{n=-\infty}^{\infty}\left(\sum_{m=-\infty}^{\infty}(x_n,x_m)z^{-m}\right)z^n d\mu$$

$$=\int\sum_{n=-\infty}^{\infty}\sum_{m=-\infty}^{\infty}(x_n z^n,x_m z^m)d\mu$$

$$=\sum_{n=-\infty}^{\infty}\sum_{m=-\infty}^{\infty}\int(x_n z^n,x_m z^m)d\mu$$

$$=\sum_{n=-\infty}^{\infty}\int(x_n z^n,x_n z^n)d\mu=\sum_{n=-\infty}^{\infty}\|x_n\|^2\ .$$

Since $\|f\|^2 = \sum\limits_{n=-\infty}^{\infty} \|x_n\|^2$, it follows that $f \sim \sum\limits_{-\infty}^{\infty} x_n e_n$ implies $f = \sum\limits_{-\infty}^{\infty} x_n e_n$ in $\mathscr{L}^2(\mathscr{K})$. Thus $\mathscr{L}^2(\mathscr{K})$ is isomorphic, in a natural way, to the direct sum of countably many copies of $\mathscr{K}$, indexed by the set of all integers.

Let U be the operator defined by multiplication by e_1 on $\mathscr{L}^2(\mathscr{K})$; the above remarks show that U is the bilateral shift whose multiplicity is equal to the dimension of $\mathscr{K}$. We define $\mathscr{H}^2(\mathscr{K})$ as the subspace of $\mathscr{L}^2(\mathscr{K})$ consisting of all functions $f \in \mathscr{L}^2(\mathscr{K})$ whose negative Fourier coefficients are 0; i.e.,

$$\mathscr{H}^2(\mathscr{K}) = \left\{ f \in \mathscr{L}^2(\mathscr{K}) : f = \sum_{n=0}^{\infty} x_n e_n \right\}.$$

Let $S = U | \mathscr{H}^2(\mathscr{K})$; then S is the unilateral shift with multiplicity the dimension of $\mathscr{K}$.

The Hilbert space $\mathscr{L}^2(\mathscr{K})$ is a collection of $\mathscr{K}$-valued functions. We shall also find it useful to study operator-valued functions defined on C. A function F from C into $\mathscr{B}(\mathscr{K})$ is said to be *measurable* if, for every $x \in \mathscr{K}$, the $\mathscr{K}$-valued function $z \to F(z)x$ is measurable. Given such an F let $\|F\|_\infty$ denote the essential supremum of $\|F(z)\|$ on C. If $\|F\|_\infty$ is finite, then we define an operator $\hat{F}$ on $\mathscr{L}^2(\mathscr{K})$ by $(\hat{F}f)(z) = F(z)f(z)$ for $f \in \mathscr{L}^2(\mathscr{K})$ and $z \in C$. Then $\|\hat{F}\| \le \|F\|_\infty$, since

$$\|\hat{F}f\|^2 = \int \|F(z)f(z)\|^2 \, d\mu \le \int \|F(z)\|^2 \cdot \|f(z)\|^2 \, d\mu \le \|F\|_\infty^2 \cdot \|f\|^2$$

for all $f \in \mathscr{L}^2(\mathscr{K})$. Let $\mathscr{F}$ denote the collection of all (equivalence classes modulo sets of measure 0 of) measurable functions F from C to $\mathscr{B}(\mathscr{K})$ such that $\|F\|_\infty < \infty$. Let $\hat{\mathscr{F}}$ denote the collection $\{\hat{F} : F \in \mathscr{F}\}$ of operators on $\mathscr{L}^2(\mathscr{K})$. (Note that $F_1(z) = F_2(z)$ a.e. implies $\hat{F}_1 = \hat{F}_2$.) If I denotes the identity operator on $\mathscr{K}$, then the bilateral shift U is equal to $\hat{F}$ with $F(z) = zI$.

Define addition, multiplication, and adjoints in $\mathscr{F}$ pointwise; then $\mathscr{F}$, with norm $\|\cdot\|_\infty$, is a normed algebra with involution. Note that $\hat{\mathscr{F}}$, is a subalgebra of $\mathscr{B}(\mathscr{L}^2(\mathscr{K}))$.

Theorem 3.17. *The mapping from $\mathscr{F}$ onto $\hat{\mathscr{F}}$ defined by $F \to \hat{F}$ is an adjoint-preserving algebra isomorphism. In particular, $\hat{F}$ is respectively* (i) *normal*, (ii) *Hermitian*, (iii) *unitary*, *or* (iv) *a projection if and only if for almost every z, $F(z)$ has the corresponding property.*

Proof. It is very easy to see that this mapping is an adjoint-preserving algebra homomorphism. To show that the mapping is one-to-one suppose that F_1 and F_2 are in $\mathscr{F}$ and that $\hat{F}_1 = \hat{F}_2$. Let $\{x_n\}$ be a countable dense subset of $\mathscr{K}$, and, for each n, let x_n denote the function in $\mathscr{L}^2(\mathscr{K})$ which is identically equal to x_n. Then

$$F_1(z)x_n = (\hat{F}_1 x_n)(z) = (\hat{F}_2 x_n)(z) = F_2(z)x_n;$$

(each equality is a. e.). Thus there is a set E of measure 0 such that, whenever z is in the complement of E, $F_1(z)x_n = F_2(z)x_n$ for all n. For z in the complement of E, then, $F_1(z)x = F_2(z)x$ for all $x \in \mathcal{K}$. Therefore, F_1 and F_2 are in the same equivalence class in $\mathcal{F}$. ☐

It is easy to see that $\hat{\mathcal{F}}$ is contained in the commutant of U; the converse is proved below, (Corollary 3.19). We first prove a stronger result which will be needed later. For this we need to recall a definition.

Definition. If V is a partial isometry on a Hilbert space $\mathcal{H}$, (i.e., $\|Vx\| = \|x\|$ for all x in (null $V)^{\perp}$), the *initial space* of V is the subspace (null $V)^{\perp}$, and the *final space* of V is $V\mathcal{H}$.

Partial isometries are an interesting class of operators, (cf. Propositions 1.1 and 3.6). They are also useful in the study of invariant subspaces of shifts.

We also require some notation for certain subspaces of $\mathcal{L}^2(\mathcal{K})$. If $\mathcal{N}$ is a subspace of $\mathcal{K}$, let

$$\mathcal{L}^2(\mathcal{N}) = \{f \in \mathcal{L}^2(\mathcal{K}) : f(z) \in \mathcal{N} \text{ a.e.}\}.$$

Then $\mathcal{L}^2(\mathcal{N})$ is a subspace of $\mathcal{L}^2(\mathcal{K})$, and the orthocomplement of $\mathcal{L}^2(\mathcal{N})$ is $\mathcal{L}^2(\mathcal{K} \ominus \mathcal{N})$.

Theorem 3.18. *Let $\mathcal{N}$ be a subspace of $\mathcal{K}$ and let A be a partial isometry on $\mathcal{L}^2(\mathcal{K})$ with initial space $\mathcal{L}^2(\mathcal{N})$. If A commutes with U, then there exists an $F \in \mathcal{F}$ such that $A = \hat{F}$ and such that, for almost every z, $F(z)$ is a partial isometry with initial space $\mathcal{N}$.*

Proof. Choose an orthonormal basis $\{y_n\}$ for $\mathcal{N}$. For each n, let y_n denote the function in $\mathcal{L}^2(\mathcal{N})$ which is identically y_n and choose a fixed function f_n in the equivalence class $A y_n$ of $\mathcal{K}$-valued functions. For each $z \in C$ define $F(z)y_n = f_n(z)$. We first show that $F(z)$ can be extended to a partial isometry for almost every z. For this let $g \in \mathcal{L}^\infty$. Then the operator $\hat{G}$ with $G(z) = g(z)I$, (I the identity on $\mathcal{K}$), commutes with A; (this follows from the proof of Theorem 3.2). Thus

$$\int g(z)(F(z)y_m, F(z)y_n)d\mu = \int ((\hat{G}A y_m)(z), (A y_n)(z))d\mu$$
$$= (\hat{G}A y_m, A y_n) = (A \hat{G} y_m, A y_n).$$

Since A is isometric on $\mathcal{L}^2(\mathcal{N})$, this is equal to

$$(\hat{G} y_m, y_n) = \int (g(z) y_m(z), y_n(z))d\mu$$
$$= \int (y_m, y_n)g(z)d\mu = (y_m, y_n)\int g(z)d\mu.$$

Let δ_{mn} denote the Kronecker delta; then

$$\int g(z)[(F(z)y_m, F(z)y_n) - \delta_{mn}]d\mu = 0$$

for all $g \in \mathscr{L}^\infty$. Thus $(F(z) y_m, F(z) y_n) = \delta_{mn}$ a.e. It follows that $F(z)$ has a unique extension to an isometry from $\mathscr{N}$ into $\mathscr{K}$ for almost every z. If we define $F(z)x = 0$ for $x \in \mathscr{N}^\perp$, then $F(z)$ is a partial isometry a.e. Since $F(z) y_n$ is measurable, it follows that $F(z)x$ is measurable for every $x \in \mathscr{K}$. Thus $F \in \mathscr{F}$.

We need only show that $A = \hat{F}$. Note that $Af = \hat{F}f$ for all $f \in \mathscr{L}^2(\mathscr{K})$ of the form $x e_0$. If $f = \sum\limits_{n=-\infty}^{\infty} x_n e_n$, then $f = \sum\limits_{n=-\infty}^{\infty} U^n(x_n e_0)$, and

$$Af = \sum_{n=-\infty}^{\infty} A \, U^n(x_n e_0) = \sum_{n=-\infty}^{\infty} U^n A(x_n e_0) = \sum_{n=-\infty}^{\infty} U^n \hat{F}(x_n e_0)$$

$$= \sum_{n=-\infty}^{\infty} \hat{F} \, U^n(x_n e_0) = \hat{F}f. \quad \square$$

Corollary 3.19. *The commutant of U is $\hat{\mathscr{F}}$.*

Proof. Theorem 3.18 applies in particular to unitary operators A; (in which case $\mathscr{N} = \mathscr{K}$). The present result then follows from the standard fact that every operator in the commutant of U is a linear combination of unitary operators in the commutant. (To prove this assume, without loss of generality, that A commutes with U and $\|A\| \le 1$. If we write $A = H + iK$, with H and K Hermitian, then H and K each commute with U and have norm at most 1. If $(1-H^2)^{\frac{1}{2}}$ and $(1-K^2)^{\frac{1}{2}}$ denote the unique positive square roots of $1-H^2$ and $1-K^2$, (which exist by Corollary 1.8), then $H \pm i(1-H^2)^{\frac{1}{2}}$ and $\pm(1-K^2)^{\frac{1}{2}} + iK$ are four unitary operators whose sum is $2A$.) $\quad \square$

Definition. A function $F \in \mathscr{F}$ is *analytic* if $\mathscr{H}^2(\mathscr{K}) \in Lat \hat{F}$. Let $\mathscr{F}_0$ denote the set of all analytic elements of $\mathscr{F}$.

Corollary 3.20. *The commutant of S is $\{\hat{F} | \mathscr{H}^2(\mathscr{K}) : F \in \mathscr{F}_0\}$.*

Proof. The proof is essentially the same as the proof of Theorem 3.4. If $AS = SA$, extend A to an operator $\hat{A}$ on $\mathscr{L}^2(\mathscr{K})$ by defining

$$\hat{A} \, U^n(x e_0) = U^n A(x e_0)$$

for all $x \in \mathscr{K}$ and all $n < 0$. Then $\hat{A} U = U \hat{A}$ and the result follows from Corollary 3.19. $\quad \square$

3.4 Invariant Subspaces of Shifts

We shall obtain results about the invariant subspaces of shifts of arbitrary multiplicity that generalize those of Section 3.2. We first discuss reducing subspaces.

Theorem 3.21. *A subspace $\mathscr{M}$ of $\mathscr{L}^2(\mathscr{K})$ reduces U if and only if there exists a $P \in \mathscr{F}$ such that $P(z)$ is a projection for almost all z and such that*

$$\mathscr{M} = \{f: f(z) \in P(z)\mathscr{K} \text{ a.e.}\}.$$

Proof. It follows from Theorem 3.19 that a subspace $\mathscr{M}$ reduces U if and only if the projection onto $\mathscr{M}$ has the form $\hat{P}$ for some $P \in \mathscr{F}$. Then $P(z)$ is a projection a.e. by Theorem 3.17, and the result follows. □

If $\mathscr{N}$ is a subspace of $\mathscr{K}$, we use $\mathscr{H}^2(\mathscr{N})$ to denote the subspace $\{f \in \mathscr{H}^2(\mathscr{K}): f(z) \in \mathscr{N} \text{ a.e.}\}$ of $\mathscr{H}^2(\mathscr{K})$.

Theorem 3.22. *A subspace $\mathscr{M}$ of $\mathscr{H}^2(\mathscr{K})$ reduces S if and only if there exists a subspace $\mathscr{N}$ of $\mathscr{K}$ such that $\mathscr{M} = \mathscr{H}^2(\mathscr{N})$.*

Proof. Obviously each such subspace reduces S. If $\mathscr{M}$ reduces S, and if P is the projection onto $\mathscr{M}$, then P commutes with $1 - SS^*$. Thus the range of $1 - SS^*$, i.e., $\{xe_0 : x \in \mathscr{K}\}$, is invariant under P. Let

$$\mathscr{N} = \{y \in \mathscr{K}: P(xe_0) = ye_0 \text{ for some } x \in \mathscr{K}\}.$$

Then $\mathscr{N} \subset \mathscr{M}$ and thus $\mathscr{H}^2(\mathscr{N}) \subset \mathscr{M}$. For the other inclusion suppose that $f = \sum_{n=0}^{\infty} x_n e_n$ is in $\mathscr{M}$. Then, for each $n > 0$,

$$P(x_n e_n) = P S^n(x_n e_0) = S^n P(x_n e_0) \in \mathscr{H}^2(\mathscr{N}).$$

Thus $f \in \mathscr{H}^2(\mathscr{N})$. □

Lemma 3.23. *Every invariant subspace of U has a unique decomposition of the form $\mathscr{M}_1 \oplus \mathscr{M}_2$ with $\mathscr{M}_1$ reducing for U, $\mathscr{M}_2 \in \text{Lat } U$, and $\bigcap_{n=0}^{\infty} U^n \mathscr{M}_2 = \{0\}$.*

Proof. Let $\mathscr{M} \in \text{Lat } U$. Set $\mathscr{M}_1 = \bigcap_{n=0}^{\infty} U^n \mathscr{M}$ and $\mathscr{M}_2 = \mathscr{M} \ominus \mathscr{M}_1$. Clearly $\mathscr{M}_1 \in \text{Lat } U$. Moreover, since $U\mathscr{M} \subset \mathscr{M}$, $\mathscr{M}_1 = \bigcap_{n=1}^{\infty} U^n \mathscr{M}$, and therefore

$$U^* \mathscr{M}_1 = U^{-1} \mathscr{M}_1 = U^{-1}\left(\bigcap_{n=1}^{\infty} U^n \mathscr{M}\right) = \bigcap_{n=0}^{\infty} U^n \mathscr{M} = \mathscr{M}_1.$$

Thus $\mathscr{M}_1$ reduces U and $\mathscr{M}_2 \in \text{Lat } U$. Now if $f \in \bigcap_{n=0}^{\infty} U^n \mathscr{M}_2$, then f is in both $\bigcap_{n=0}^{\infty} U^n \mathscr{M}$ and $\mathscr{M}_2$, which implies that $f = 0$. □

Lemma 3.24. *If $\mathscr{M} \in \text{Lat } U$ and U has multiplicity α, then the dimension of $\mathscr{M} \ominus U \mathscr{M}$ is at most α.*

Proof. If α is not finite, then, since we are only considering separable spaces, the result is trivial. If α is finite let $\mathcal{M}_0 = \mathcal{M} \ominus U \mathcal{M}$. The subspaces $\{U^n \mathcal{M}_0\}_{n=-\infty}^{\infty}$ are mutually orthogonal; (this follows from a direct computation, using the definition of $\mathcal{M}_0$ and the fact that U is unitary).

Let $\{f_j\}_{j=1}^m$ be any orthonormal set in $\mathcal{M}_0$. Then $\{U^{-n}f_j\}, j=1,\ldots,m$, $n=0,\pm 1,\ldots$, is also an orthonormal set. Let P be the projection of $\mathcal{L}^2(\mathcal{K})$ onto $\mathcal{K}$ given by $P\left(\sum_{n=-\infty}^{\infty} x_n e_n\right) = x_0$, and let $f_j = \sum_{n=-\infty}^{\infty} x_{n_j} e_n$. Then

$$m = \sum_{j=1}^m \|f_j\|^2 = \sum_{j=1}^m \sum_{n=-\infty}^{\infty} \|x_{n_j}\|^2.$$

Note that $x_{n_j} = P U^{-n} f_j$ for each n and j. Thus $m = \sum_{j=1}^m \sum_{n=-\infty}^{\infty} \|P U^{-n} f_j\|^2$. Since $\{U^{-n} f_j\}$ is an orthonormal set, and P is a projection of rank α, it follows that $m \le \alpha$. Since this is so whenever $\{f_j\}_{j=1}^m$ is an orthonormal subset of $\mathcal{M}_0$, the dimension of $\mathcal{M}_0$ is at most α. $\square$

Theorem 3.25. *A subspace $\mathcal{M}$ of $\mathcal{L}^2(\mathcal{K})$ is in $Lat\,U$ if and only if $\mathcal{M}$ can be written in the form*

$$\mathcal{M}_1 \oplus \hat{V} \mathcal{H}^2(\mathcal{N}),$$

where $\mathcal{M}_1$ reduces U, $\mathcal{N}$ is a subspace of $\mathcal{K}$, and V is a function in $\mathcal{F}$ such that $V(z)$ is a partial isometry with initial space $\mathcal{N}$ for almost every z. Furthermore,

(i) *$\mathcal{M}_1$ is uniquely determined by $\mathcal{M}$ and*

(ii) *if $V_1 \in \mathcal{F}$ has the property that $V_1(z)$ is a partial isometry with initial space $\mathcal{N}_1$ (for almost all z), and if $\hat{V} \mathcal{H}^2(\mathcal{N}) = \hat{V}_1 \mathcal{H}^2(\mathcal{N}_1)$, then there is a partial isometry W with initial space $\mathcal{N}_1$ and final space $\mathcal{N}$ such that $\hat{V}_1 = \hat{V} \hat{W}$, where $W(z) = W$ for all z.*

Proof. Clearly each such $\mathcal{M}$ is in $Lat\,U$. If $\mathcal{M} \in Lat\,U$, then $\mathcal{M} = \mathcal{M}_1 \oplus \mathcal{M}_2$ as in Lemma 3.23. We must show that $\mathcal{M}_2$ has the form $\hat{V} \mathcal{H}^2(\mathcal{N})$ as above. Let $\mathcal{M}_0 = \mathcal{M}_2 \ominus U \mathcal{M}_2$. Then it is easily verified that $\mathcal{M}_2 = \sum_{n=0}^{\infty} \oplus U^n \mathcal{M}_0$. By Lemma 3.24 the dimension of $\mathcal{M}_0$ is less than or equal to the dimension of $\mathcal{K}$; let $\mathcal{N}$ be any subspace of $\mathcal{K}$ whose dimension is equal to the dimension of $\mathcal{M}_0$. Now let J be an isometry of $\mathcal{N}$ onto $\mathcal{M}_0$, and define a partial isometry V on $\mathcal{L}^2(\mathcal{K})$ by setting $Vf = 0$ for $f \in \mathcal{L}^2(\mathcal{K} \ominus \mathcal{N})$ and

$$V\left(\sum_{n=-\infty}^{\infty} x_n e_n\right) = \sum_{n=-\infty}^{\infty} U^n J x_n$$

for $\sum_{n=-\infty}^{\infty} x_n e_n$ in $\mathcal{L}^2(\mathcal{N})$. Then V commutes with U and $V\mathcal{H}^2(\mathcal{N}) = \mathcal{M}_2$; Theorem 3.18 implies that $V = \hat{V}$ with V having the stated properties.

Statement (i) is part of Lemma 3.23. Suppose that $\mathcal{M}_2 = \hat{V}\mathcal{H}^2(\mathcal{N})$ $= \hat{V}_1\mathcal{H}^2(\mathcal{N}_1)$ as in (ii). Then

$$\mathcal{M}_2 \ominus U\mathcal{M}_2 = \hat{V}(\mathcal{H}^2(\mathcal{N}) \ominus U\mathcal{H}^2(\mathcal{N}))$$
$$= \hat{V}_1(\mathcal{H}^2(\mathcal{N}_1) \ominus U\mathcal{H}^2(\mathcal{N}_1)).$$

This clearly implies that $\mathcal{N}_1$ and $\mathcal{N}$ have the same dimension. Let W be the isometry of $\mathcal{N}_1$ onto $\mathcal{N}$ such that $\hat{V}_1(x e_0) = \hat{V}((Wx)e_0)$ for x in $\mathcal{N}_1$. Extend W to $\mathcal{K}$ by defining $Wx = 0$ for $x \in \mathcal{N}_1^\perp$. Now let W be defined by $W(z) = W$ for all z. To see that $\hat{V}_1 = \hat{V}\hat{W}$, first note that both of these operators are 0 on $\mathcal{L}^2(\mathcal{K} \ominus \mathcal{N}_1)$. If $f = \sum_{n=-\infty}^{\infty} x_n e_n$ with $\{x_n\} \subset \mathcal{N}_1$, then $\hat{V}\hat{W}f = \sum_{n=-\infty}^{\infty} \hat{V}\hat{W}(x_n e_n)$, and

$$\hat{V}\hat{W}(x_n e_n) = \hat{V}\hat{W}U^n(x_n e_0)$$
$$= U^n\hat{V}((Wx_n)e_0)$$
$$= U^n\hat{V}_1(x_n e_0) = \hat{V}_1(x_n e_n). \quad \square$$

Corollary 3.26. *A subspace $\mathcal{M}$ of $\mathcal{H}^2(\mathcal{K})$ is in $Lat\,S$ if and only if $\mathcal{M} = \hat{V}\mathcal{H}^2(\mathcal{N})$, where $\mathcal{N}$ is a subspace of $\mathcal{K}$ and V is an element of $\mathcal{F}_0$ such that $V(z)$ is a.e. a partial isometry with initial space $\mathcal{N}$. Also, $\hat{V}$ and $\mathcal{N}$ are determined by $\mathcal{M}$ as in statement (ii) of Theorem 3.25.*

Proof. Suppose that $\mathcal{M} \in Lat\,S$. Then $\mathcal{M}$, as a subspace of $\mathcal{L}^2(\mathcal{K})$, belongs to $Lat\,U$. Thus $\mathcal{M}$ has the form exhibited in Theorem 3.25. Then $\mathcal{M}_1$ is $\{0\}$, since if $f = \sum_{n=0}^{\infty} x_n e_n$ were in $\mathcal{M}_1$ with $x_k \neq 0$, then $U^{-k-1}f$ would not be in $\mathcal{H}^2(\mathcal{K})$, and thus could not be in $\mathcal{M}_1$. Therefore $\mathcal{M} = \hat{V}\mathcal{H}^2(\mathcal{N})$ for some V and $\mathcal{N}$. Now the range of $\hat{V}$ is contained in $\mathcal{H}^2(\mathcal{K})$, and therefore $V \in \mathcal{F}_0$. $\quad \square$

3.5 Parts of Shifts

Definition. An operator A is a *part* of the operator B if there exists an $\mathcal{M} \in Lat\,B$ such that A is (unitarily equivalent to) $B|\mathcal{M}$.

It turns out that parts of the adjoint of the unilateral shift of multiplicity $\aleph_0$ form a remarkably rich class of operators. In fact, this class contains all operators with norm less than one, (Corollary 3.30). Note that if $A = S^*|\mathcal{M}$ with S a unilateral shift, then $\|A\| \leq 1$, and if A is similar to a part of S^* then $r(A) \leq 1$.

We first consider finite-dimensional parts of the adjoint of the unilateral shift of multiplicity 1.

Theorem 3.27. *Let S be the unilateral shift of multiplicity 1, and let A be an operator on a finite-dimensional Hilbert space. Then A is similar to a part of S^* if and only if A is cyclic, (or, equivalently, the characteristic and minimal polynomials of A coincide), and $r(A) < 1$.*

Proof. Recall that the minimal and characteristic polynomials of A are equal if and only if every eigenspace of A is one-dimensional. By Theorem 3.8, then, every operator similar to a finite-dimensional part of S^* satisfies the stated conditions; (if $\mathcal{M}$ is finite-dimensional, then $r(S^*|\mathcal{M}) < 1$, for otherwise S^* would have an eigenvalue of modulus 1).

Now suppose that $p(\lambda) = (\lambda - \lambda_1)^{n(1)} \ldots (\lambda - \lambda_m)^{n(m)}$ is the characteristic and the minimal polynomial of A, and that $|\lambda_j| < 1$ for each j. Then, by Theorem 3.8, if $\mathcal{M} = \bigvee_{j=1}^{m} \text{null}(S^* - \lambda_j)^{n(j)}$, then $S^*|\mathcal{M}$ has the same Jordan canonical form as A. $\quad\square$

The operators on finite-dimensional spaces which are similar to parts of S^*, where S has any multiplicity, can be characterized in a manner similar to the above; (see Proposition 3.13).

We now return to operators on infinite-dimensional spaces.

Theorem 3.28. *If A is an operator on a separable space such that $r(A) < 1$, then A is similar to a part of S^*, where S is the unilateral shift of multiplicity $\aleph_0$.*

Proof. Since $r(A) = \lim_{n \to \infty} \|A^n\|^{1/n} < 1$, the series $\sum_{n=0}^{\infty} \|A^n\|^2$ converges (by the root test). Let $\mathcal{K}$ be the Hilbert space on which A is defined. Then map $\mathcal{K}$ into $\mathcal{H}^2(\mathcal{K})$ by $Vx = \sum_{n=0}^{\infty} (A^n x) e_n$ for all $x \in \mathcal{K}$. Since
$$\|Vx\|^2 = \sum_{n=0}^{\infty} \|A^n x\|^2 \leq \left(\sum_{n=0}^{\infty} \|A^n\|^2 \right) \|x\|^2, \ V \text{ is bounded. Also } \|Vx\| \geq \|x\|,$$
and thus the range, $\mathcal{M}$, of V is closed. Now $VAx = S^* Vx$ for all $x \in \mathcal{K}$. Hence $\mathcal{M} \in \text{Lat } S^*$, and $V^{-1}(S^*|\mathcal{M})V = A$. $\quad\square$

With a little more care a necessary and sufficient condition that A be a part of S^* can be obtained. Clearly it is necessary that $\|A\| \leq 1$ and that $\{A^n x\} \to 0$ for all x.

Theorem 3.29. *Let A be an operator such that $\|A\| \leq 1$ and $\{A^n x\} \to 0$ for all x, and let α be the dimension of the closure of the range of $1 - A^*A$. Then A is a part of S^*, where S is the unilateral shift of multiplicity α.*

Proof. Since $\|A\| \leq 1$, the operator $1 - A^*A$ is positive. Let B denote its unique positive square root, (Corollary 1.8). Let $\mathcal{K}$ denote the closure of the range of B. Then the dimension of $\mathcal{K}$ is α; let S be the unilateral shift on $\mathcal{H}^2(\mathcal{K})$. Consider the mapping V from the space $\mathcal{H}$, on which A is defined, given by $Vx = \sum_{n=0}^{\infty} (BA^n x)e_n$ for $x \in \mathcal{H}$. Then for each $k > 0$,

$$\left\| \sum_{n=0}^{k} (BA^n x)e_n \right\|^2 = \sum_{n=0}^{k} \|BA^n x\|^2 = \sum_{n=0}^{k} ((1 - A^*A)A^n x, A^n x)$$

$$= \sum_{n=0}^{k} (\|A^n x\|^2 - \|A^{n+1} x\|^2)$$

$$= \|x\|^2 - \|A^{k+1} x\|^2.$$

Thus

$$\|Vx\|^2 = \lim_{k \to \infty} \left\| \sum_{n=0}^{k} (BA^n x)e_n \right\|^2$$

$$= \|x\|^2 - \lim_{k \to \infty} \|A^{k+1} x\|^2 = \|x\|^2,$$

and V is an isometry of $\mathcal{H}$ into $\mathcal{H}^2(\mathcal{K})$. Note that $VAx = S^*Vx$ for all $x \in \mathcal{H}$. Thus the subspace $V\mathcal{H}$ of $\mathcal{H}^2(\mathcal{K})$ is in $Lat\, S^*$, and $V^{-1}(S^*|V\mathcal{H})V = A$. $\square$

Corollary 3.30. *Let S denote the unilateral shift of multiplicity $\aleph_0$. Then every operator of norm less than 1 (on a separable space) is a part of S^*.*

Proof. Let $\|A\| < 1$, and let the dimension of $[\text{null}(1 - A^*A)]^{\perp}$ equal α. Then A is a part of the adjoint of the shift of multiplicity α, by Theorem 3.29. If $\alpha = \aleph_0$ we are done; if α is finite, then the result follows immediately from the obvious fact that the adjoint of the shift of multiplicity α is a part of the adjoint of the shift of multiplicity $\aleph_0$. $\square$

It follows from Corollary 3.30 that the invariant subspace problem can be re-formulated. If A is any operator, let $B = A/(2\|A\|)$. Then B is unitarily equivalent to $S^*|\mathcal{M}$, where S is the unilateral shift of multiplicity $\aleph_0$ and $\mathcal{M} \in Lat\, S^*$, and A and B have the same invariant subspaces. Now B has a non-trivial invariant subspace if and only if S^* has an invariant subspace $\mathcal{N}$ such that $\{0\} \subsetneqq \mathcal{N} \subsetneqq \mathcal{M}$. This is so if and only if $\mathcal{N}^{\perp} \in Lat\, S$ and $\mathcal{M}^{\perp} \subsetneqq \mathcal{N}^{\perp} \subsetneqq \mathcal{H}^2(\mathcal{K})$. Thus the invariant subspace problem is equivalent to the question: if $\mathcal{L} \in Lat\, S$ and if the dimension of $\mathcal{H}^2(\mathcal{K}) \ominus \mathcal{L}$ is greater than 1, must S have an invariant subspace properly contained between $\mathcal{L}$ and $\mathcal{H}^2(\mathcal{K})$?

Corollary 3.31. *If A is an operator on a space of dimension greater than 1 such that $\|A\| \le 1$, $\{A^n x\} \to 0$ for all x, and the dimension of the range of $1 - A^* A$ is 1, then A has a non-trivial invariant subspace.*

Proof. This follows immediately from Theorem 3.29, Corollary 3.16, and remarks similar to the above. □

In general, given a unilateral shift S and $\mathcal{M} \in Lat\, S$, the problem of determining whether or not there is an $\mathcal{N} \in Lat\, S$ such that $\mathcal{M} \subsetneqq \mathcal{N} \subsetneqq \mathscr{H}^2(\mathscr{K})$ is equivalent to a factorization problem for operator-valued analytic functions.

Theorem 3.32. *Let $\mathcal{M}_1 = \hat{V}_1 \mathscr{H}^2(\mathscr{N}_1)$ and $\mathcal{M}_2 = \hat{V}_2 \mathscr{H}^2(\mathscr{N}_2)$ be invariant subspaces of S as described in Corollary 3.26. Then $\mathcal{M}_1 \subset \mathcal{M}_2$ if and only if there exists $W \in \mathscr{F}_0$ such that $W(z)$ is a. e. a partial isometry with initial space $\mathscr{N}_1$ and final space contained in $\mathscr{N}_2$ satisfying $V_1(z) = V_2(z) W(z)$.*

Proof. Obviously the existence of such a W implies $\mathcal{M}_1 \subset \mathcal{M}_2$. Suppose that $\mathcal{M}_1 \subset \mathcal{M}_2$. For each z let $W(z) = V_2^*(z) V_1(z)$. Then $\hat{W} = \hat{V}_2^* \hat{V}_1$, (as operators on $\mathscr{L}^2(\mathscr{K})$), by Theorem 3.17. Moreover, $W \in \mathscr{F}_0$ and $\hat{W}$ is a partial isometry with initial space $\mathscr{L}^2(\mathscr{N}_1)$, since the initial space of $\hat{V}_2^*$ contains the final space of $\hat{V}_1$. Theorem 3.18, and the uniqueness given by Theorem 3.17, now give the result. □

Thus the invariant subspace problem is equivalent to the question: if $V \in \mathscr{F}_0$ is a partial isometry a. e. such that $\hat{V} \mathscr{H}^2(\mathscr{K})$ has co-dimension greater than 1, must there exist a factoring $V = V_2 W$ as in the above theorem with W non-constant? This problem can be reduced to the study of those functions V such that $V(z)$ is unitary a. e., (Proposition 3.15). The factorization question has been answered affirmatively in the case where $\mathscr{K}$ is finite-dimensional, thus proving the existence of invariant subspaces for certain operators. We shall not attack this problem directly here, (see Helson [1] for such an approach), but shall instead obtain the result as a corollary of a more general theorem on existence of invariant subspaces; (see Corollary 6.18).

Theorem 3.33. *If S is a unilateral shift (of arbitrary multiplicity) and $\mathcal{M}$ is a cyclic invariant subspace of S, then $S|\mathcal{M}$ is unitarily equivalent to the unilateral shift of multiplicity 1.*

Proof. By Corollary 3.26, $S|\mathcal{M}$ is unitarily equivalent to $S_1 = S|\mathscr{H}^2(\mathscr{N})$, where $\mathscr{N}$ is a subspace of $\mathscr{K}$ as described in that corollary. If the dimension of $\mathscr{N}$ is denoted by d, then S_1 is a unilateral shift of multiplicity d, and we need only show that $d = 1$. Since, by hypothesis, $\mathscr{H}^2(\mathscr{N})$ is a cyclic subspace for S, there exists $x \in \mathscr{H}^2(\mathscr{N})$ such that $\bigvee_{n=0}^{\infty} \{S_1^n x\}$

$= \mathcal{H}^2(\mathcal{N})$. But $\bigvee_{n=0}^{\infty} \{S_1^n x\} \ominus \bigvee_{n=1}^{\infty} \{S_1^n x\}$ is at most one-dimensional. Since the subspace $\bigvee_{n=1}^{\infty} \{S_1^n x\}$ is contained in the range of S, and the range of S has co-dimension d, we must have $d=1$. $\quad\square$

3.6 Additional Propositions

Proposition 3.1. (A function in $\mathcal{H}^2$ is said to be *outer* if it is a cyclic vector for S.) Every function other than 0 in $\mathcal{H}^2$ can be written as the product of an inner function and an outer function. This factorization is unique up to constant multipliers.

Proposition 3.2. If α is any cardinal number less than or equal to $\aleph_0$, then there is an analytic Toeplitz operator which is unitarily equivalent to the unilateral shift of multiplicity α.

Proposition 3.3. If $\phi \in \mathcal{H}^{\infty}$, then the norm of the analytic Toeplitz operator corresponding to ϕ is the essential supremum of ϕ on C.

Proposition 3.4. (i) An operator A on $\mathcal{H}$ is unitarily equivalent to an operator in $\mathcal{L}^{\infty}$, (i.e., an operator of the form M_{ϕ} on $\mathcal{L}^2(C, \mu)$), if and only if there is an orthonormal basis $\{e_n\}_{n=-\infty}^{\infty}$ for $\mathcal{H}$ such that $(A e_n, e_m) = (A e_{n+1}, e_{m+1})$ for all m and n.

(ii) An operator A on $\mathcal{H}$ is unitarily equivalent to an analytic Toeplitz operator if and only if there is an orthonormal basis $\{e_n\}_{n=0}^{\infty}$ of $\mathcal{H}$ such that $(A e_n, e_m) = (A e_{n+1}, e_{m+1})$ for all m and n and such that $(A e_n, e_m) = 0$ when $m < n$.

Proposition 3.5. The only compact analytic Toeplitz operator is 0.

Proposition 3.6. If $\|A\| \leq 1$, then

$$\begin{pmatrix} A & (1-AA^*)^{\frac{1}{2}} \\ 0 & 0 \end{pmatrix}$$

is a partial isometry. If A and B are invertible and $\|A\| \leq 1$, $\|B\| \leq 1$, then A and B are unitarily equivalent if and only if

$$\begin{pmatrix} A & (1-AA^*)^{\frac{1}{2}} \\ 0 & 0 \end{pmatrix} \quad \text{and} \quad \begin{pmatrix} B & (1-BB^*)^{\frac{1}{2}} \\ 0 & 0 \end{pmatrix}$$

are unitarily equivalent.

Proposition 3.7. If V is any isometry, then there is a unilateral shift S (of some multiplicity) and a unitary operator W such that V is unitarily equivalent to $S \oplus W$.

Proposition 3.8. Let $\{x_n\}$ be an orthonormal basis for $\mathcal{K}$. If $f \in \mathcal{L}^2(\mathcal{K})$, then the function f_n defined by $f_n(z) = (f(z), x_n)$ is in $\mathcal{L}^2$ for each integer n, and $f(z) = \sum_n f_n(z) x_n$, (in the sense of pointwise convergence a.e. with respect to the norm in $\mathcal{K}$). Moreover, $\|f\|^2 = \sum_n \|f_n\|^2$.

Proposition 3.9. If S is any unilateral shift, and if A is an operator such that $Lat\,S \subset Lat\,A$, then it follows (from considering adjoints) that $AS = SA$.

Proposition 3.10. Let S be any unilateral shift.

(i) If $\lambda \in \Pi_0(S^*)$, then $\bigvee_{n=1}^{\infty} \text{null}(S^* - \lambda)^n = \mathcal{H}$.

(ii) If E is a subset of the unit disc D, then $\bigvee_{\lambda \in E} \text{null}(S^* - \lambda) = \mathcal{H}$ if and only if there is a countable subset $\{\lambda_j\}$ of E such that $\sum_{j=1}^{\infty} (1 - |\lambda_j|) = \infty$. In particular, $\bigvee_{\lambda \in E} \text{null}(S^* - \lambda) = \mathcal{H}$ whenever E is uncountable.

Proposition 3.11. Every unitary operator is the product of two bilateral shifts. (One proof uses the fact that every unitary operator is the direct sum of infinitely many unitary operators.)

Proposition 3.12. If A is quasinilpotent and $\varepsilon > 0$, then there exists an operator of norm less than ε which is similar to A. If A is a compact operator and $\|A^n\| \leq M$ for all n, then A is similar to a contraction.

Proposition 3.13. Let S be the unilateral shift of multiplicity n, and let A be an operator on a finite-dimensional space whose distinct eigenvalues are $\lambda_1, \ldots, \lambda_m$. Then A is similar to a part of S^* if and only if $|\lambda_j| < 1$ and n is greater than or equal to the dimension of the nullspace of $A - \lambda_j$ for every j.

Proposition 3.14. If $V \in \mathscr{F}_0$ and x and y are any fixed vectors in $\mathcal{K}$, then the function $z \to (V(z)x, y)$ is in $\mathscr{H}^\infty$. It follows that two functions in $\mathscr{F}_0$ which agree on a set of positive measure are equal a.e.

Proposition 3.15. If every $V \in \mathscr{F}_0$ such that V is unitary a.e. and $\hat{V} \mathscr{H}^2(\mathcal{K})$ has co-dimension greater than 1 can be factored as in Theorem 3.32 with W non-constant, then every operator has a non-trivial invariant subspace.

Proposition 3.16. Theorem 3.33 follows easily from Proposition 3.7.

3.7 Notes and Remarks

The basic properties of functions of class $\mathcal{H}^p$ were developed early in the twentieth century in the work of Hardy, (for whom the "$\mathcal{H}$" stands), Littlewood, Fatou, F. Riesz, M. Riesz, and others: an excellent account of this work is given in Duren [1]. Privalov [1] and Goluzin [1] contain a great deal of classical material, and Hoffman [1], Porcelli [1] and de Branges-Rovnyak [1] present more functional-analytic approaches.

In 1949 Beurling [1] appeared, and this paper led to a tremendous amount of work along the lines presented in Chapter 3 and also stimulated renewed interest in the classical work on the subject. Beurling [1] discussed inner and outer functions and proved the fundamental Theorem 3.9; the proof we present is a simplification due to Srinivasan (cf. Helson [1]) of the proof found by Helson and Lowdenslager [1]. Corollary 3.10 is due to F. and M. Riesz [1]; our proof is from Halmos [3]. Corollary 3.16 follows trivially, of course, from the factorization theory for inner functions; the independent proof presented in the text appears to be new.

The results of Sections 3.3 and 3.4 generalizing Beurling's work to shifts of multiplicity greater than 1 are due to Lax [1] in the finite-multiplicity case and Halmos [8] and Helson and Lowdenslager [1] in the general case. The geometric approach to Theorem 3.25 presented above is due to Halmos [8]; our exposition of these results has been strongly influenced by Fillmore [1]. The proof of Lemma 3.24 above was found by I. Halperin (cf. Sz.-Nagy-Foiaş [2]) and is much easier than Halmos' original one. Theorem 3.28 was discovered by Rota [1], and Theorem 3.29 was independently obtained by Foias [1] and de Branges and Rovnyak [2]. Caradus [1] shows that there are many universal operators in the sense of Corollary 3.30.

A more analytic treatment of shifts of higher multiplicity, as well as numerous related results and references to applications to prediction theory, can be found in Helson [1] and Sz.-Nagy-Foias [1].

Propositions 3.3, 3.4, and 3.5 are from Brown-Halmos [1], and Proposition 3.6 is from Halmos-McLauglin [1]. Proposition 3.7 is due to von Neumann [3] and, independently, to Wold [1] who applies it to problems in statistics. A proof similar to that of Lemma 3.23 in the text is in Halmos [8]. Proposition 3.9 was observed by Sarason [4], while Proposition 3.11 is from Halmos-Kakutani [1]. The first assertion of Proposition 3.12 is from Rota [1], and the second is from Sz.-Nagy [1]. A nice discussion of Proposition 3.15 is in Fillmore ([1], pp. 43—44).

Some additional results on invariant subspaces of shifts can be found in Ahern-Clark ([1], [2]), Douglas-Shapiro-Shields [1], and Sherman ([1], [2]).

Chapter 4. Examples of Invariant Subspace Lattices

If A is any operator, then $Lat\,A$ is a complete lattice of subspaces, (Section 0.2). The problem of "determining $Lat\,A$" for specific operators A is not a very clearly posed problem. For example, Corollary 3.26 and Theorem 3.30 allow us to "determine $Lat\,A$" in some sense for any operator A. However, the description of $Lat\,A$ so obtained does not, in general, even help in deciding whether or not A has any non-trivial invariant subspaces.

An adequate description of $Lat\,A$ must enable us to decide whether any two given elements of $Lat\,A$ are comparable, and to determine $\bigvee_{n=0}^{\infty} \{A^n x\}$ for each vector x. We will determine $Lat\,A$ in this sense for certain particular operators; such results can have interesting analytic implications, (e.g., see Section 4.4). Theorem 3.9 and the theory of inner functions discussed in Section 3.1 give such a description of the invariant subspaces of the unilateral shift of multiplicity 1. It is likely that the problem of completely characterizing invariant subspace lattices is too difficult to be solved in general.

4.1 Preliminaries

Definition. An abstract lattice $\mathscr{L}$ is *attainable* if there exists an operator A on a separable, infinite-dimensional complex Hilbert space such that $Lat\,A$ is order-isomorphic to $\mathscr{L}$; we use the notation $Lat\,A \approx \mathscr{L}$. The notation $Lat\,A$ stands for the collection of subspaces invariant under A; we sometimes abuse this notation by letting it denote the abstract lattice which is order-isomorphic to $Lat\,A$. The meaning should be clear from the context.

Note that the invariant subspace problem can be re-phrased: is the two-element totally ordered lattice, (i.e., the ordinal number 2), attainable?

New attainable lattices can often be obtained from given ones, and many of the results of this chapter indicate techniques for doing this.

Theorem 4.1. *If $\mathscr{L}$ is an attainable lattice, so is the dual of $\mathscr{L}$; (i.e., the lattice $\mathscr{L}^*$ which is equal to $\mathscr{L}$ as a set and which has the ordering defined by: $x \leqq y$ if and only if $x \geqq y$ in $\mathscr{L}$).*

Proof. Suppose that $\mathscr{L} \approx Lat\,A$. We have observed (Proposition 0.1) that $Lat\,A^* = \{\mathscr{M} : \mathscr{M}^\perp \in Lat\,A\}$. Since $\mathscr{M}^\perp \subset \mathscr{N}^\perp$ if and only if $\mathscr{M} \supset \mathscr{N}$, it follows that $Lat\,A^*$ is the dual of $\mathscr{L}$. $\quad\square$

Definition. *A segment $[x,y]$ in a lattice $\mathscr{L}$ is a sublattice of the form $[x,y] = \{z \in \mathscr{L} : x \leqq z \leqq y\}$, (where $x \leqq y$).*

Theorem 4.2. *Let $\mathscr{M}$ and $\mathscr{N}$ be in $Lat\,A$, with $\mathscr{M} \subset \mathscr{N}$ and $\mathscr{N} \ominus \mathscr{M}$ infinite-dimensional. Then $[\mathscr{M}, \mathscr{N}]$ is attainable.*

Proof. Let P denote the projection onto $\mathscr{N} \ominus \mathscr{M}$, and let B denote the compression of A to $\mathscr{N} \ominus \mathscr{M}$, i.e., $B = PA|(\mathscr{N} \ominus \mathscr{M})$. We show that $\mathscr{K} \in Lat\,B$ if and only if $\mathscr{K} \oplus \mathscr{M} \in Lat\,A$ and $\mathscr{K} \oplus \mathscr{M} \subset \mathscr{N}$. Suppose that $\mathscr{K} \in Lat\,B$ and that $k \in \mathscr{K}$, $m \in \mathscr{M}$. Then $A(k+m) = Ak + Am$. Now $Am \in \mathscr{M}$, and $Ak = PAk + (1-P)Ak = Bk + (1-P)Ak$. Since $Ak \in \mathscr{N}$ it follows that $(1-P)Ak \in \mathscr{M}$, and thus $A(k+m) \in \mathscr{K} \oplus \mathscr{M}$. Therefore $\mathscr{K} \in Lat\,B$ implies $\mathscr{K} \oplus \mathscr{M} \in Lat\,A$.

It is trivial to see that $\mathscr{K} \oplus \mathscr{M} \in Lat\,A$ and $\mathscr{K} \subset \mathscr{N}$ implies that $\mathscr{K} \in Lat\,B$. Thus $Lat\,B$ is order-isomorphic to the segment $[\mathscr{M}, \mathscr{N}]$ of $Lat\,A$. $\quad\square$

Definition. The operator A is *unicellular* if $Lat\,A$ is totally ordered.

Theorem 4.2 can be used to give an example of a unicellular operator.

Example 4.3. *If S is the unilateral shift of multiplicity 1, if ϕ_1 is the inner function defined by $\phi_1(z) = \exp\{(z+1)/(z-1)\}$, and if P is the projection onto $(\phi_1 \mathscr{H}^2)^\perp$, then the operator $PS|(\phi_1 \mathscr{H}^2)^\perp$ is unicellular. In fact, $Lat(PS|(\phi_1 \mathscr{H}^2)^\perp) \approx [0,1]$.*

Proof. It follows from the theory of inner functions, (see the discussion in Section 3.1), that the inner functions ϕ such that ϕ_1/ϕ is inner are the functions of the form $\phi(z) = \exp\{a(z+1)/(z-1)\}$ for $a \in [0,1]$. Thus, by Corollary 3.12 and the proof of Theorem 4.2, the invariant subspaces of $PS|(\phi_1 \mathscr{H}^2)^\perp$ are precisely the subspaces of the form $P(\phi \mathscr{H}^2)$ with $\phi(z) = \exp\{a(z+1)/(z-1)\}$ for some $a \in [0,1]$. It follows that $Lat(PS|(\phi_1 \mathscr{H}^2)^\perp) \approx [0,1]$. $\quad\square$

There is one easily proven general result on unicellular operators.

Theorem 4.4. *If A is unicellular, then the set of cyclic vectors of A is a residual set; (i.e., the set of non-cyclic vectors is a set of first category).*

Proof. We shall need the trivial fact that a proper closed subspace is nowhere dense. If $\{\mathscr{M}_\alpha\}_{\alpha \in \Lambda}$ is the family of proper invariant subspaces

of A, then clearly the set of non-cyclic vectors of A is equal to $\bigcup_{\alpha \in \Lambda} \mathcal{M}_\alpha$.
Let $\mathcal{S} = \bigcup_{\alpha \in \Lambda} \mathcal{M}_\alpha$. Since $Lat\, A$ is totally ordered, $\mathcal{S}$ is a linear manifold
in $\mathcal{H}$. If the closure of $\mathcal{S}$ is not $\mathcal{H}$, then we are done: $\mathcal{S}$ is nowhere
dense. If the closure of $\mathcal{S}$ is $\mathcal{H}$, let $\{x_i\}_{i=1}^\infty$ be a countable dense subset
of $\mathcal{S}$. For each i choose an $\mathcal{M}_{\alpha_i}$ containing x_i. Then $\mathcal{S} = \bigcup_{i=1}^\infty \mathcal{M}_{\alpha_i}$, for
if $\mathcal{M}_\alpha \supset \mathcal{M}_{\alpha_i}$ for all i, then $\mathcal{M}_\alpha$ contains $\{x_i\}_{i=1}^\infty$ and $\mathcal{M}_\alpha = \mathcal{H}$. The set $\mathcal{S}$
is therefore the countable union of the closed nowhere dense sets $\mathcal{M}_{\alpha_i}$.   ☐

Corollary 4.5. *If A is unicellular, then every invariant subspace of A is cyclic.*

Proof. Let $\mathcal{M} \in Lat\, A$. Then $A|\mathcal{M}$ is unicellular, and $A|\mathcal{M}$ has a cyclic vector by Theorem 4.4.   ☐

4.2 Algebraic Operators

We first consider two results about operators on finite-dimensional spaces.

Theorem 4.6. *If A is an operator on a finite-dimensional complex Hilbert space, then $Lat\, A$ is self-dual; (i.e., order-isomorphic to its dual).*

Proof. By the proof of Theorem 4.1 it suffices to show that $Lat\, A$
and $Lat\, A^*$ are isomorphic. Choose an orthonormal basis $\{e_i\}$ for $\mathcal{H}$.
Then the matrix of A^* with respect to $\{e_i\}$ is the conjugate transpose
of the matrix of A. Let A^T denote the operator corresponding to the
transpose of A. It is a standard fact in linear algebra, (verified by considering Jordan canonical forms), that A and A^T are similar. Hence
$Lat\, A \approx Lat\, A^T$, and we need only show that $Lat\, A^T \approx Lat\, A^*$. If
$\mathcal{M} \in Lat\, A^T$, and if $\mathcal{M}^* = \{\sum \alpha_i e_i : \sum \bar{\alpha}_i e_i \in \mathcal{M}\}$, then clearly $\mathcal{M}^* \in Lat\, A^*$.
It is obvious that the correspondence $\mathcal{M} \leftrightarrow \mathcal{M}^*$ is an order-isomor-
phism between $Lat\, A^T$ and $Lat\, A^*$.   ☐

Note that invariant subspace lattices need not be self-dual on infinite-dimensional spaces: the lattice of Corollary 4.13 below is not
self-dual.

Theorem 4.7. *An operator on a finite-dimensional space is unicellular if and only if it is cyclic and its spectrum is a singleton.*

Proof. Let A be unicellular. Then A is cyclic, by Theorem 4.4, and
if $\sigma(A)$ contained two points then A would have two non-comparable
invariant subspaces, (the corresponding eigenspaces). For the converse
first note that if A is cyclic and $\sigma(A) = \{\lambda\}$, then the Jordan canonical

form of A is

$$J = \begin{pmatrix} \lambda & 1 & & & \\ & \lambda & \ddots & & \\ & & \ddots & \ddots & \\ & & & \ddots & 1 \\ & & & & \lambda \end{pmatrix},$$

with respect to some basis $\{e_i\}_{i=1}^n$.

The invariant subspaces of J are the same as the invariant subspaces of $J - \lambda$. We shall show that every non-trivial invariant subspace of $J - \lambda$ has the form $\mathcal{M}_k = \bigvee_{n=1}^k \{e_n\}$ for some k. Since the span of any subspaces of this form is again of this form, it suffices to prove that every cyclic subspace is an $\mathcal{M}_k$. Let $f = \sum_{i=1}^k \alpha_i e_i$ with $\alpha_k \neq 0$, and let $\mathcal{M} = \bigvee_{n=0}^\infty \{(J - \lambda)^n f\}$. Then $(J - \lambda)^{(k-1)} f = \alpha_k e_1$, and thus $e_1 \in \mathcal{M}$. If $k \geq 2$ then $(J - \lambda)^{(k-2)} f = \alpha_k e_2 + \alpha_{k-1} e_1$, and thus $e_2 \in \mathcal{M}$. Continuing in this manner shows that the span of $\{(J - \lambda)^n f : n \leq k\}$ is $\mathcal{M}_k$, and thus $\mathcal{M} = \mathcal{M}_k$. $\square$

Operators on finite-dimensional spaces are *algebraic*, in the sense that whenever A is an operator on a finite-dimensional space there exists a non-zero polynomial p such that $p(A) = 0$; (this follows from the fact that the space of operators is finite-dimensional). On infinite-dimensional spaces, of course, there are many operators which are not algebraic. The algebraic operators on infinite-dimensional spaces can be characterized in terms of their invariant subspaces.

Theorem 4.8. *An operator is algebraic if and only if the union of its finite-dimensional invariant subspaces is $\mathcal{H}$.*

Proof. If A is an operator and p is a non-zero polynomial such that $p(A) = 0$, then obviously the dimension of $\bigvee_{n=0}^\infty \{A^n x\}$ is less than or equal to the degree of p. Hence every vector x is contained in a finite-dimensional invariant subspace of A.

Conversely, suppose that the union of the finite-dimensional members of $Lat\,A$ is $\mathcal{H}$. Then, for each $x \in \mathcal{H}$, $\bigvee_{n=0}^\infty \{A^n x\}$ is finite-dimensional, and thus there exists an integer k such that $A^k x$ is a linear combination of $\{x, Ax, \dots, A^{k-1} x\}$. Thus for each x there is a non-zero polynomial p_x such that $p_x(A) x = 0$. Let $\mathscr{F}_k$ denote the set of all vectors x such that $p_x(A) x = 0$ for some non-zero polynomial p_x of degree less than or equal to k. Then $\bigcup_{k=1}^\infty \mathscr{F}_k = \mathcal{H}$.

If $\{x_n\}$ is a sequence in $\mathscr{F}_k$ and $\{x_n\} \to x$, then $x \in \mathscr{F}_k$. For suppose that, for each n, p_{x_n} is a non-zero polynomial of degree at most k such that $p_{x_n}(A) x_n = 0$. Divide p_{x_n} by its largest coefficient; we can thus assume that each coefficient of p_{x_n} has absolute value at most 1 and that at least one coefficient has absolute value equal to 1. Then a subsequence of $\{p_{x_n}\}$ converges coefficient-wise to a polynomial p of degree at most k; p is not 0, since at least one of its coefficients has modulus 1. Re-label so that $\{p_{x_n}\}$ is such a subsequence. Then $\{p_{x_n}(A)\}$ converges to $p(A)$ uniformly, and it follows that $p(A) x = 0$.

Thus each $\mathscr{F}_k$ is closed, and by the Baire category theorem there exists some k such that $\mathscr{F}_k$ contains an open ball, say $\{x : \|x - x_0\| < \varepsilon\}$. If $y = x - x_0$, then $p_x(A) p_{x_0}(A) y = 0$. Therefore whenever $\|y\| < \varepsilon$ there exists a non-zero polynomial p_y of degree at most $2k$ such that $p_y(A) y = 0$. Since $p_y(A)(\alpha y) = \alpha(p_y(A) y)$, it follows that every vector is in $\mathscr{F}_{2k}$.

The rest of the proof reduces to finite-dimensional linear algebra. For each vector $x \neq 0$ let n_x denote the degree of the lowest-degree non-zero polynomial p_x such that $p_x(A) x = 0$, and let $n = \max\{n_x\}$. Then $n \leq 2k$. Choose any x such that $n = n_x$, and let p be a polynomial of degree n such that $p(A) x = 0$ and the leading coefficient of p is 1. We claim that $p(A) y = 0$ for all y. Given y, let $\mathscr{M} = \bigvee_{j=0}^{\infty} \{A^j x, A^j y\}$. Then $\mathscr{M}$ is an invariant subspace of A, and the dimension of $\mathscr{M}$ is at most $2n$. Let q denote the minimal polynomial of $A|\mathscr{M}$. Since the set of polynomials r such that $r(A) x = 0$ is an ideal (in the ring of polynomials), it follows that p divides q. Moreover, the degree of q is equal to the degree of p, (since there must exist a vector $z \in \mathscr{M}$ such that no polynomial in A of degree less than the degree of q annihilates z—this is a basic fact that is used in deriving the Jordan canonical form theorem). Thus p is a multiple of q, and $p(A|\mathscr{M}) = 0$. For any y, then, $p(A) y = 0$. □

4.3 Lattices of Normal Operators

The problem of determining $Lat\,A$ is often quite difficult even in the case where A is normal. The bilateral shift, for example, has a rather complex invariant subspace lattice, as was shown in Chapter 3. We shall determine the lattices of some completely normal operators. First we consider an easy general result: the lattices of normal operators are similar to the lattices of finite-dimensional operators in one respect.

Theorem 4.9. *If A is normal, then $Lat\,A$ is self-dual.*

Proof. By Theorem 1.6 we can assume that A is an operator of the form M_ϕ on the space $\mathscr{L}^2(X, \mu)$. We must show that $Lat\,M_\phi \approx Lat\,M_\phi^*$.

For $\mathscr{M} \in Lat\, M_\phi$ define $\mathscr{M}^* = \{\bar{f} : f \in \mathscr{M}\}$, where $\bar{f}(x) = \overline{f(x)}$ for each x. Then the fact that $M_\phi^* = M_{\bar\phi}$ implies that $\mathscr{M}^* \in Lat\, M_\phi^*$. The map $\mathscr{M} \to \mathscr{M}^*$ is therefore an order-isomorphism between $Lat\, M_\phi$ and $Lat\, M_\phi^*$. □

Example 4.10. *Let $\{\lambda_n\}_{n=1}^\infty$ be a sequence of distinct complex numbers which converges to 0, and let $\{e_n\}_{n=1}^\infty$ be an orthonormal basis for $\mathscr{H}$. If A is the operator defined by $Ae_n = \lambda_n e_n$, then $Lat\, A$ is isomorphic to the set of all subsets of the natural numbers.*

Proof. We show that $Lat\, A$ consists of all subspaces of the form $\bigvee_{n \in \mathscr{S}} \{e_n\}$ for $\mathscr{S}$ a subset of the natural numbers. Clearly every subspace of this form is in $Lat\, A$. If $\mathscr{M} \in Lat\, A$ then, since A is compact and normal, Corollary 1.24 implies that $\mathscr{M}$ reduces A. Let P denote the orthogonal projection onto $\mathscr{M}$; then $PA = AP$. Thus $APe_n = PAe_n = \lambda_n Pe_n$, and, since the eigenspace of A corresponding to λ_n is one-dimensional, it follows that for each n there exists a γ_n such that $Pe_n = \gamma_n e_n$. Since $P^2 = P$, each γ_n is either 0 or 1. If $\mathscr{S} = \{n : \gamma_n = 1\}$ then $\mathscr{M} = \bigvee_{n \in \mathscr{S}} \{e_n\}$. □

This example is one of a class of examples. Let (X, μ) be a finite measure space and let $\mathscr{L}^\infty$ denote the algebra of multiplication operators on $\mathscr{L}^2(X, \mu)$ as in Section 1.6. Let $\mathscr{G}$ denote the lattice of equivalence classes of measurable subsets of X, where E and F are equivalent if $\mu(E \backslash F) = \mu(F \backslash E) = 0$, ordered by inclusion.

Example 4.11. *If the smallest weakly closed algebra containing the operator A is $\mathscr{L}^\infty$, then $Lat\, A \approx \mathscr{G}$. In particular, in the case where $X = [0, 1]$, μ is ordinary Lebesgue measure and M_x is multiplication by the function $\phi(x) = x$, it follows that $Lat\, M_x \approx \mathscr{G}$.*

Proof. The first assertion is immediate: since the weakly closed algebra generated by A contains A^*, every invariant subspace is reducing, and if P is a projection that commutes with A, then P commutes with $\mathscr{L}^\infty$ and hence is multiplication by a characteristic function χ_E (by Theorem 1.20). The correspondence between invariant subspaces and measurable subsets defined by $\{\chi_E f : f \in \mathscr{L}^2(X, \mu)\} \to E$ is obviously an isomorphism between $Lat\, A$ and $\mathscr{G}$.

The second assertion is an easy consequence of the first. By the Weierstrass theorem every continuous function ϕ is a uniform limit of polynomials, and $\| M_p - M_\phi \|$ is $\sup_{x \in [0,1]} |p(x) - \phi(x)|$. Thus, whenever ϕ is continuous, M_ϕ is a uniform limit of a sequence of polynomials in M_x. If χ_E is the characteristic function of the measurable set E, and if $\varepsilon > 0$, choose a compact set K and an open set U with $K \subset E \subset U$ such that $\mu(U\setminus K) < \varepsilon$. By Urysohn's Lemma there exists a continuous function ϕ mapping $[0,1]$ into $[0,1]$ such that ϕ is 1 on K and 0 on $[0,1]\setminus U$. If $f \in \mathscr{L}^2(0,1)$, then

$$\|\phi f - \chi_E f\|^2 = \int_K |\phi f - \chi_E f|^2 \, d\mu + \int_{[0,1]\setminus U} |\phi f - \chi_E f|^2 \, d\mu + \int_{U\setminus K} |\phi f - \chi_E f|^2 \, d\mu$$

$$= 0 + 0 + \int_{U\setminus K} |\phi - \chi_E|^2 |f|^2 \, d\mu \leq \int_{U\setminus K} |f|^2 \, d\mu.$$

The absolute continuity of the integral implies that $\int_{U\setminus K} |f|^2 \, d\mu$ approaches 0 with ε.

Thus M_{χ_E} is the strong limit of a sequence of polynomials in M_x. Since every $\psi \in \mathscr{L}^\infty(0,1)$ can be uniformly approximated by a sequence of linear combinations of characteristic functions, it follows that M_x generates $\mathscr{L}^\infty$. □

In Chapter 7 we shall see that, whenever $\mathscr{L}^2(X,\mu)$ is separable, the algebra $\mathscr{L}^\infty$ is generated by a single operator. The space $\mathscr{L}^2(X,\mu)$ is separable if and only if the measure space (X,μ) is separable, (in the sense that the metric space corresponding to the symmetric-difference metric is a separable metric space—see Halmos([1], p. 177). This gives a number of examples of lattices of completely normal operators. On the other hand, the fact that every separable non-atomic normalized measure algebra is isomorphic to the measure algebra of $[0,1]$ (Halmos [1], p. 173) shows that, up to lattice isomorphism, all these lattices are direct products of sublattices of Example 4.10 and the lattice $\mathscr{G}$ associated with $[0,1]$.

The theory of spectral multiplicity (see Halmos [2]) gives a description of the reducing subspaces of arbitrary normal operators, and hence gives a determination of the invariant subspace lattices of all completely normal operators. The operators which generate $\mathscr{L}^\infty$ are of uniform multiplicity 1.

4.4 Two Unicellular Operators

Definition. Let $\mathscr{H}$ have an orthonormal basis $\{e_n\}_{n=0}^\infty$ and let $\{w_n\}_{n=1}^\infty$ be a sequence of non-zero complex numbers such that $\{|w_n|\}_{n=1}^\infty$ is monotone decreasing and is in ℓ^2. Then the operator A such that $Ae_0 = 0$ and $Ae_n = w_n e_{n-1}$ for $n > 0$ is the *Donoghue operator* with weight sequence $\{w_n\}_{n=1}^\infty$.

Given a Donoghue operator A with respect to the basis $\{e_n\}$ let $\mathscr{M}_n = \bigvee_{k=0}^n \{e_k\}$. Then obviously $\mathscr{M}_n \in Lat\, A$ for all n.

Theorem 4.12. *If A is a Donoghue operator and $\mathcal{M}$ is a non-trivial invariant subspace of A, then $\mathcal{M} = \mathcal{M}_n$ for some n.*

Proof. Note first that if $x = \sum_{i=0}^{k} \alpha_i e_i$ with $\alpha_k \neq 0$, it follows (from a computation such as that of the proof of Theorem 4.7) that $\bigvee_{n=0}^{\infty} \{A^n x\} = \mathcal{M}_k$. Also, since the span of any number of $\mathcal{M}_n$'s is again an $\mathcal{M}_n$, (unless it is $\mathcal{H}$), it suffices to show that every cyclic subspace is an $\mathcal{M}_n$. Suppose, then, that $x = \sum_{k=0}^{\infty} \alpha_k e_k$ with infinitely many α_k non-zero. We must show that $\bigvee_{n=0}^{\infty} \{A^n x\} = \mathcal{H}$, or, that $e_k \in \bigvee_{n=0}^{\infty} \{A^n x\}$ for all k. Let $\mathcal{M} = \bigvee_{n=0}^{\infty} \{A^n x\}$.

We first show that $e_0 \in \mathcal{M}$. For this note that

$$A^n x = \sum_{k=n}^{\infty} \alpha_k w_k w_{k-1} \cdots w_{k-n+1} e_{k-n}.$$

If $\alpha_n \neq 0$ then

$$\frac{1}{\alpha_n w_n \cdots w_1} A^n x = e_0 + \sum_{k=n+1}^{\infty} \frac{\alpha_k}{\alpha_n} \frac{w_k w_{k-1} \cdots w_{k-n+1}}{w_n \cdots w_1} e_{k-n}.$$

Now, since $\{|w_n|\}$ is monotone decreasing, $|w_{k-j-1}/w_{n-j}| \leq 1$ for $k \geq n+1$. Thus

$$\left\| \sum_{k=n+1}^{\infty} \frac{\alpha_k}{\alpha_n} \frac{w_k \cdots w_{k-n+1}}{w_n \cdots w_1} e_{k-n} \right\|^2 \leq \sum_{k=n+1}^{\infty} \left| \frac{\alpha_k}{\alpha_n} \right|^2 \left| \frac{w_k}{w_1} \right|^2.$$

Now we show that for all $\varepsilon > 0$ there exists an n such that

$$\left\| \frac{1}{\alpha_n w_n \cdots w_1} A^n x - e_0 \right\|^2 < \varepsilon.$$

Given ε, choose N such that $\sum_{k=N}^{\infty} |w_k/w_1|^2 < \varepsilon$, (recall that $\{w_k\} \in \ell^2$). Then choose $M \geq N$ such that $|\alpha_M| = \max_{k \geq N} \{|\alpha_k|\}$. It follows that

$$\left\| \frac{1}{\alpha_M w_M \cdots w_1} A^M x - e_0 \right\|^2 \leq \sum_{k=M+1}^{\infty} \left| \frac{\alpha_k}{\alpha_M} \right|^2 \left| \frac{w_k}{w_1} \right|^2 \leq \sum_{k=M+1}^{\infty} \left| \frac{w_k}{w_1} \right|^2 < \varepsilon.$$

Thus $e_0 \in \mathcal{M}$.

For each n, therefore,

$$A^n x - \alpha_n w_n \cdots w_1 e_0 = \sum_{k=n+1}^{\infty} \alpha_k w_k \cdots w_{k-n+1} e_{k-n} \quad \text{is in } \mathcal{M}.$$

Whenever α_{n+1} is non-zero, the vector

$$e_1 + \sum_{k=n+2}^{\infty} \frac{\alpha_k w_k \cdots w_{k-n+1}}{\alpha_{n+1} w_{n+1} \cdots w_2} e_{k-n}$$

is in $\mathcal{M}$, and therefore a computation such as that given above shows that $e_1 \in \mathcal{M}$. It is clear that we can continue in this manner to get $e_k \in \mathcal{M}$ for all k. ☐

Corollary 4.13. *The lattices $\omega + 1$ and $1 + {}^*\omega$ are attainable, (where ω denotes the order type of the positive integers and ${}^*\omega$ denotes the order type of the negative integers).*

Proof. If A is a Donoghue operator, then $\operatorname{Lat} A = \{\{0\}, \mathcal{M}_1, \mathcal{M}_2, \ldots, \mathcal{H}\}$, and thus $\operatorname{Lat} A \approx \omega + 1$. It follows that $\operatorname{Lat} A^*$ is the dual of $\omega + 1$, which is $1 + {}^*\omega$. ☐

Definition. *A Volterra-type integral operator* on $\mathscr{L}^2(0,1)$ is an operator K of the form $(Kf)(x) = \int_0^x k(x,y) f(y) dy$, where k is any square-integrable function on the unit square. The *Volterra operator* is the particular Volterra-type operator obtained when k is the constant function 1.

For each $\alpha \in [0,1]$ let

$$\mathcal{M}_\alpha = \{f \in \mathscr{L}^2(0,1): f = 0 \text{ a.e. on } [0,\alpha]\}.$$

It is obvious that $\{\mathcal{M}_\alpha : \alpha \in [0,1]\} \subset \operatorname{Lat} K$ for every Volterra-type integral operator K.

Theorem 4.14. *If V is the Volterra operator, then*

$$\operatorname{Lat} V = \{\mathcal{M}_\alpha : \alpha \in [0,1]\}.$$

Proof. We prove this result by relating the operator V to the unicellular compression of the unilateral shift discussed in Example 4.3. The operator V is compact and quasinilpotent (Proposition 4.11); hence the function $f(z) = (1-z)(1+z)^{-1}$ is analytic on $\eta(\sigma(V))$. Since f is also one-to-one, it follows from Theorem 2.14 that the operator $W = f(V) = (1-V)(1+V)^{-1}$ has the same invariant subspaces as V. We shall show that W is unitarily equivalent to $PS|(\phi_1 \mathcal{H}^2)^\perp$, where S is the unilateral shift, ϕ_1 is the inner function $\phi_1(z) = \exp\{(z+1)(z-1)^{-1}\}$, and P is the projection onto $(\phi_1 \mathcal{H}^2)^\perp$.

To exhibit this unitary equivalence it will be useful to regard $\mathscr{L}^2(0,1)$ as the subspace of $\mathscr{L}^2(0,\infty)$ consisting of all functions in $\mathscr{L}^2(0,\infty)$ whose support is contained in $[0,1]$. Let Q be the orthogonal projection of $\mathscr{L}^2(0,\infty)$ onto $\mathscr{L}^2(0,1)$, (i.e. Q is multiplication by $\chi_{[0,1)}$). Define the operator K on $\mathscr{L}^2(0,\infty)$ by

$$(Kf)(x) = \int_0^x f(t) e^{t-x} dt.$$

We first prove that $(1 + V)^{-1} = Q(1 - K)|\mathscr{L}^2(0, 1)$. For this let $f \in \mathscr{L}^2(0, 1)$; we must show that $(1 + V)Q(1 - K)f = f$.

Now, for $s \in [0, 1]$,

$$(QKf)(s) = \chi_{[0, 1]} \int_0^s f(t)e^{t - s} dt = \int_0^s f(t)e^{t - s} dt,$$

and it follows that

$$[(1 + V)Q(1 - K)f](s) = f(s) + \int_0^s f(t) dt - \int_0^s f(t)e^{t - s} dt - \int_0^s \int_0^x f(t)e^{t - x} dt\, dx.$$

By Fubini's theorem,

$$\int_0^s \int_0^x f(t)e^{t - x} dt\, dx = \int_0^s f(t)\left[\int_t^s e^{t - x} dx\right] dt = \int_0^s [f(t) - f(t)e^{t - s}] dt.$$

Thus

$$[(1 + V)Q(1 - K)f](s) = f(s), \quad \text{and} \quad (1 + V)^{-1} = Q(1 - K)|\mathscr{L}^2(0, 1).$$

It follows that

$$W = (1 - V)(1 + V)^{-1} = 2(1 + V)^{-1} - 1 = Q(1 - 2K)|\mathscr{L}^2(0, 1).$$

We shall show that $1 - 2K$ is unitarily equivalent to S.

We construct an isometry of $\mathscr{L}^2(0, \infty)$ onto $\mathscr{H}^2$ as follows. First, for $f \in \mathscr{L}^2(0, \infty)$, define the function $U_1 f$ with domain the right half-plane, (i.e., $\{w: \text{Re}(w) > 0\}$), by $(U_1 f)(w) = (2\pi)^{-\frac{1}{2}} \int_0^\infty f(t)e^{-wt} dt$. One version of the Paley-Wiener Theorem (see Hoffman [1], p. 131) states that this map is an isometry of $\mathscr{L}^2(0, \infty)$ onto H^2 of the right half-plane. Now the map U_2 of H^2 of the right half-plane onto $\mathscr{H}^2$ defined by $(U_2 g)(z) = 2\sqrt{\pi}(1 - z)^{-1} g((1 + z)(1 - z)^{-1})$ is also an isometry, (Hoffman [1], p. 106). Thus the composition $U_2 U_1$ is an isometry of $\mathscr{L}^2(0, \infty)$ onto $\mathscr{H}^2$. In other words, the linear transformation U of $\mathscr{L}^2(0, \infty)$ into $\mathscr{H}^2$ defined by

$$(Uf)(z) = \frac{\sqrt{2}}{1 - z} \int_0^\infty f(t)e^{-\frac{1 + z}{1 - z}t} dt$$

is a surjective isometry.

We claim that $SU = U(1 - 2K)$, where S is the unilateral shift. To see this let $f \in \mathscr{L}^2(0, \infty)$. Then $U(1 - 2K)(f)(z) = (Uf)(z) - 2(UKf)(z)$, and

$$(UKf)(z) = \frac{\sqrt{2}}{1-z} \int\limits_0^\infty \left[\int\limits_0^t f(s) e^{s-t} ds \right] e^{-\frac{1+z}{1-z}t} dt$$

$$= \frac{\sqrt{2}}{1-z} \int\limits_0^\infty \left[\int\limits_s^\infty e^{\left(-\frac{1+z}{1-z}-1\right)t} dt \right] f(s) e^s ds$$

$$= \frac{\sqrt{2}}{1-z} \int\limits_0^\infty \frac{1-z}{2} e^{\left(-\frac{1+z}{1-z}-1\right)s} f(s) e^s ds$$

$$= \frac{\sqrt{2}}{2} \int\limits_0^\infty f(s) e^{-\frac{1+z}{1-z}s} ds.$$

Therefore

$$U(1-2K)(f)(z) = (Uf)(z) - \sqrt{2} \int\limits_0^\infty f(s) e^{-\frac{1+z}{1-z}s} ds$$

$$= (Uf)(z) - (1-z)(Uf)(z)$$

$$= z(Uf)(z).$$

It follows that $SU = U(1-2K)$, or $U^{-1}SU = 1-2K$.

Now the operator W with $Lat\, W = Lat\, V$ has been shown to be a compression of $1-2K$; it follows that W is unitarily equivalent to a compression of S. For $\alpha \in [0,1]$ let $\mathscr{M}_\alpha = \{ f \in \mathscr{L}^2(0,\infty) : f = 0$ a.e. on $[0,\alpha]\}$. The fact that $Lat\, W = \{\mathscr{M}_\alpha : \alpha \in [0,1]\}$ will follow from Example 4.3 if we show that $U\mathscr{M}_\alpha = \phi_\alpha \mathscr{H}^2$, where ϕ_α is the inner function $\phi_\alpha(z) = e^{\alpha(z+1)(z-1)^{-1}}$. Suppose that $f \in \mathscr{L}^2(0,\infty)$ and $f = 0$ a.e. on $[0,\alpha]$. Then

$$\frac{(Uf)(z)}{\phi_\alpha} = e^{-\frac{z+1}{z-1}\alpha} \cdot \frac{\sqrt{2}}{1-z} \int\limits_0^\infty f(t) e^{-\frac{1+z}{1-z}t} dt$$

$$= \frac{\sqrt{2}}{1-z} \int\limits_\alpha^\infty f(t) e^{-\frac{1+z}{1-z}(t-\alpha)} dt$$

$$= \frac{\sqrt{2}}{1-z} \int\limits_0^\infty f(t+\alpha) e^{-\frac{1+z}{1-z}t} dt.$$

Thus $(Uf/\phi_\alpha)\in\mathcal{H}^2$, $Uf\in\phi_\alpha\mathcal{H}^2$, and we conclude that $U\mathcal{M}_\alpha\in\phi_\alpha\mathcal{H}^2$ for each α. For the reverse inclusion let $\phi_\alpha g\in\phi_\alpha\mathcal{H}^2$. Then $g=Uf$ for some $f\in\mathcal{L}^2(0,\infty)$; i.e., $g(z)=\sqrt{2}(1-z)^{-1}\int_0^\infty f(t)e^{-(1+z)(1-z)^{-1}t}\,dt$. It follows that

$$\phi_\alpha(z)g(z) = \frac{\sqrt{2}}{1-z}\int_0^\infty f(t)e^{-\frac{1+z}{1-z}(t+\alpha)}\,dt$$

$$= \frac{\sqrt{2}}{1-z}\int_\alpha^\infty f(t-\alpha)e^{-\frac{1+z}{1-z}t}\,dt.$$

If we define

$$h(t) = \begin{cases} f(t-\alpha) & \text{for } t\geq\alpha \\ 0 & \text{for } t<\alpha \end{cases},$$

then $h\in\mathcal{M}_\alpha$ and $\phi_\alpha g=Uh$. $\square$

The computational drudgery involved in the above proof is amply rewarded by the fact that the following important classical theorem is an easy corollary. Recall that the *convolution* of two functions f and g in $\mathcal{L}^1(0,1)$ is defined by $(f*g)(x)=\int_0^x f(x-t)g(t)dt$; convolution is associative and commutative.

Theorem 4.15 *(The Titchmarsh Convolution Theorem). If f and g are in $\mathcal{L}^1(0,1)$, 0 is in the support of f (i.e., there does not exist any $\alpha>0$ such that $f=0$ a.e. on $[0,\alpha]$), and if $f*g=0$ a.e., then $g=0$ a.e.*

Proof. The connection between this result and the Volterra operator stems from the obvious relation $Vf=u*f$, where u is the constant function 1. First consider the case where f and g are continuous. Then, if $u^{(n)}$ denotes $u*u*\ldots*u$ (n factors), the relation $V^nf=u^{(n)}*f$ gives $0=f*g=u^{(n)}*f*g=(V^nf)*g$. Thus the continuous function V^nf*g is 0 everywhere. In particular, $0=(V^nf*g)(1)=\int_0^1 (V^nf)(t)g(1-t)dt$. Thus $\overline{g(1-t)}$ is orthogonal, in $\mathcal{L}^2(0,1)$, to V^nf for all n. It follows, since $\bigvee_{n=0}^\infty \{V^nf\}=\mathcal{L}^2(0,1)$ by Theorem 4.14, that $\overline{g(1-t)}=0$ for all $t\in[0,1]$, and hence that $g=0$.

The general case follows easily from the case of continuous functions. For if f and g are in $\mathcal{L}^1(0,1)$ and $f*g=0$, then

$$(u*f)*(u*g) = u*f*u*g = u^{(2)}*f*g=0,$$

and $u*f$ and $u*g$ are continuous. Moreover, if 0 is in the support of f, then 0 is in the support of $u*f$. Thus the continuous case implies that $u*g=0$, and therefore $g=0$ a.e. $\square$

4.5 Direct Products of Attainable Lattices

Definition. If $\{\mathscr{L}_1, \mathscr{L}_2, \mathscr{L}_3, ...\}$ is a finite or denumerable collection of abstract lattices, their *direct product*, denoted $\mathsf{X}\mathscr{L}_n$, is the lattice whose elements are the elements of the Cartesian product of the $\{\mathscr{L}_n\}$ and whose partial ordering is defined coordinate-wise. That is, $(a_1, a_2, ...)$ $\leqq (b_1, b_2, ...)$ if and only if $a_n \leqq b_n$ for all n.

Note that $\{\mathscr{M} \oplus \mathscr{N} : \mathscr{M} \in Lat\, A, \ \mathscr{N} \in Lat\, B\}$, partially ordered by inclusion, is order-isomorphic to $Lat\, A \times Lat\, B$. Obviously

$$Lat(A \oplus B) \supset \{\mathscr{M} \oplus \mathscr{N} : \mathscr{M} \in Lat\, A, \mathscr{N} \in Lat\, B\}.$$

Definition. If $Lat(A \oplus B) = \{\mathscr{M} \oplus \mathscr{N} : \mathscr{M} \in Lat\, A, \mathscr{N} \in Lat\, B\}$, then we shall say that the invariant subspaces of $A \oplus B$ *split*. When this is the case, then, $Lat(A \oplus B) \approx Lat\, A \times Lat\, B$.

Theorem 4.16. *If $\eta(\sigma(A)) \cap \eta(\sigma(B)) = \emptyset$, then the invariant subspaces of $A \oplus B$ split.*

Proof. First, $\sigma(A \oplus B) = \sigma(A) \cup \sigma(B)$. It is easily verified that $\eta(\sigma(A \oplus B)) \supset \eta(\sigma(A)) \cup \eta(\sigma(B))$. In fact, $\eta(\sigma(A \oplus B)) = \eta(\sigma(A)) \cup \eta(\sigma(B))$; the second inclusion follows immediately from a property of the extended complex plane (or 2-sphere) known as unicoherence (see Wilder [1], p. 47 and p. 60). Given this relation, let U_1 and U_2 be disjoint open sets containing $\eta(\sigma(A))$ and $\eta(\sigma(B))$ respectively. Let f_1 denote the characteristic function of U_1 and f_2 the characteristic function of U_2. Then f_1 and f_2 are analytic on $\eta(\sigma(A \oplus B))$. Also $f_1(A \oplus B) = f_1(A) \oplus f_1(B)$ $= 1 \oplus 0$, and $f_2(A \oplus B) = 0 \oplus 1$, by the basic properties of the functional calculus.

We must show that $\mathscr{K} \in Lat(A \oplus B)$ implies that $\mathscr{K} = \mathscr{M} \oplus \mathscr{N}$. Given $\mathscr{K} \in Lat(A \oplus B)$ let $\mathscr{M} \oplus \{0\} = (1 \oplus 0)\mathscr{K}$ and $\{0\} \oplus \mathscr{N} = (0 \oplus 1)\mathscr{K}$. Obviously $\mathscr{K} \subset \mathscr{M} \oplus \mathscr{N}$. Moreover, by Corollary 2.13, $f_1(A \oplus B)\mathscr{K} \subset \mathscr{K}$ and $f_2(A \oplus B)\mathscr{K} \subset \mathscr{K}$. Thus $\mathscr{M} \oplus \mathscr{N} \subset \mathscr{K}$, and it follows that $\mathscr{K} = \mathscr{M} \oplus \mathscr{N}$. Clearly $\mathscr{M} \in Lat\, A$ and $\mathscr{N} \in Lat\, B$. $\square$

Corollary 4.17. *If $\mathscr{L}_1$ and $\mathscr{L}_2$ are attainable, then $\mathscr{L}_1 \times \mathscr{L}_2$ is attainable.*

Proof. Suppose that $Lat\, A_1 \approx \mathscr{L}_1$ and $Lat\, A_2 \approx \mathscr{L}_2$. Choose any two disjoint closed disks D_1 and D_2. Then multiply and translate A_1 and A_2 by appropriate non-zero scalars $\alpha_1, \alpha_2, \beta_1, \beta_2$ so that $\sigma(\alpha_j A + \beta_j) \subset D_j$.

Then $\eta(\alpha_j A_j + \beta_j) \subset D_j$, and Theorem 4.16 implies that

$$Lat[(\alpha_1 A_1 + \beta_1) \oplus (\alpha_2 A_2 + \beta_2)] \approx \mathscr{L}_1 \times \mathscr{L}_2. \quad \square$$

It is clear that Corollary 4.17 can be extended to show that the direct product of a finite number of attainable lattices is attainable. With a little more care this can be extended to the case of denumerably many attainable lattices.

Theorem 4.18. *If $\{\mathscr{L}_j\}_{j=1}^{\infty}$ is a collection of attainable lattices, then $\underset{j=1}{\overset{\infty}{\times}} \{\mathscr{L}_j\}$ is attainable.*

Proof. Suppose that $Lat A_j \approx \mathscr{L}_j$ for each j.

Let $\{D_j\}$ be a sequence of disjoint closed disks in the complex plane such that the centre of D_j is at the point $1/j$ on the real axis. By multiplying and translating A_j by appropriate scalars we can assume that $\|A_j\| \leq 1/j$ and that $\sigma(A_j)$ is contained in the interior of D_j. Let $A = \sum_{j=1}^{\infty} \oplus A_j$; we must show that $\mathscr{M} \in Lat A$ implies $\mathscr{M} = \sum_{j=1}^{\infty} \oplus \mathscr{M}_j$ with $\mathscr{M}_j \in Lat A_j$.

In order to apply Theorem 4.16 we must describe $\eta(\sigma(A))$. We claim that $\sigma(A) = \left(\bigcup_{j=1}^{\infty} \sigma(A_j)\right) \cup \{0\}$. Obviously $\sigma(A) \supset \left(\bigcup_{j=1}^{\infty} \sigma(A_j)\right) \cup \{0\}$. Now if $\lambda \notin \left(\bigcup_{j=1}^{\infty} \sigma(A_j)\right) \cup \{0\}$, choose j_0 such that $1/j_0 < |\lambda|$. Then $|\lambda| > \left\| \sum_{j=j_0}^{\infty} \oplus A_j \right\|$, and thus $\left(\sum_{j=j_0}^{\infty} \oplus A_j - \lambda\right)$ is invertible. Since

$$\sigma\left(\sum_{j=1}^{j_0-1} \oplus A_j\right) = \bigcup_{j=1}^{j_0-1} \sigma(A_j),$$

$A - \lambda$ is invertible, proving the claim. Then it is obvious that $\eta(\sigma(A)) \subset \left(\bigcup_{j=1}^{\infty} D_j\right) \cup \{0\}$. It follows that

$$\eta(\sigma(A)) = \left(\bigcup_{j=1}^{\infty} \eta(\sigma(A_j))\right) \cup \{0\}.$$

Now let $\mathscr{M} \in Lat A$, and let f_j denote the characteristic function of D_j. Then, by the above, f_j is analytic on $\eta(\sigma(A))$, and by Corollary 2.13, $f_j(A)\mathscr{M}$ is contained in $\mathscr{M}$. Also, $f_j(A)\mathscr{M}$ has the form

$$\{0\} \oplus \cdots \oplus \{0\} \oplus \mathscr{M}_j \oplus \{0\} \oplus \cdots,$$

where $\mathscr{M}_j$ is in the jth coordinate space. It follows that $\mathscr{M} = \sum_{j=1}^{\infty} \oplus \mathscr{M}_j$, and it is obvious that $\mathscr{M}_j \in Lat A_j$ for each j. $\quad \square$

This theorem allows us to produce many examples of attainable lattices from the few previously mentioned ones. Also note that the proof of the theorem shows that $\bigtimes_{j=1}^{\infty} \mathscr{L}_j$ is attainable if each $\mathscr{L}_j$ is isomorphic to the lattice of invariant subspaces of an operator on a finite-dimensional space. There are other questions about direct products of attainable lattices however. For example, can a lattice of the form $\mathscr{L}_1 \times \mathscr{L}_2$ be the invariant subspace lattice of a compact operator, or of a quasinilpotent operator? These and some other such questions are not answered by Theorem 4.16; (if $\sigma(A) \cap \sigma(B) = \emptyset$, then $A \oplus B$ cannot be compact or quasinilpotent). Thus there is some interest in finding other sufficient conditions that the invariant subspaces of $A \oplus B$ split.

Definition. In an abstract lattice $\mathscr{L}$, y is a *cover* for x if $x < y$ and there is no $z \in \mathscr{L}$ such that $x < z < y$. The element y is an *atom* if it is a cover for 0.

Theorem 4.19. *If $Lat\, A$ has no covers, and if B is a nilpotent operator, then the invariant subspaces of $A \oplus B$ split.*

Proof. Let $\mathscr{M} \in Lat(A \oplus B)$. It suffices to show that whenever $x \oplus y$ is in $\mathscr{M}$ so is $x \oplus 0$. By hypothesis $B^n = 0$ for some positive integer n. Thus if $x \oplus y \in \mathscr{M}$, and $x \neq 0$, $(A \oplus B)^n(x \oplus y) = A^n x \oplus 0$ is in $\mathscr{M}$. It follows that $\bigvee_{k=n}^{\infty} \{A^k x\} \oplus \{0\} \subset \mathscr{M}$. Now $\bigvee_{k=n}^{\infty} \{A^k x\} = \bigvee_{k=0}^{\infty} \{A^k x\}$; for if not, then, since $\bigvee_{k=n}^{\infty} \{A^k x\}$ has finite co-dimension in $\bigvee_{k=0}^{\infty} \{A^k x\}$, the subspace $\bigvee_{k=n}^{\infty} \{A^k x\}$ would have a cover in $Lat\, A$; (if $\mathscr{K}$ is a one-dimensional invariant subspace of the compression of A to $\left(\bigvee_{k=0}^{\infty} \{A^k x\} \right) \ominus \left(\bigvee_{k=n}^{\infty} \{A^k x\} \right)$ then, as in the proof of Theorem 4.2, $\bigvee_{k=n}^{\infty} \{A^k x\} \oplus \mathscr{K} \in Lat\, A$). Thus $x \in \bigvee_{k=n}^{\infty} \{A^k x\}$, and $x \oplus 0 \in \mathscr{M}$. $\square$

Example 4.20. *If M_x denotes multiplication by the independent variable on $\mathscr{L}^2(0,1)$, and if B is any nilpotent operator, then the invariant subspaces of $M_x \oplus B$ split.*

Proof. It follows from Example 4.11 that $Lat\, M_x$ has no covers, and thus the result follows from Theorem 4.19. $\square$

Example 4.21. *If V denotes the Volterra operator, and if B is any nilpotent operator, then the invariant subspaces of $V \oplus B$ split.*

Proof. Theorem 4.14 shows that Theorem 4.19 applies here. ⊔
There is another result similar to Theorem 4.19.

Theorem 4.22. *If Lat A has no covers, and if the weakly closed algebra generated by A^2 contains A, then the invariant subspaces of $A \oplus (-A)$ split.*

Proof. Let $\mathcal{M} \in Lat(A \oplus (-A))$ and suppose that $x \oplus y \in \mathcal{M}$. We must show that $x \oplus 0 \in \mathcal{M}$. Note that $[A \oplus (-A)]^2 = A^2 \oplus A^2$, and thus the weakly closed algebra generated by $A \oplus (-A)$ contains $A \oplus A$. It follows that $Ax \oplus Ay \in \mathcal{M}$, and therefore $(2Ax) \oplus 0 \in \mathcal{M}$. Since *Lat A* has no covers it follows, as in the proof of Theorem 4.19, that $\bigvee_{k=1}^{\infty} \{A^k x\} = \bigvee_{k=0}^{\infty} \{A^k x\}$. Hence $x \oplus 0 \in \mathcal{M}$. ⊔

Theorem 4.22 can be used to give other examples of operators whose invariant subspaces split; (e.g., see Example 5.16).

We have seen (Corollary 0.14) that $\sigma(A) \cap \sigma(B) = \emptyset$ implies that every operator commuting with $A \oplus B$ has the form $C \oplus D$. In particular, if $A \oplus B$ satisfies the hypotheses of Theorem 4.16, then this is the case. Note also that the statement that every projection commuting with $A \oplus B$ has the form $C \oplus D$ is equivalent to the statement that the reducing subspaces of $A \oplus B$ split.

Theorem 4.23. *If the invariant subspaces of $A \oplus B$ split, then every operator commuting with $A \oplus B$ has the form $C \oplus D$, where C commutes with A and D commutes with B.*

Proof. Suppose that $\begin{pmatrix} C & E \\ F & D \end{pmatrix}$ commutes with $\begin{pmatrix} A & 0 \\ 0 & B \end{pmatrix}$; it suffices to show that $E=0$ and $F=0$. Since

$$\begin{pmatrix} C & E \\ F & D \end{pmatrix} \begin{pmatrix} A & 0 \\ 0 & B \end{pmatrix} = \begin{pmatrix} A & 0 \\ 0 & B \end{pmatrix} \begin{pmatrix} C & E \\ F & D \end{pmatrix},$$

it follows that $AE = EB$ and $BF = FA$. Now consider the subspaces $\mathcal{M} = \{x \oplus Fx : x \in \mathcal{H}\}$ and $\mathcal{N} = \{Ex \oplus x : x \in \mathcal{H}\}$. Then $(A \oplus B)(x \oplus Fx) = (Ax \oplus BFx) = (Ax \oplus FAx)$; hence $\mathcal{M} \in Lat(A \oplus B)$. Similarly $\mathcal{N} \in Lat(A \oplus B)$. Obviously $0 \oplus y \in \mathcal{M}$ implies $y=0$. Hence, since $\mathcal{M}$ splits, F must be 0. Similarly, the fact that $\mathcal{N}$ splits implies that $E=0$. ⊔

4.6 Attainable Ordinal Sums

There are other ways of piecing together known attainable lattices to produce new ones.

Definition. If $\mathscr{L}_1$ and $\mathscr{L}_2$, are partially ordered sets, then the *ordinal sum* of $\mathscr{L}_1$ and $\mathscr{L}_2$, denoted $\mathscr{L}_1 + \mathscr{L}_2$, is the disjoint union of $\mathscr{L}_1$ and $\mathscr{L}_2$ with the partial ordering defined by: $a \leq b$ if $a \in \mathscr{L}_1$ and $b \in \mathscr{L}_2$ or if a and b are both in $\mathscr{L}_j$ and $a \leq b$ in $\mathscr{L}_j$. If $\mathscr{L}$ is a complete lattice, let $\mathscr{L}^-$ denote the partially ordered subset of $\mathscr{L}$ consisting of all elements except the largest.

Theorem 4.24. *Suppose that $A \in \mathscr{B}(\mathscr{H})$ and there exists a subspace $\mathscr{M}$ of $\mathscr{H}$ of Hilbert space dimension m, $(m \leq \aleph_0)$, such that $Ax + y$ is cyclic for A whenever $x \in \mathscr{H}$ and y is any non-zero vector in $\mathscr{M}$. Then $(Lat\, A)^- + \mathscr{L}$ is attainable whenever $\mathscr{L}$ is the invariant subspace lattice of a nilpotent operator on a Hilbert space of dimension less than or equal to m.*

Proof. Let $\mathscr{K}$ have dimension less than or equal to m, let $B \in \mathscr{B}(\mathscr{K})$ be nilpotent, and let C be any one-to-one operator mapping $\mathscr{K}$ into $\mathscr{M}$. We shall show that $Lat\begin{pmatrix} A & C \\ 0 & B \end{pmatrix} \approx (Lat\, A)^- + Lat\, B$. This will follow if we show that $\mathscr{N} \in Lat\begin{pmatrix} A & C \\ 0 & B \end{pmatrix}$ implies that $\mathscr{N}$ either has the form $\mathscr{N}_1 \oplus \{0\}$ with $\mathscr{N}_1 \in Lat\, A$ or has the form $\mathscr{H} \oplus \mathscr{N}_2$ with $\mathscr{N}_2 \in Lat\, B$.

Suppose, then, that $\mathscr{N} \in Lat\begin{pmatrix} A & C \\ 0 & B \end{pmatrix}$. If $\mathscr{N} \subset \mathscr{H} \oplus \{0\}$ then obviously $\mathscr{N} = \mathscr{N}_1 \oplus \{0\}$ with $\mathscr{N}_1 \in Lat\, A$. If $\mathscr{N} \not\subset \mathscr{H} \oplus \{0\}$ then $\mathscr{N}$ contains some vector $x \oplus y$ with $y \neq 0$. Note that

$$\begin{pmatrix} A & C \\ 0 & B \end{pmatrix}^n \begin{pmatrix} x \\ y \end{pmatrix} = \begin{pmatrix} z \\ B^n y \end{pmatrix}$$

for some z. Since some power of B is 0, by hypothesis, it follows that $\mathscr{N}$ contains a vector $x \oplus y$ with $y \neq 0$ and $By = 0$. Then

$$\begin{pmatrix} A & C \\ 0 & B \end{pmatrix} \begin{pmatrix} x \\ y \end{pmatrix} = \begin{pmatrix} Ax + Cy \\ 0 \end{pmatrix}$$

is in $\mathscr{N}$, and it follows that $\bigvee_{n=0}^{\infty} \{A^n(Ax + Cy)\} \oplus \{0\} \subset \mathscr{N}$. Since the range of C is contained in $\mathscr{M}$, and since Cy is not 0, the vector $Ax + Cy$ is cyclic for A. Thus $\mathscr{H} \oplus \{0\} \subset \mathscr{N}$. Now clearly if $\mathscr{N}_2 = \{y : 0 \oplus y \in \mathscr{N}\}$, then $\mathscr{N}_2 \in Lat\, B$ and $\mathscr{N} = \mathscr{H} \oplus \mathscr{N}_2$. ☐

To use Theorem 4.24 to construct new attainable lattices we must find appropriate subspaces $\mathscr{M}$ for given operators A.

Example 4.25. *If $\mathscr{L}$ is the lattice of a nilpotent operator, then $\omega + \mathscr{L}$ is attainable.*

Proof. It suffices to show that if A is a Donoghue operator (as in Theorem 4.12), then A satisfies the hypotheses of Theorem 4.24. Let A be the Donoghue operator with weights $\{w_n\}_{n=1}$ relative to the basis $\{e_n\}_{n=0}^{\infty}$ of $\mathscr{H}$. For each natural number n let

$$a_{nj} = \begin{cases} w_j & \text{if } j=p^n \text{ for some prime } p \\ 0 & \text{if } j \neq p^n \text{ for any prime } p \end{cases},$$

and define $x_n = \sum_{j=1}^{\infty} a_{nj} e_{j-1}$. Note that $(x_n, x_m) = 0$ if $n \neq m$. Let $\mathscr{M} = \bigvee_{n=1}^{\infty} \{x_n\}$; we claim that $Ax + y$ is cyclic for A whenever $x \in \mathscr{H}$ and y is a non-zero vector in $\mathscr{M}$. To see this, let $x = \sum_{j=0}^{\infty} \alpha_j e_j$ and $y = \sum_{j=1}^{\infty} \beta_j x_j$. Then $Ax + y = \sum_{j=1}^{\infty} \alpha_j w_j e_{j-1} + \sum_{j=1}^{\infty} \beta_j x_j$. If $Ax + y$ is not cyclic for A, then $(Ax+y, e_j) = 0$ for j sufficiently large. It follows that $\alpha_j w_j = - \sum_{n=1}^{\infty} (\beta_n x_n, e_{j-1})$ for j sufficiently large. At least one β_n is non-zero; suppose that $\beta_m \neq 0$. Now if $j = p^m$ for a prime p, then $\sum_{n=1}^{\infty} (\beta_n x_n, e_{j-1}) = \beta_m w_j$. It follows that, for infinitely many j, $\alpha_j w_j = -\beta_m w_j$ or $\alpha_j = -\beta_m$; this contradicts the square-summability of $\{\alpha_j\}$. Thus $\mathscr{M}$ satisfies the hypothesis of Theorem 4.24. $\square$

Other examples can be constructed using the Volterra operator.

Example 4.26. *If $\mathscr{L}$ is the lattice of a nilpotent operator, then $[0,1) + \mathscr{L}$ is attainable.*

Proof. Let V denote the Volterra operator on $\mathscr{L}^2(0,1)$, (as in Theorem 4.14). We construct a subspace $\mathscr{M}$ satisfying the hypotheses of Theorem 4.24. Let $\{[a_n, b_n]\}_{n=1}^{\infty}$ be a sequence of disjoint non-trivial closed subintervals of $[0,1]$ such that $\{a_n\}$ and $\{b_n\}$ are monotone decreasing sequences converging to 0. For each n let F_n be a function which vanishes on the complement of $[a_n, b_n]$, which is bounded by 1, and which is continuous but not absolutely continuous on $[a_n, b_n]$; (e.g., F_n could be a Cantor function (see Titchmarsh [1]) on $[a_n, b_n]$). For each prime number p define $G_p = \sum_{k=1}^{\infty} F_{p^k}$, and let $\mathscr{M} = \bigvee_p \{G_p\}$.

Note that $(G_p, G_q) = 0$ when $p \neq q$. Now if $f \in \mathscr{L}^2(0,1)$ and $g = \sum_{p \text{ prime}} \alpha_p G_p$ with $\alpha_{p_0} \neq 0$, we claim that $Vf + g$ is cyclic for V. If not, then $Vf + g = 0$ a.e. on $[0, \alpha]$ for some $\alpha \in (0,1)$. Then $g(x) = - \int_0^x f(t)\,dt$

a.e. on $[0,\alpha]$. Choose k so large that $[a_n,b_n]\subset[0,\alpha]$ for $n=p_0^k$. Then if $n=p_0^k$, $g(x)=\alpha_{p_0}F_n(x)$ a.e. on $[a_n,b_n]$, and thus $Vf=-\alpha_{p_0}F_n$ a.e. on $[a_n,b_n]$. But both Vf and F_n are continuous on $[a_n,b_n]$, and therefore $F_n(x)=(1/\alpha_{p_0})(Vf)(x)$ for all $x\in[a_n,b_n]$. This contradicts the fact that Vf is absolutely continuous on $[a_n,b_n]$ and F_n is not. $\square$

Particular attainable lattices obtained from Examples 4.25 and 4.26 include $\omega+n$ and $[0,1]+n$ for any natural number $n\geq1$, (by using a Jordan block on a finite-dimensional space as the nilpotent operator), and, also, $\omega+\mathscr{L}$ and $[0,1)+\mathscr{L}$ where $\mathscr{L}$ is the lattice of all subspaces of a separable Hilbert space, (by using 0 as the nilpotent operator).

4.7 Transitive Lattices

The general problem of characterizing attainable lattices is too difficult. This suggests that it might be sensible to consider the converse problem: what are some examples of lattices that are not attainable? An attainable lattice must be complete, have at most c elements, and have the property that its chains have countable order-dense subsets, (this follows easily from the separability of $\mathscr{H}$). Beyond this nothing is known. If we vary the question to ask for concrete lattices (i.e., lattices of subspaces) that are not invariant subspace lattices of an operator, then examples are easy to construct. For instance, let $\mathscr{M}$ be a finite-dimensional subspace of $\mathscr{H}$ of dimension greater than 1. Then the subspace lattice $\{\{0\},\mathscr{M},\mathscr{H}\}$ is not the lattice of invariant subspaces of any operator; (if A had this lattice then $A|\mathscr{M}$ would be an operator on a finite-dimensional space with $\Pi_0(A|\mathscr{M})=\emptyset$). Many similar examples can be given, using the known facts (e.g., $Lat\,A$ is self-dual) that are true about invariant subspace lattices of operators on finite-dimensional spaces; (cf. Section 10.2).

If $\mathscr{F}$ is a complete lattice of subspaces, it may be the case that the only operators which leave invariant all the members of $\mathscr{F}$ are the multiples of the identity. Clearly if $\mathscr{F}$ is such a lattice and $\mathscr{F}$ is also attainable, then $\mathscr{F}$ is the lattice of all subspaces.

Definition. A *transitive* lattice is a complete lattice $\mathscr{F}$ of subspaces of $\mathscr{H}$ which contains $\{0\}$ and $\mathscr{H}$ and which has the property that $\mathscr{F}\subset Lat\,A$ implies A is a multiple of the identity.

Let $\mathscr{F}$ denote the lattice of all subspaces of $\mathscr{H}$, and suppose that $\mathscr{F}\subset Lat\,A$. Then every one-dimensional subspace is invariant under A, and thus to each $x\in\mathscr{H}$ there corresponds a complex number λ_x such that $Ax=\lambda_x x$. Now $A(x+y)=\lambda_{x+y}(x+y)$, and thus $\lambda_{x+y}x+\lambda_{x+y}y=\lambda_x x+\lambda_y y$, or $(\lambda_{x+y}-\lambda_x)x=(\lambda_y-\lambda_{x+y})y$. If x and y are linearly independent, then, $\lambda_x = \lambda_{x+y} = \lambda_y$. Hence λ_x is a constant independent of x. In other words, the lattice of all subspaces of $\mathcal{H}$ is transitive.

There are more interesting transitive lattices.

Example 4.27. *Let $\mathcal{H}$ have orthonormal basis $\{e_n\}_{n=0}^\infty$, and let*

$$\mathscr{F} = \{\mathcal{M} : \mathcal{M} \text{ is a subspace of } \mathcal{H} \text{ and } \sum_{n=0}^\infty \alpha_n e_n \in \mathcal{M} \text{ implies } \sum_{n=0}^\infty \bar{\alpha}_n e_n \in \mathcal{M}\}.$$

Then $\mathscr{F}$ is a transitive lattice.

Proof. Suppose that $\mathscr{F} \subset Lat\, A$. Then, for each n, the one-dimensional space spanned by e_n is in $Lat\, A$, and thus there exists a complex number λ_n such that $Ae_n = \lambda_n e_n$. Now, for any m and n, the one-dimensional space spanned by $e_n + e_m$ is in $\mathscr{F}$, and it follows that $A(e_n + e_m) = \lambda(e_n + e_m)$ for some λ. Hence $\lambda_n = \lambda = \lambda_m$. $\square$

The lattice of the above example has many elements and it is thus not surprising that it is transitive. It is perhaps surprising that there exist finite transitive lattices.

Theorem 4.28. *There is a six-element transitive lattice.*

Proof. Let $\mathcal{H}$ be a Hilbert space with orthonormal basis $\{e_n\}_{n=-\infty}^\infty$. We construct a transitive lattice of subspaces of the space $\mathcal{K} = \mathcal{H} \oplus \mathcal{H}$. Let T be the bilateral weighted shift on $\mathcal{H}$ defined by $Te_n = w_n e_{n+1}$, where

$$w_n = \begin{cases} 1, & \text{for } n \leq 0 \\ \exp((-1)^n n!), & \text{for } n > 0 \end{cases}.$$

Then T is an unbounded linear transformation. Let $\mathscr{D}$ denote the set of all vectors $x = \sum_{i=-\infty}^\infty \alpha_i e_i$ in $\mathcal{H}$ such that $\sum_{i=-\infty}^\infty |\alpha_i w_i|^2 < \infty$. It is easily seen that T is a closed linear transformation on the domain $\mathscr{D}$; i.e., the set $\mathcal{M}_T = \{x \oplus Tx : x \in \mathscr{D}\}$ is a closed subspace of $\mathcal{H} \oplus \mathcal{H} = \mathcal{K}$. Let $\mathcal{M}_{x0} = \{x \oplus 0 : x \in \mathcal{H}\}$, $\mathcal{M}_{0x} = \{0 \oplus x : x \in \mathcal{H}\}$, and $\mathcal{M}_{xx} = \{x \oplus x : x \in \mathcal{H}\}$. We claim that

$$\mathscr{F} = \{\{0\}, \mathcal{K}, \mathcal{M}_{x0}, \mathcal{M}_{0x}, \mathcal{M}_{xx}, \mathcal{M}_T\}$$

is a transitive lattice of subspaces of $\mathcal{K}$.

We must first show that $\mathscr{F}$ is a subspace lattice, i.e., that $\mathscr{F}$ is closed under intersections and spans. In fact, the span of any two non-trivial elements of $\mathscr{F}$ is $\mathcal{K}$ and the intersection of any two non-trivial elements is $\{0\}$. Some of these relations are immediately obvious; we prove the others. First $\mathcal{M}_{x0} \cap \mathcal{M}_T = \{0\}$, since the nullspace of T is $\{0\}$, and $\mathcal{M}_{x0} \vee \mathcal{M}_T = \mathcal{K}$, since the closure of the range of T is $\mathcal{H}$. Now suppose

that $x \oplus y \in \mathcal{M}_{xx} \cap \mathcal{M}_T$. Then $y = Tx = x$, and $(T-1)x = 0$. If $x = \sum\limits_{i=-\infty}^{\infty} \alpha_i e_i$
then $\alpha_{n+1} = w_n \alpha_n$ for all n, and thus $\alpha_n = \alpha_0$ for all negative n. Therefore
$\alpha_0 = 0$, and it follows that $\alpha_n = 0$ for all n. Hence $\mathcal{M}_{xx} \cap \mathcal{M}_T = \{0\}$. Also
$\mathcal{M}_{xx} \vee \mathcal{M}_T$ contains all vectors of the form $0 \oplus (T-1)e_n$, and if
$x = \sum\limits_{n=-\infty}^{\infty} \alpha_n e_n$ is orthogonal to $(T-1)e_n$ for all n it follows that
$w_n \alpha_{n+1} - \alpha_n = 0$ for all n, and this implies that $\alpha_n = 0$ for all n. Hence
$\mathcal{M}_{xx} \vee \mathcal{M}_T$ contains $0 \oplus \mathcal{H}$, and $\mathcal{M}_{xx} \vee \mathcal{M}_T = \mathcal{K}$. The set $\mathcal{F}$ is therefore a complete lattice.

We must now show that $\mathcal{F}$ is transitive. Suppose that $A \in \mathcal{B}(\mathcal{K})$
and $\text{Lat } A \supset \mathcal{F}$. Since $\text{Lat } A \supset \{\mathcal{M}_{x0}, \mathcal{M}_{0x}\}$, $A = A_1 \oplus A_2$ with A_1 and
A_2 in $\mathcal{B}(\mathcal{H})$. Since $\mathcal{M}_{xx} \in \text{Lat } A$, $(A_1 \oplus A_2)(x \oplus x) = (A_1 x \oplus A_2 x) \in \mathcal{M}_{xx}$ for
all $x \in \mathcal{H}$, and it follows that $A_1 = A_2$. Thus A has the form $B \oplus B$,
with $B \in \mathcal{B}(\mathcal{H})$. Now if $x \oplus Tx \in \mathcal{M}_T$, then $(B \oplus B)(x \oplus Tx) = Bx \oplus BTx$,
and $\mathcal{M}_T \in \text{Lat } A$ implies that $TBx = BTx$. Thus $\mathcal{D}$ is an invariant linear
manifold of B and $BT = TB$ on $\mathcal{D}$. We claim that the only operators B
with this property are multiples of the identity.

Suppose that B is not a multiple of the identity on $\mathcal{H}$. Note that
$x \in \mathcal{D}$ implies $(Tx, e_n) = w_{n-1}(x, e_{n-1})$ for each n. Since (BTe_{m-1}, e_n)
$= (TBe_{m-1}, e_n)$ for every m and n, it follows that $w_{m-1}(Be_m, e_n)$
$= w_{n-1}(Be_{m-1}, e_{n-1})$. In particular, $(Be_m, e_m) = (Be_{m-1}, e_{m-1})$ for all m,
and (Be_m, e_m) is a constant independent of m. Since B is not a multiple
of the identity there must exist distinct integers m and n such that
$(Be_m, e_n) \neq 0$. The above computation, repeated k times, gives

$$(Be_{m+k}, e_{n+k}) = \frac{w_n w_{n+1} \cdots w_{n+k-1}}{w_m w_{m+1} \cdots w_{m+k-1}} (Be_m, e_n).$$

Now observe that, for each fixed integer r,

$$\limsup_{j \to \infty} (w_j w_{j+1} \cdots w_{j+r}) = \infty \quad \text{and}$$

$$\liminf_{j \to \infty} (w_j w_{j+1} \cdots w_{j+r}) = 0.$$

It follows, using the first of these equations if $n > m$ and the second
if $m > n$, that

$$\{(Be_{m+k}, e_{n+k}) : k \text{ a positive integer}\}$$

is unbounded, which contradicts the fact that B is a bounded operator. $\square$

There are other known facts about transitive lattices (e.g., Propositions 4.9 and 4.10) but the general problem seems to be as difficult as
the problem of attainable lattices.

4.8 Additional Propositions

Proposition 4.1. If every invariant linear manifold of A is invariant under B, and if $AB = BA$, then B is a polynomial in A; (this is related to the proof of Theorem 4.8).

Proposition 4.2. If $\eta(\sigma(A)) \cap \eta(\sigma(B)) = \emptyset$, then the only closed linear transformation C such that the domain of C is invariant under A and $CA = BC$ is $C = 0$; (this is related in spirit, though not in proof, to Corollary 0.13).

Proposition 4.3. If $\mathscr{H}$ has orthonormal basis $\{e_n\}_{n=0}^{\infty}$, and if

$$
w_j = \begin{cases}
\frac{1}{2}, & \text{for } j = 1 \\
1, & \text{for } j \text{ a multiple of } 3 \\
\prod_{m=1}^{j-1} w_m, & \text{for } j \text{ greater than 1 and not a multiple of 3,}
\end{cases}
$$

then the weighted shift A defined by $Ae_0 = 0$, $Ae_j = w_j e_{j-1}$ for $j > 0$ has lattice $\omega + 1$. The same result holds for weighted shifts whose weight sequence $\{w_j\}$ is monotone decreasing and belongs to ℓ^p for some $p > 0$.

Proposition 4.4. The commutant of a Donoghue operator A is equal to the uniformly closed algebra generated by A and 1.

Proposition 4.5. Define $(Wf)(x) = \pi^{-\frac{1}{2}} \int_0^x (f(t)(x-t)^{-\frac{1}{2}}) dt$ for $f \in \mathscr{L}^2(0,1)$. Then W is a compact operator and W^2 is the Volterra operator.

Proposition 4.6. The ordinal number 2 is attainable if and only if the ordinal number n is attainable for each natural number n greater than 2.

Proposition 4.7. The lattice $([0,1] \times m) + n$ is attainable for each pair of natural numbers m and n.

Proposition 4.8. If $\mathscr{M}_0$ is a non-trivial subspace of $\mathscr{H}$, and if $\mathscr{F} = \{\mathscr{M} : \mathscr{M} \supset \mathscr{M}_0 \text{ or } \mathscr{M} \subset \mathscr{M}_0\}$, then there does not exist a bounded linear operator A such that $Lat\, A = \mathscr{F}$.

Proposition 4.9. There is no transitive lattice with only four elements.

Proposition 4.10. On a finite-dimensional space of dimension greater than 2 every transitive lattice has at least seven elements.

Proposition 4.11. Every Volterra-type integral operator is compact and quasinilpotent.

Proposition 4.12. In Theorem 4.19 the hypothesis that B is nilpotent can be relaxed to the requirement that B be algebraic.

Proposition 4.13. If A is a Donoghue operator and B is the direct sum of a finite number of copies of A, then every non-trivial invariant subspace of B contains an eigenvector of B.

Proposition 4.14. If x_0 is a cyclic vector for A, and if $\alpha \in \mathbb{C}$ such that $\|\alpha A\| < 1$, then, for each positive integer n, the vector $x_n = (1 - \alpha A)^n x_0$ is a cyclic vector for A. Also, $\bigvee_{n=0}^{\infty} \{x_n\} = \mathscr{H}$. Thus an operator which has a cyclic vector has a spanning set of cyclic vectors.

Proposition 4.15. An operator on a space of finite dimension n is unicellular if and only if it is the sum of a multiple of the identity and a nilpotent operator of index n.

4.9 Notes and Remarks

Beurling [1] aroused interest in the determination of the invariant subspaces of particular operators by his work on the shift which is described in Chapter 3. Dixmier [3] found the invariant subspaces of the Volterra operator on real $\mathscr{L}^1(0,1)$, and Donoghue [1] and Brodski [2] independently discovered the invariant subspaces of the Volterra operator on complex $\mathscr{L}^2(0,1)$.

The desirability of determining attainable lattices was pointed out by P.R. Halmos, who raised a number of specific questions about attainability in conversations with many mathematicians.

Theorem 4.4 was discovered independently by Gohberg-Krein [2] and Rosenthal [1]; the proof presented is from Rosenthal [1]. Theorem 4.6 was independently discovered by Halmos (unpublished) and Brickman-Fillmore [1], and Theorem 4.8 is a reformulation of a result of Kaplansky [1]. Theorem 4.9 is an unpublished observation of David Topping. Theorem 4.12 was found by Donoghue [1] in the case $w_n = 2^{-n}$, and the generalization presented in the text was independently discovered by Nikolskii [1], Stephen Parrott (unpublished) and Allen Shields (unpublished). The first assertion of Proposition 4.3 appeared in Rosenthal [7], while more general results that include both Theorem 4.12 and Proposition 4.3 were found at about the same time by Nikolskii [2].

Part of the history of Theorem 4.14 was discussed above. Another treatment is in Kalisch [1]; the proof in the text is due to Sarason [2], and the functional-analytic proof of the Titchmarsh convolution theorem (Titchmarsh [2]) is due to Kalisch [2]. Other interesting proofs of Theorem 4.14 can be found in Gohberg-Krein [2] and Ringrose [5]. Theorems 4.16 and 4.18 are from Crimmins-Rosenthal [1], and Theorems 4.19—4.24 and Examples 4.25 and 4.26 are from Rosenthal [3].

The study of transitive lattices was initiated by Halmos [13]. Example 4.27 was found by McLaughlin (cf. Halmos [12]). Theorem 4.28, due to Harrison-Radjavi-Rosenthal [1], is a modification of a result of Halmos [12]. A penetrating study of transitive lattices on finite-dimensional spaces is presented in Harrison [2]. The related concept of reflexive lattices is discussed in Halmos [10] and Arveson [3].

Proposition 4.1 is a generalization by Fillmore [2] of the corresponding result in the finite-dimensional case obtained in Brickman-Fillmore [1]. Proposition 4.5 is classical, and can be found in Hille-Phillips [1]. Propositions 4.2, 4.6, and 4.7 are from Rosenthal [3], and Proposition 4.4 was discovered by Nordgren [1]. Proposition 4.8 is in Rosenthal [4] and Proposition 4.10 in Halmos [12]. Proposition 4.12 was observed by Peter Fillmore (unpublished). Proposition 4.13 is from Rosenthal [7] (cf. Nordgren-Radjavi-Rosenthal [1]), and much deeper results on direct sums of Donoghue operators are contained in Nordgren [2]. Proposition 4.14 is due to Geher [1].

Two other attainable lattices have recently been discovered. Harrison [1] has shown that the lattice $\omega + \omega + 1$ is attainable, and Sarason [5] has proven that the lattice of closed subsets of $[0,1]$ is also attainable.

Foiaş-Williams [1] shows that there exist operators in the commutant of the Volterra operator which are unicellular and have spectra containing more than one point; thus part of Theorem 4.7 does not extend to the infinite-dimensional case. Some analogues of Theorem 4.7 for certain operators on infinite-dimensional spaces are discussed in Sz.-Nagy-Foiaş [1] (Chapter IX) and Brodskii [1] (Chapter III).

Chapter 5. Compact Operators

Compact operators are often more easily studied than arbitrary bounded operators. In this chapter we show that compact (in fact, polynomially compact) operators have non-trivial invariant subspaces; much more general results are obtained in Chapter 8. We also show that the properties of normality and quasinilpotence for compact operators are determined by their invariant subspaces, and give examples of attainable lattices that are not attainable by compact operators.

5.1 Existence of Invariant Subspaces

We prove a slightly more general theorem than the existence of invariant subspaces for polynomially compact operators. For this it will be helpful to consider another class of operators.

Definition. The operator A is *quasitriangular* if there exists a sequence $\{P_n\}$ of projections of finite rank which converges to the identity in the strong topology such that $\{\|P_n A P_n - A P_n\|\} \to 0$.

The definition of quasitriangularity says, roughly, that A has a sequence of "approximately invariant" finite-dimensional subspaces. Several results about quasitriangular operators will be needed to prove the invariant subspace theorem given below.

Lemma 5.1. *If $\{Q_n\}$ is any sequence of projections, then the set $\mathfrak{A} = \{A \in \mathscr{B}(\mathscr{H}) : \{\|Q_n A Q_n - A Q_n\|\} \to 0\}$ is a uniformly closed subalgebra of $\mathscr{B}(\mathscr{H})$.*

Proof. It is clear that $\mathfrak{A}$ is a linear manifold in $\mathscr{B}(\mathscr{H})$. If A and B are in $\mathfrak{A}$, then

$$Q_n A B Q_n - A B Q_n = (Q_n A Q_n - A Q_n) B Q_n + (A - Q_n A)(Q_n B Q_n - B Q_n),$$

and it follows that AB is in $\mathfrak{A}$. Hence $\mathfrak{A}$ is an algebra. Now, if $\{A_k\} \subset \mathfrak{A}$ and $\{A_k\}$ converges uniformly to A, then

$$\|Q_n A Q_n - A Q_n\| \leq \|Q_n A_k Q_n - A_k Q_n\| + \|Q_n(A - A_k)Q_n\| + \|(A_k - A)Q_n\|,$$

and therefore $A \in \mathfrak{A}$. $\quad\square$

Lemma 5.2. *If A is quasitriangular, then there exists a sequence $\{Q_n\}$ of projections of finite rank which converges weakly to an operator other than 0 or 1 and which has the property that $\{\|Q_n A Q_n - A Q_n\|\} \to 0$.*

Proof. Let $\{P_n\}$ be a sequence of projections of finite rank which strongly converges to 1 such that $\{\|P_n A P_n - A P_n\|\} \to 0$. The sequence $\{Q_n\}$ is constructed as follows. Let x and y be any two mutually orthogonal unit vectors, and define the linear functional ρ on $\mathscr{B}(\mathscr{H})$ by $\rho(B) = \frac{1}{2}(Bx, x) + \frac{1}{2}(By, y)$ for $B \in \mathscr{B}(\mathscr{H})$. Note that ρ is weakly continuous. Since $\{P_n\}$ converges strongly to 1, $\rho(P_n) \geq \frac{3}{4}$ for n sufficiently large. Suppose that P is a projection of rank 1: i.e., there is a unit vector e such that $Px = (x, e)e$ for $x \in \mathscr{H}$. Then $\rho(P) = \frac{1}{2}[|(x, e)|^2 + |(y, e)|^2] \leq \frac{1}{2}$, (by Bessel's inequality). Thus $\rho(P) \leq \frac{1}{2}$ for every projection P of rank 1.

Now consider $P_n A P_n$ for $\rho(P_n) \geq \frac{3}{4}$. Since $(P_n A P_n)|(P_n \mathscr{H})$ is an operator on a finite-dimensional space it can be put in triangular form. In other words, there exists a chain $0 = Q_n^0 < Q_n^1 < \cdots < Q_n^{N_n} = P_n$ of projections whose ranges are invariant under $P_n A P_n$ such that $Q_n^{j+1} - Q_n^j$ is a projection of rank 1 for each j. Now $\rho(Q_n^j) \leq \frac{1}{2}$, and $\rho(Q_n^{j+1}) - \rho(Q_n^j) = \rho(Q_n^{j+1} - Q_n^j) \leq \frac{1}{2}$ for each j. Hence there is at least one Q_n^j, say $Q_n^{j_n}$, such that $\frac{1}{4} \leq \rho(Q_n^{j_n}) \leq \frac{3}{4}$. The unit ball of $\mathscr{B}(\mathscr{H})$ is weakly sequentially compact, (cf. Proposition 0.2); hence some subsequence $\{Q_n\}$ of $\{Q_n^{j_n}\}$ converges weakly to an operator Q. The weak continuity of ρ implies that $\frac{1}{4} \leq \rho(Q) \leq \frac{3}{4}$, so that Q is neither 0 nor 1.

To complete the proof we need only show that $\{\|Q_n A Q_n - A Q_n\|\} \to 0$. For each n there is an integer k with $k \geq n$ and $P_k \geq Q_n$ such that $Q_n \mathscr{H}$ is invariant under $P_k A P_k$. Then

$$\|Q_n A Q_n - A Q_n\| = \|Q_n P_k A P_k Q_n - A P_k Q_n\| = \|P_k A P_k Q_n - A P_k Q_n\|$$
$$\leq \|P_k A P_k - A P_k\|.$$

It follows that $\{\|Q_n A Q_n - A Q_n\|\} \to 0$. □

The *uniformly closed algebra generated by A* is the smallest subalgebra of $\mathscr{B}(\mathscr{H})$ which contains A and 1 and which is closed in the norm topology.

Theorem 5.3. *If A is quasitriangular and the uniformly closed algebra generated by A contains a compact operator different from 0, then A has a non-trivial invariant subspace.*

Proof. Let C denote the compact operator in the algebra. If C has a non-trivial nullspace, then, since C commutes with A, A has a nontrivial invariant subspace. We can assume, therefore, that the nullspace of C is $\{0\}$. In this case choose a sequence $\{Q_n\}$ converging weakly to some Q as in Lemma 5.2, and let $\mathscr{M} = \{x : Qx = x\}$.

We show that $\mathscr{M}$ is a non-trivial invariant subspace of A. If $x \in \mathscr{M}$ then $\|x - Q_n x\|^2 = ((1 - Q_n)x, x)$, and $\{((1 - Q_n)x, x)\} \to 0$. Thus $\{Q_n x\}$

converges in norm to x, and $\{AQ_nx\} \to Ax$ in norm. Hence $\{Q_nAQ_nx\}$ converges weakly to QAx. Also, it follows that $\{Q_nAQ_nx\} \to Ax$, since $\{\|Q_nAQ_nx - AQ_nx\|\} \to 0$ by Lemma 5.2. Therefore $QAx = Ax$, and $Ax \in \mathcal{M}$. This shows that $\mathcal{M} \in \operatorname{Lat} A$. Since $Q \ne 1$, $\mathcal{M} \ne \mathcal{H}$. The compact operator C is needed to show that $\mathcal{M} \ne \{0\}$. We show, in fact, that $\mathcal{M} \supset CQ\mathcal{H}$.

To verify this first note that $\{\|Q_nCQ_n - CQ_n\|\} \to 0$ by Lemma 5.1. For each $x \in \mathcal{H}$, $\{Q_nx\}$ converges weakly to Qx, and the compactness of C implies that $\{CQ_nx\}$ converges to CQx in norm. Hence $\{Q_nCQ_nx\}$ converges to $QCQx$ weakly, and, as in the preceding paragraph, we conclude that $QCQx = CQx$. Hence $CQ\mathcal{H} \subset \mathcal{M}$.   □

To get more natural invariant subspace theorems from Theorem 5.3 we need to find easily stated sufficient conditions that an operator be quasitriangular.

Theorem 5.4. *If A has a cyclic vector e such that $\liminf\{\|A^ne\|^{1/n}\} = 0$, then A is quasitriangular.*

Proof. We can assume that $\|e\| = 1$. Apply the Gram-Schmidt orthogonalization procedure to the sequence $\{e, Ae, A^2e, \ldots\}$ to get an orthonormal basis $\{e_n\}_{n=0}^\infty$ with $e_0 = e$. Then the matrix of A with respect to $\{e_n\}_{n=0}^\infty$ is almost upper triangular: the only non-zero elements below the main diagonal are those on the first subdiagonal. Let $a_n = (Ae_n, e_{n+1})$ for each n. Note that $a_n \ne 0$ for all n, since e is cyclic.

Now $A^ne = f + (A^ne, e_n)e_n$, where $f \in \bigvee_{j=0}^{n-1}\{e_j\}$, so that $A^{n+1}e = Af + (A^ne, e_n)Ae_n$. Hence $(A^{n+1}e, e_{n+1}) = (A^ne, e_n)(Ae_n, e_{n+1}) = a_n(A^ne, e_n)$, or $a_n = (A^{n+1}e, e_{n+1})/(A^ne, e_n)$. Thus $\prod_{n=0}^{k-1} a_n = (A^ke, e_k)$, and $\left|\prod_{n=0}^{k-1} a_n\right|^{1/k} = |(A^ke, e_k)|^{1/k} \le \|A^ke\|^{1/k}$. Hence the sequence $\left\{\left|\prod_{n=0}^{k-1} a_n\right|^{1/k}\right\}_{k=1}^\infty$ converges to 0, and it follows that some subsequence $\{a_{n_j}\}$ of $\{a_n\}$ converges to 0. If P_j denotes the projection onto $\bigvee_{n=0}^{n_j}\{e_n\}$, then $\|P_jAP_j - AP_j\| = |a_{n_j}|$, and A is quasitriangular.   □

Corollary 5.5. *If A is a quasi-nilpotent operator such that the uniformly closed algebra generated by A contains a compact operator other than 0, then A has a non-trivial invariant subspace.*

Proof. Let e be any unit vector. If e is not cyclic, then, by definition, the invariant subspace generated by e is proper. If e is cyclic, then A has an invariant subspace by the above theorem, since $\|A^ne\| \le \|A^n\|$.   □

Corollary 5.6. *Every polynomially compact operator has a non-trivial invariant subspace.*

Proof. Suppose that $p(A)=K$ where K is compact and p is a non-zero polynomial. If K has point spectrum, then so does A, and *Lat A* is non-trivial. Thus, by the Fredholm alternative, we can assume that $\sigma(K)=\{0\}$. In this case $\sigma(A)$ is finite, (by the spectral mapping theorem). If $\sigma(A)$ has more than one point, then A has invariant subspaces by Theorem 2.10. If $\sigma(A)=\{\lambda\}$ then $A-\lambda$ has a non-trivial invariant subspace by Corollary 5.5. □

Corollary 5.7. *Every compact operator has a non-trivial invariant subspace.*

A much stronger result than Corollaries 5.5 and 5.6 is given in Corollary 8.24, which is obtained using techniques very different from the above.

5.2 Normality and *Lat A*

We have seen (Corollary 1.24) that a polynomially compact operator which is normal has the property that all its invariant subspaces are reducing. Below we establish the converse; i.e., if every invariant subspace of a polynomially compact operator is reducing, then the operator is normal.

Lemma 5.8. *If $\{\mathscr{M}_\alpha\}$ is a chain of subspaces of $\mathscr{H}$, and if $\mathscr{N}=\bigcap\mathscr{M}_\alpha$, then there is a countable subfamily $\{\mathscr{M}_{\alpha_i}\}$ such that $\mathscr{N}=\bigcap\mathscr{M}_{\alpha_i}$ and $\mathscr{M}_{\alpha_{i+1}}\subset\mathscr{M}_{\alpha_i}$ for each i.*

Proof. Taking set-theoretic complements gives $\mathscr{N}'=\bigcup\mathscr{M}'_\alpha$. Since $\mathscr{N}'$ has the Lindelöf property, there is a countable subcover $\{\mathscr{M}'_{\alpha_i}\}$, and then $\mathscr{N}=\bigcap\mathscr{M}_{\alpha_i}$. Now, for each i, discard $\mathscr{M}_{\alpha_{i+1}}$ if $\mathscr{M}_{\alpha_{i+1}}\supset\mathscr{M}_{\alpha_i}$. □

Theorem 5.9. *If A is polynomially compact, and if every invariant subspace of A is reducing, then A is normal.*

Proof. We show that A is diagonable; i.e., that $\mathscr{H}$ has an orthonormal basis $\{e_n\}$ such that each e_n is an eigenvector of A and of A^*.

Let p be a non-zero polynomial such that $p(A)$ is a compact operator K. Zorn's lemma trivially implies that there exists a maximal orthonormal set of common eigenvectors of A and A^*; let $\mathscr{E}$ be such a maximal set; ($\mathscr{E}$ could conceivably be $\emptyset$, in which case define $\mathscr{E}^\perp=\mathscr{H}$). We must show that $\mathscr{E}^\perp=\{0\}$. Suppose that $\mathscr{E}^\perp\neq\{0\}$.

If $K|\mathscr{E}^\perp=0$ then, by the spectral mapping theorem, A has an eigenvector $e\in\mathscr{E}^\perp$ of norm 1. Since the one-dimensional space spanned by

e reduces A by hypothesis, it follows that $\mathscr{E} \cup \{e\}$ is a larger ortho-normal set of common eigenvectors of A and A^*. Thus $K|\mathscr{E}^{\perp} \neq 0$.

Now let $\mathscr{F}$ denote the family of subspaces

$$\{\mathscr{M} : \mathscr{M} \subset \mathscr{E}^{\perp}, \; \mathscr{M} \text{ reduces } A, \text{ and } \|K|\mathscr{M}\| = \|K|\mathscr{E}^{\perp}\|\}.$$

The family $\mathscr{F}$ is non-empty: $\mathscr{E}^{\perp} \in \mathscr{F}$. Let $\mathscr{N}$ be the intersection of the subspaces in a maximal chain in $\mathscr{F}$, and use Lemma 5.8 to write $\mathscr{N} = \bigcap_{i=1}^{\infty} \mathscr{M}_i$ with $\mathscr{M}_i \in \mathscr{F}$ and $\mathscr{M}_{i+1} \subset \mathscr{M}_i$ for all i. Then $\mathscr{N}$ reduces A, and the proof will be complete if it is shown that the dimension of $\mathscr{N}$ is 1, (for this would contradict the maximality of $\mathscr{E}$).

A compact operator attains its norm; thus for each i there is an $x_i \in \mathscr{M}_i$ such that $\|x_i\| = 1$ and $\|Kx_i\| = \|K|\mathscr{E}^{\perp}\|$. Choose a subsequence $\{x_{i_n}\}$ of $\{x_i\}$ that converges weakly to some x. Then $\{Kx_{i_n}\} \to Kx$, and it follows that $\|Kx\| = \|K|\mathscr{E}^{\perp}\|$. Since $\|x\| \leq 1$, $\|K|\mathscr{N}\| = \|K|\mathscr{E}^{\perp}\|$, and $\mathscr{N} \in \mathscr{F}$.

If the dimension of $\mathscr{N}$ were greater than 1, then $A|\mathscr{N}$ would have a non-trivial invariant subspace $\mathscr{L}$ by Corollary 5.6. Then $\mathscr{L}$ reduces A by hypothesis. Since $\|K|\mathscr{N}\| = \max\{\|K|\mathscr{L}\|, \|K|(\mathscr{L}^{\perp} \cap \mathscr{N})\|\}$, at least one of $\mathscr{L}$ and $\mathscr{L}^{\perp} \cap \mathscr{N}$ is in $\mathscr{F}$, contradicting the definition of $\mathscr{N}$. □

5.3 Spectrum and $Lat\,A$

In this section we show that the spectrum of a compact operator can be computed from any maximal chain of invariant subspaces. This leads to an interesting characterization of compact quasinilpotent operators.

Definition. If $\mathscr{F}$ is a chain of subspaces and $\mathscr{M} \in \mathscr{F}$, then $\mathscr{M}_-$ denotes the span of all subspaces in $\mathscr{F}$ which are properly contained in $\mathscr{M}$. The chain $\mathscr{F}$ is *complete* if $\{0\}$ and $\mathscr{H}$ are in $\mathscr{F}$ and the span and intersection of any subfamily of $\mathscr{F}$ are in $\mathscr{F}$; it is *maximal* if the only chain $\mathscr{F}_0$ of subspaces such that $\mathscr{F}_0 \supset \mathscr{F}$ is $\mathscr{F}_0 = \mathscr{F}$.

Lemma 5.10. *A chain $\mathscr{F}$ of subspaces is maximal if and only if $\mathscr{F}$ is complete and $\mathscr{M} \in \mathscr{F}$ implies that the dimension of $\mathscr{M} \ominus \mathscr{M}_-$ is at most 1.*

Proof. If $\mathscr{F}$ is maximal, then it is clear that $\mathscr{F}$ is complete. Also if the dimension of $\mathscr{M} \ominus \mathscr{M}_-$ were greater than 1, and if $\mathscr{K}$ were a non-trivial subspace of $\mathscr{M} \ominus \mathscr{M}_-$, then $\mathscr{M}_- \oplus \mathscr{K}$ would be comparable with every element of $\mathscr{F}$, contradicting the maximality of $\mathscr{F}$.

Conversely, if $\mathscr{F}$ has the stated properties, then $\mathscr{F}$ is maximal. For if $\mathscr{L}$ is a subspace not in $\mathscr{F}$ such that $\mathscr{F} \cup \{\mathscr{L}\}$ is a chain, let $\mathscr{M} = \bigcap \{\mathscr{K} : \mathscr{K} \in \mathscr{F} \text{ and } \mathscr{K} \supset \mathscr{L}\}$ and let $\mathscr{N} = \bigvee \{\mathscr{K} : \mathscr{K} \in \mathscr{F} \text{ and }$

$\mathcal{K} \subset \mathcal{L}$}. Clearly $\mathcal{M}_- = \mathcal{N}$, and, since $\dim(\mathcal{M} \ominus \mathcal{M}_-) = 1$ and $\mathcal{M}_- \subset \mathcal{L} \subset \mathcal{M}$, it follows that $\mathcal{L} = \mathcal{M}_-$ or $\mathcal{L} = \mathcal{M}$. Hence $\mathcal{L} \in \mathcal{F}$. ☐

Theorem 5.11. *If A is a polynomially compact operator, then there exists a maximal subspace chain $\mathcal{F}$ such that $\mathcal{F} \subset Lat\,A$.*

Proof. Zorn's Lemma implies that the set of all chains contained in *Lat A* has a maximal element, $\mathcal{F}$. To show that $\mathcal{F}$ is a maximal subspace chain we use Lemma 5.10.

Obviously $\mathcal{F}$ is complete. Let $\mathcal{M} \in \mathcal{F}$. If the dimension of $\mathcal{M} \ominus \mathcal{M}_-$ is greater than 1, let P denote the projection onto $\mathcal{M} \ominus \mathcal{M}_-$. Then $PA|(\mathcal{M} \ominus \mathcal{M}_-)$ is polynomially compact, and thus has a non-trivial invariant subspace $\mathcal{K}$, by Corollary 5.6. Then $\mathcal{M}_- \oplus \mathcal{K} \in Lat\,A$ and $\mathcal{F} \cup \{\mathcal{M}_- \oplus \mathcal{K}\}$ is a larger invariant subspace chain than $\mathcal{F}$. ☐

Definition. Let $\mathcal{F}$ be a maximal subspace chain contained in *Lat A*. For each $\mathcal{M}$ in $\mathcal{F}$ other than $\{0\}$ define $\alpha_{\mathcal{M}} = 0$ if $\mathcal{M}_- = \mathcal{M}$, and define $\alpha_{\mathcal{M}}$ as the number α such that $PAP = \alpha P$, where P is the projection onto $\mathcal{M} \ominus \mathcal{M}_-$, if $\mathcal{M}_- \neq \mathcal{M}$. (Since $\mathcal{M}_- \neq \mathcal{M}$ implies $\dim(\mathcal{M} \ominus \mathcal{M}_-) = 1$, it is clear that PAP is a multiple of P in this case.) Then $\{\alpha_{\mathcal{M}} : \mathcal{M} \in \mathcal{F}\}$ is the set of *diagonal coefficients of A relative to* $\mathcal{F}$.

Note that, in the case where A is an operator on a finite-dimensional space, the set of diagonal coefficients relative to $\mathcal{F}$ is the diagonal of the upper triangular matrix representation of A associated with $\mathcal{F}$. This makes the following very plausible.

Theorem 5.12. *If A is compact and $\mathcal{F}$ is any maximal chain in Lat A, then $\sigma(A)$ is the union of the set of diagonal coefficients of A relative to $\mathcal{F}$ and $\{0\}$.*

Proof. Since $\sigma(A) = \{0\} \cup \Pi_0(A)$ for every compact operator A, we need only show that $\{\alpha_{\mathcal{M}} : \mathcal{M} \in \mathcal{F}\} \subset \sigma(A)$ and that $(\Pi_0(A) \setminus \{0\}) \subset \{\alpha_{\mathcal{M}} : \mathcal{M} \in \mathcal{F}\}$. If $\alpha_{\mathcal{M}} \neq 0$, then $\mathcal{M}_- \neq \mathcal{M}$, and, by the definition of $\alpha_{\mathcal{M}}$, $(A - \alpha_{\mathcal{M}}) \mathcal{M} \subset \mathcal{M}_-$. Thus $(A|\mathcal{M}) - \alpha_{\mathcal{M}}$ maps $\mathcal{M}$ into a proper subspace of $\mathcal{M}$, and $\alpha_{\mathcal{M}} \in \sigma(A|\mathcal{M})$. Since $A|\mathcal{M}$ is compact, $\alpha_{\mathcal{M}} \in \Pi_0(A|\mathcal{M}) \subset \Pi_0(A)$.

For the other inclusion suppose that $\alpha \in \Pi_0(A) \setminus \{0\}$, and let $\mathcal{S} = \{x : Ax = \alpha x$ and $\|x\| = 1\}$. Then $\mathcal{S}$ is the unit sphere in a finite-dimensional space. Let $\mathcal{M} = \bigcap\{\mathcal{K} : \mathcal{K} \in \mathcal{F}$ and $\mathcal{K} \cap \mathcal{S} \neq \emptyset\}$; we show that $\alpha = \alpha_{\mathcal{M}}$.

The fact that $\mathcal{S}$ is compact implies, since $\mathcal{F}$ is totally ordered, that $\mathcal{M} \cap \mathcal{S} \neq \emptyset$. Hence $\alpha \in \Pi_0(A|\mathcal{M})$. If $\mathcal{M}$ is finite-dimensional, then clearly $\mathcal{M}_- \neq \mathcal{M}$ and $\mathcal{M}_- \cap \mathcal{S} = \emptyset$. Then $\alpha \notin \sigma(A|\mathcal{M}_-)$ and $\alpha_{\mathcal{M}} = \alpha$. If $\mathcal{M}$ is infinite-dimensional, then it requires a longer argument to show that $\alpha_{\mathcal{M}} = \alpha$. In this case $\sigma(A|\mathcal{M})$ contains the two distinct points α and 0; by the Riesz decomposition theorem (Theorem 2.10), $\mathcal{M}$ is the direct sum of two non-trivial invariant subspaces $\mathcal{M}_1$ and $\mathcal{M}_2$ of $A|\mathcal{M}$ such that $\sigma(A|\mathcal{M}_1)=\{\alpha\}$ and $0\in\sigma(A|\mathcal{M}_2)$. Moreover, the compactness of $A|\mathcal{M}$ implies that $\eta(\sigma(A|\mathcal{M}))=\sigma(A|\mathcal{M})$; hence, by Theorem 4.16 (which obviously applies to direct sums whether or not they are orthogonal), every $\mathcal{K}\in Lat(A|\mathcal{M})$ splits into a direct sum $\mathcal{K}_1+\mathcal{K}_2$ with $\mathcal{K}_1\subset\mathcal{M}_1$ and $\mathcal{K}_2\subset\mathcal{M}_2$. If $\mathcal{K}$ is a proper subspace of $\mathcal{M}$ and $\mathcal{K}\in\mathcal{F}$, then $\mathcal{K}\cap\mathcal{M}_1=\{0\}$ (for otherwise $\mathcal{K}\cap\mathcal{S}$ would not be empty). Thus every $\mathcal{K}\in\mathcal{F}$ which is a proper subspace of $\mathcal{M}$ is a subspace of $\mathcal{M}_2$, and $\mathcal{M}_-\subset\mathcal{M}_2$. It follows that $\mathcal{M}_-=\mathcal{M}_2$, the dimension of $\mathcal{M}_1$ is 1, and $\alpha_{\mathcal{M}}=\alpha$. ◻

Definition. A subspace chain $\mathcal{F}$ is *continuous* if $\mathcal{M}_-=\mathcal{M}$ for all $\mathcal{M}$ in $\mathcal{F}$ other than $\{0\}$.

Corollary 5.13. *If A is compact and $Lat\,A$ contains a continuous maximal chain, then A is quasinilpotent.*

Proof. The diagonal coefficients of A relative to a continuous maximal chain are all 0; hence the result follows immediately from Theorem 5.12. ◻

As one application of Corollary 5.13 we note that it implies, without any computation, that every Volterra-type integral operator (see Section 4.4) is quasinilpotent.

5.4 Lattices of Compact Operators

Theorem 5.14. *If $\mathcal{L}_1$ and $\mathcal{L}_2$ are totally ordered complete lattices, and if each lattice has an atom, then $\mathcal{L}_1\times\mathcal{L}_2$ is not attainable by a compact operator.*

Proof. Suppose that A is compact and $Lat\,A\approx\mathcal{L}_1\times\mathcal{L}_2$. For $i=1,2$, let ℓ_i be the largest element of $\mathcal{L}_i$, and let $\mathcal{H}_1$ and $\mathcal{H}_2$ be the invariant subspaces of A corresponding to $\ell_1\times0$ and $0\times\ell_2$ respectively. Then $Lat(A|\mathcal{H}_i)\approx\mathcal{L}_i$, and, since $\mathcal{L}_i$ is totally ordered, $\sigma(A|\mathcal{H}_i)$ is connected by the Riesz decomposition theorem, (Theorem 2.10). Since $A|\mathcal{H}_i$ is compact, it follows that $\sigma(A|\mathcal{H}_i)=\{0\}$.

Now if $\mathcal{N}_i$ is the atom in $Lat(A|\mathcal{H}_i)$, the invariant subspace theorem for compact operators (Corollary 5.7) implies that $\dim\mathcal{N}_i=1$. Thus $A|\mathcal{N}_i$ has an eigenvector; the corresponding eigenvalue must be 0, and therefore $A|\mathcal{N}_i$ has non-trivial nullspace. Then the dimension of the nullspace of A is at least 2, and the 0 element in $\mathcal{L}_1\times\mathcal{L}_2$ has uncountably many covers; (all the one-dimensional subspaces of the nullspace of A). This contradicts the fact that 0 has only two covers in $\mathcal{L}_1\times\mathcal{L}_2$. ◻

Example 5.15. *The lattice* $(\omega+1)\times(\omega+1)$ *is attained by a polynomially compact operator but not by a compact operator.*

Proof. If A is a Donoghue operator, then $\sigma(A)=\{0\}$, and Theorems 4.12 and 4.16 imply that $Lat(A\oplus(A+1))\approx(\omega+1)\times(\omega+1)$. Also $[A\oplus(A+1)]^2-[A\oplus(A+1)]=(A^2-A)\oplus(A^2+A)$ is compact. Theorem 5.14 implies that $(\omega+1)\times(\omega+1)$ is not attainable by a compact operator. $\square$

Some direct products of totally ordered lattices are attainable by compact operators. One such example can be obtained from Example 4.21 by choosing a nilpotent operator with a totally-ordered lattice on a finite-dimensional space for B. Another such example is the following.

Example 5.16. *There is a compact operator K such that*

$$Lat\,K \approx [0,1]\times[0,1].$$

Proof. Define the operator W on $\mathscr{L}^2(0,1)$ by

$$Wf(x) = \frac{1}{\sqrt{\pi}}\int_0^x \frac{f(t)}{\sqrt{x-t}}\,dt.$$

Then W^2 is the Volterra operator V and W is compact (Proposition 4.5). If $\{\mathscr{M}_\alpha\}$ is the family of subspaces invariant under V (as in Theorem 4.14), then clearly $\{\mathscr{M}_\alpha\}\subset Lat\,W$. But $W^2=V$ implies that $Lat\,W\subset Lat\,V$, and thus $Lat\,W=\{\mathscr{M}_\alpha\}$ and $Lat\,W\approx[0,1]$. Now apply Theorem 4.22 to the operator $K=W\oplus(-W)$; we need only verify the hypothesis that the weakly closed algebra generated by V includes W. For this we quote a more general result due to Sarason [3]: every operator which commutes with V is in the weakly closed algebra generated by V. $\square$

There are other attainable lattices that are not attained by any compact operator.

Theorem 5.17. *If A is polynomially compact, and if the supremum of the set of atoms in $Lat\,A$ is $\mathscr{H}$, (i.e., the largest element of $Lat\,A$), then the dual of $Lat\,A$ has at least one atom.*

Proof. It follows from the existence of invariant subspaces for polynomially compact operators, (Corollary 5.6), that each atom in $Lat\,A$ corresponds to a one-dimensional invariant subspace of A. We show that A^* has an eigenvector; this is enough since $Lat\,A^*$ is the dual of $Lat\,A$.

Let p be a non-zero polynomial such that $p(A)$ is compact. Then, since every eigenvector of A is an eigenvector of $p(A)$, the eigenvectors of $p(A)$ span the space $\mathscr{H}$. If $p(A)=0$, then A^* is algebraic too, and it follows that A^* has an eigenvector. If the nullspace of $p(A)$ is not $\mathscr{H}$, then $p(A)$ must have a non-zero eigenvalue λ. Then $\bar{\lambda}$ is an eigenvalue of $[p(A)]^*$, since $[p(A)]^*$ is compact, and the spectral mapping theorem implies that A^* has an eigenvalue. □

Theorem 5.18. *If S is a unilateral shift, then there does not exist a polynomially compact operator A such that* $Lat\,A \approx Lat\,S$.

Proof. It follows from the proof of Theorem 3.1 that the adjoint of the unilateral shift of multiplicity 1 has a spanning set of eigenvectors; hence this is the case for every unilateral shift. Now if $Lat\,A \approx Lat\,S$ and A is polynomially compact, then $Lat\,A^* \approx Lat\,S^*$. By Theorem 5.17, then, $Lat\,A$ has an atom. But $Lat\,S$ has no atoms. □

We have exhibited several examples of attainable lattices not attained by compact operators. Are there lattices which are attained only by compact operators? No, since $Lat(A+\lambda) = Lat\,A$ for each complex number λ. Are there lattices, other than transitive ones, which are attained only by translates of compact operators? For abstract lattices the answer is unknown. The following example answers the question affirmatively for concrete lattices.

Example 5.19. *If A is a Donoghue operator and if* $Lat(A \oplus A) \subset Lat\,B$, *then B is a translate of a compact operator.*

Proof. Since the co-ordinate spaces and the diagonal space $\{x \oplus x : x \in \mathscr{H}\}$ are in $Lat\,B$, the operator B has the form $C \oplus C$. Now $\{x \oplus Ax : x \in \mathscr{H}\}$ is in $Lat(A \oplus A)$ and hence also in $Lat(C \oplus C)$. It follows that $AC = CA$. Therefore, by Proposition 4.4, C is in the uniformly closed algebra generated by A and 1. Now if $\{p_n(A)\} \Rightarrow C$, and $p_n(z)$ has constant term λ_n, then $(p_n(A)e_0, e_0) = \lambda_n$, and $\{\lambda_n\} \to (Ce_0, e_0)$. Thus the sequence $\{p_n(A) - \lambda_n\}$ of compact operators converges to $C - (Ce_0, e_0)$, and C is the translate of a compact operator. □

5.5 Additional Propositions

Proposition 5.1. The direct sum of countably many quasitriangular operators is quasitriangular.

Proposition 5.2. There exist operators that are not quasitriangular.

Proposition 5.3. If $\sigma(A)$ is finite, then A is quasitriangular.

Proposition 5.4. If A has the property that $A|\mathscr{M}$ has a non-trivial reducing subspace whenever $\mathscr{M}$ is a reducing subspace of A of dimension at least 2, then each finite-dimensional eigenspace of A is reducing.

Proposition 5.5. Suppose that A is polynomially compact, *Lat A* $\approx 3 + \mathscr{L}$ for some complete lattice $\mathscr{L}$, ("3" denotes the ordinal number), and n is an integer greater than 1. Then A has an nth root if and only if A is invertible.

Proposition 5.6. If $\{\|A^n x\|^{1/n}\} \to 0$ for all $x \in \mathscr{H}$, then A is quasi-nilpotent.

Proposition 5.7. The operator A has a non-trivial invariant subspace if and only if the equation $XAX = AX$ has a solution other than 0 and 1.

Proposition 5.8. If A is compact and if $A|\mathscr{M}$ has a non-trivial reducing subspace whenever $\mathscr{M}$ reduces A and has dimension greater than 1, then A is normal.

5.6 Notes and Remarks

The basic techniques used in proving the existence theorems of Section 5.1 were independently discovered by von Neumann (unpublished) and Aronszajn-Smith [1], who used them to prove that compact operators have non-trivial invariant subspaces. Refining the basic techniques to obtain the results of Section 5.1 proved to be much more difficult than might be supposed. The breakthrough was obtained by Bernstein-Robinson [1], who proved Corollary 5.6 using Robinson's theory of non-standard analysis (see Robinson [1] for such a proof). Halmos [11] translated the Bernstein-Robinson proof into standard analysis, and Arveson-Feldman [1] obtained Corollary 5.5 by modifying Halmos' approach. For treatments of these results on Banach spaces (where the same theorems hold but proofs are more difficult) see Aronszajn-Smith [1], Bernstein [1], Gillespie [1], Kitano [2], Meyer-Nieberg ([1], [2]), Apostol [2] and Hsu [1].

The concept of quasitriangularity was isolated and studied by Halmos [9], who derived several conditions equivalent to quasitriangularity, proved Propositions 5.1 and 5.2, and found some other related results. Proposition 5.3 is due to Douglas-Pearcy [2]; it has been generalized by Halmos [14] to the theorem that every operator whose spectrum has capacity 0 is quasitriangular. Other results on quasitriangular operators can be found in Douglas-Pearcy [4], Deckard-Douglas-Pearcy [1], Meyer-Nieberg [2] and Deddens [2]. Some of these results are generalized to the case of operators in von Neumann algebras (projections in an ideal replacing projections of finite rank) in Olsen ([2], [3]). Olsen [1] proves that a polynomially compact operator is the sum of a compact operator and an algebraic operator.

Theorem 5.3 is improved in Meyer-Nieberg [2] and Pearcy-Salinas [1], where it is shown that every quasitriangular operator A with the property that the uniformy closed algebra generated by the rational functions of A contains a compact operator different from 0 has a non-trivial invariant subspace. Apostol-Foiaş-Voiculescu ([1], [2]) include a remarkable result: if $\Pi_0(A^*) = \emptyset$, then A is quasitriangular. This implies, in particular, that the hypothesis of quasitriangularity can be omitted in Theorem 5.3 and in the stronger result quoted above.

Theorem 5.9 was proven by Ando [2] for compact operators, by Saito [1] for operators some power of which is compact, and by Rosenthal [2] as presented in the text.

The beautiful results of Section 5.3 are due to Ringrose [1], although our proofs are somewhat different, and the results of Section 5.4 are due to Rosenthal ([3], [4]).

Proposition 5.4 is from Rosenthal [2], Proposition 5.5 from Rosenthal [7], and Proposition 5.6 from Colojoara-Foiaş [1]. Proposition 5.7 is implicitly contained in Aronszajn-Smith [1] and Proposition 5.8 is in Rosenthal [2].

Chapter 6. Existence of Invariant and Hyperinvariant Subspaces

We have seen that certain operators have non-trivial invariant subspaces: normal operators (Corollary 1.17), operators with disconnected spectra (Corollary 2.11), parts of the adjoint of the unilateral shift (Corollary 3.31), and operators closely related to compact operators (Corollaries 5.5, 5.6, and 5.7). In this chapter we show that operators that are close (in a certain sense) to normal operators with thin spectra have invariant subspaces. In particular, we obtain the result that parts of the adjoints of finite-multiplicity unilateral shifts have invariant subspaces, thereby proving the factorization theorem for isometry-valued analytic functions alluded to in Section 3.6.

Some of these existence theorems produce hyperinvariant subspaces. In Section 4 we consider several additional results on existence of hyperinvariant subspaces.

We begin by exhibiting counterexamples to two assertions related to the invariant subspace question.

6.1 Operators on Other Spaces

The invariant subspace problem is unsolved for operators on Banach spaces as well as on Hilbert spaces. It is obvious that every linear transformation on a complex vector space of dimension greater than 1 has a non-trivial invariant linear manifold. The following are counterexamples to two natural assertions in between these statements.

Example 6.1. *There is a bounded linear operator on a complex inner-product space which has no non-trivial invariant subspaces.*

Proof. Let $\mathscr{H}$ denote the vector space of all polynomials with complex coefficients, and define $(p,q) = \int_0^1 p(x)\overline{q(x)}\,dx$. Let M_x denote the operator which sends $p(x)$ into $xp(x)$ for each $p \in \mathscr{H}$. Then $\mathscr{H}$ is a dense linear manifold in $\mathscr{L}^2(0,1)$, and the operator M_x is the restriction of multiplication by the independent variable to $\mathscr{H}$. If $\mathscr{M}$ is a closed subspace of $\mathscr{H}$ invariant under M_x, and if $p \in \mathscr{M}$, $p \neq 0$, then, since $\{x \in [0,1] : p(x) = 0\}$ is a finite set and thus has measure 0, it follows from Example 4.11 that every polynomial is a limit in $\mathscr{L}^2(0,1)$ of a sequence of the form $\{q_n(x)p(x)\}$ where each q_n is a polynomial. Hence $\mathscr{M} = \mathscr{H}$. ☐

There are many other examples like the above, (e.g., a Volterra or Donoghue operator can be used instead of M_x).

Example 6.2. *There is a (not necessarily bounded) linear transformation on a Hilbert space which has no non-trivial invariant subspaces.*

Proof. Let $\mathscr{H}$ be a (separable) Hilbert space. We construct such a linear transformation by ordering the subspaces of $\mathscr{H}$ and defining a transformation which takes a vector from each subspace out of the subspace. Let ω_c denote the first ordinal number which has cardinality c (i.e., the cardinality of the continuum). The space $\mathscr{H}$ has c proper closed infinite-dimensional subspaces: let $\alpha \leftrightarrow \mathscr{M}_\alpha$ be a one-to-one correspondence between the predecessors of ω_c and this collection of subspaces.

We use transfinite induction to assign a pair (f_α, g_α) of vectors to each $\alpha < \omega_c$ in such a manner that $f_\alpha \in \mathscr{M}_\alpha$, $g_\alpha \notin \mathscr{M}_\alpha$, and $\{f_\alpha, g_\alpha : \alpha < \omega_c\}$ is a linearly independent set. Begin the construction by choosing an $f_1 \in \mathscr{M}_1$ and a $g_1 \notin \mathscr{M}_1$ such that $\{f_1, g_1\}$ is linearly independent. Now if $\alpha < \omega_c$ and (f_β, g_β) has been assigned for all $\beta < \alpha$, let $\mathscr{V}_\alpha$ denote the linear span of $\{f_\beta, g_\beta : \beta < \alpha\}$. The algebraic dimension of $\mathscr{V}_\alpha$ is less than c, and the algebraic dimension of $\mathscr{M}_\alpha$ is c. Thus there exists a vector $f_\alpha \in \mathscr{M}_\alpha$ such that $f_\alpha \notin \mathscr{V}_\alpha$. Since $\mathscr{V}_\alpha$ contains no non-empty open set, there exists a vector g_α which is not in $\mathscr{M}_\alpha \cup \mathscr{V}_\alpha$. By transfinite induction, then, there is a linearly independent set $\mathscr{S} = \{f_\alpha, g_\alpha : \alpha < \omega_c\}$ such that $f_\alpha \in \mathscr{M}_\alpha$ and $g_\alpha \notin \mathscr{M}_\alpha$ for each α.

Now extend the set $\mathscr{S}$ to an algebraic basis $\mathscr{S} \cup \mathscr{T}$ for $\mathscr{H}$. Define the linear transformation T on $\mathscr{S}$ by $T f_\alpha = g_\alpha$ and $T g_\alpha = f_{\alpha+1}$ for all $\alpha < \omega_c$. Then clearly, no matter how we define T on $\mathscr{T}$, T will have no infinite-dimensional invariant subspaces. We must also make sure that T has no finite-dimensional invariant subspaces, and for this we distinguish two cases. If $\mathscr{T}$ is infinite, define T on the linear span of $\mathscr{T}$ as any linear transformation of this span into itself which has no finite-dimensional invariant subspaces. If $\mathscr{T}$ is finite, say $\mathscr{T} = \{h_1, \ldots, h_n\}$, define $T h_i = h_{i+1}$ for $i < n$ and $T h_n = f_1$. Then, in either case, extend T by linearity to all of $\mathscr{H}$. All that remains to be shown is that T has no finite-dimensional invariant subspaces. This follows from the fact that T has no eigenvectors, (since every linear combination of the basis vectors in $\mathscr{S} \cup \mathscr{T}$ is mapped by T onto a linear combination which involves at least one new basis element). ☐

6.2 Perturbations of Normal Operators

In this section we prove the existence of invariant and hyperinvariant subspaces for certain compact perturbations of normal operators whose spectra lie on smooth Jordan arcs. The subspaces will be obtained by considering which of a class of vector-valued analytic functions have analytic continuations across arcs contained in the spectrum of the operator. The basic result is a technical one whose scope is not clear at the outset. After proving this theorem we obtain a number of corollaries which give more comprehensible existence theorems.

Definition. A *smooth* Jordan arc is a one-to-one function $z(t) = x(t) + iy(t)$ from $(0,1)$ into $\mathbb{C}$ such that d^2z/dt^2 exists everywhere in $(0,1)$. If T is a bounded linear operator, then $\sigma(T)$ *contains an exposed arc* J if there exists an open disk $\mathcal{D}$ such that $\mathcal{D} \cap \sigma(T) = J$ and J is a smooth Jordan arc.

Theorem 6.3. *Let T be a bounded linear operator such that $\sigma(T)$ contains an exposed arc C, and let k be a positive integer. Suppose that for each point $z_0 \in C$ and each closed line segment L which meets $\sigma(T)$ only in $\{z_0\}$ and which is not tangent to C, there exists a constant K such that*

$$\|(z - T)^{-1}\| \leqq \exp\{K|z - z_0|^{-k}\}$$

for all z on L other than z_0. Then T has a non-trivial hyperinvariant subspace.

Proof. By multiplying T by an appropriate complex number (which rotates $\sigma(T)$) and replacing C by a subarc if necessary, we can assume that C has a representation $y = g(x)$, $x \in (a,b)$, with g one-to-one, $|g'(x)| < \tan(\pi/5k)$ for $x \in (a,b)$ and $g''(x)$ existing everywhere in (a,b). If $\mathcal{D}$ is an open disk such that $\mathcal{D} \cap \sigma(T) = C$, then $\mathcal{D}$ is the union of disjoint Jordan regions $\mathcal{D}_1$ and $\mathcal{D}_2$ lying above and below C respectively.

We need to consider subarcs J of C such that $\bar{J} \subset C$. If J is such a subarc, with endpoints z_1 and z_2, $\operatorname{Re} z_1 < \operatorname{Re} z_2$, we construct a simple closed Jordan polygon $\Gamma_1(J)$ enclosing J as pictured on the next page. Construct $\Gamma_1(J)$ in $\mathcal{D}$, intersecting C at z_1 and z_2 only, by beginning with the angle at z_1 determined by the rays through z_1 with arguments $\pm\pi/5k$ and the angle at z_2 determined by the rays through z_2 with arguments $\pi \pm \pi/5k$. Then connect these angles by lines above and below J getting a hexagon lying in $\mathcal{D}$ as illustrated in the diagram.

Let $\Gamma_2(J)$ be the union of $\Gamma_1(J)$ and any fixed circle containing $\sigma(T) \cup \mathcal{D}$ in its interior.

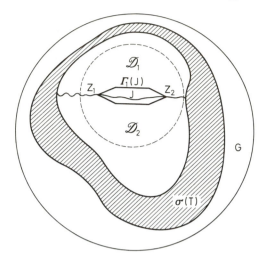

Fix an open subarc J_0 of C such that $\bar{J}_0 \subset C$, and, for each $x \in \mathcal{H}$, let R_x denote the function taking $\rho(T)$ into $\mathcal{H}$ by $R_x(z) = (z-T)^{-1}x$ for $z \in \rho(T)$. Then R_x is an analytic vector-valued function with domain $\rho(T)$. Let

$\mathcal{N} = \{x \in \mathcal{H} : R_x$ has an analytic continuation to the complement of $\bar{J}_0$ in the complex plane$\}$.

We prove that $\mathcal{N}$ is a non-trivial hyperinvariant subspace for T. It is clear that $\mathcal{N}$ is a linear manifold. If $AT = TA$, then $R_{Ax}(z) = (z-T)^{-1}Ax = A(z-T)^{-1}x = AR_x(z)$, and thus $\mathcal{N}$ is invariant under every operator which commutes with T.

We must show that $\mathcal{N}$ is closed. Let $\{x_n\} \in \mathcal{N}$ and $\{x_n\} \to x$. Let, for each n, R_n denote the analytic continuation of $(\lambda - T)^{-1}x_n$ to the complement of $\bar{J}_0$. To prove that R_x has an analytic continuation to the complement of $\bar{J}_0$ it suffices to show that R_x has an analytic continuation to the complement of $\bar{J}$ for each open subarc J such that $J \supset \bar{J}_0$. Let J be such a subarc, with endpoints λ_1 and λ_2, $\operatorname{Re}\lambda_1 < \operatorname{Re}\lambda_2$, and let G be the open annulus-like region whose boundary is $\Gamma_2(J)$. Define the function m by

$$m(z) = \begin{cases} \exp\{-(z-\lambda_1)^{-2k} - (z-\lambda_2)^{-2k}\} & \text{for } z \neq \lambda_1, \lambda_2 \\ 0 & \text{for } z = \lambda_1, \lambda_2 \end{cases}.$$

It follows from the choice of angles of $\Gamma_1(J)$ at λ_1 and λ_2 that $m(z)$ is continuous on $\bar{G}$. Thus $m(z)R_n(z)$ is analytic on G and continuous on $\bar{G}$. We show that the sequence $\{m(z)R_n(z)\}$ converges to an analytic function on G. For $z \in G$,

$$\|m(z)\,R_n(z) - m(z)\,R_m(z)\| \leq \sup_{w \in \Gamma_2(J)} \|m(w)\,(R_n(w) - R_m(w))\|$$

by the maximum modulus principle, (Proposition 2.12). Let L denote one of the lines of $\Gamma_2(J)$ with λ_1 as an endpoint. Then, by the hypothesis on the rate of growth of the resolvent of T, for $w \in L$, $w \neq \lambda_1$, we have

$$\|m(w)\,(R_n(w) - R_m(w))\| = \|m(w)\,(w - T)^{-1}(x_n - x_m)\|$$

$$\leq \|x_n - x_m\| \, |\exp\{-(w - \lambda_1)^{-2k} - (w - \lambda_2)^{-2k} + K|w - \lambda_1|^{-k}\}|$$

$$\leq \|x_n - x_m\| \, N \exp\{\mathrm{Re}(-(w - \lambda_1)^{-2k} + K|w - \lambda_1|^{-k})\},$$

where $N = \sup_{w \in L} |\exp\{-(w - \lambda_2)^{-2k}\}|$.

Since w and λ_1 are on L, $w - \lambda_1 = |w - \lambda_1| e^{i\theta}$, where θ is either $\pi/5k$ or $-(\pi/5k)$. It follows that $\exp\{\mathrm{Re}(-(w - \lambda_1)^{-2k}) + K|w - \lambda_1|^{-k}\}$ is bounded on L. Thus for $w \in L$

$$\|m(w)\,(R_n(w) - R_m(w))\| \leq M_1 \|x_n - x_m\|$$

for some constant M_1.

In exactly the same manner it can be shown that there exists a constant M_2 such that

$$\|m(w)\,(R_n(w) - R_m(w))\| \leq M_2 \|x_n - x_m\|$$

for all w on the lines of $\Gamma_2(J)$ through λ_2. Since a similar assertion is obviously true for the lines of $\Gamma_2(J)$ which do not meet either λ_1 or λ_2, it follows that the sequence $\{m(z)\,R_n(z)\}$ is a uniform Cauchy sequence on $\bar{G}$. Hence this sequence converges uniformly to some function $S(z)$ analytic on G and continuous on $\bar{G}$. For each $z \in \rho(T)$, $(m(z))^{-1} m(z) R_n(z) = (z - T)^{-1} x_n$, and thus $(m(z))^{-1} S(z) = (z - T)^{-1} x$. Hence the function $S(x)/m(z)$ is an analytic continuation of $(z - T)^{-1} x$ to the complement of $\bar{J}$. For each such J, then, $(z - T)^{-1} x$ can be analytically continued to the complement of $\bar{J}$, and it follows that $(z - T)^{-1} x$ has an analytic continuation to the complement of $\bar{J}_0$. Thus $x \in \mathcal{N}$.

All that remains to be done is to show that $\mathcal{N}$ is non-trivial. It is easily seen that $\mathcal{N}$ is not equal to $\mathcal{H}$. For suppose that $\mathcal{N} = \mathcal{H}$, i.e., $(z - T)^{-1} x$ has an analytic continuation to the complement of $\bar{J}_0$ for every x. Fix $\lambda \in C \backslash \bar{J}_0$ and a compact neighbourhood $\mathcal{S}$ of λ disjoint from $\bar{J}_0$ and contained in $\mathcal{D}$. Then $\{\|(z - T)^{-1} x\| : z \in \mathcal{S} \backslash C\}$ is bounded for each x, and the principle of uniform boundedness implies that $\{\|(z - T)^{-1}\| : z \in \mathcal{S} \backslash C\}$ is bounded. This contradicts the fact that $\lambda \in \Pi(T)$, (by Theorem 0.7). Therefore $\mathcal{N} \neq \mathcal{H}$.

It remains to be shown that $\mathcal{N}$ is not $\{0\}$.

Let z_1 and z_2 be the endpoints of J_0 and

$$m(z) = \begin{cases} \exp\{-(z - z_1)^{-2k} - (z - z_2)^{-2k}\} & \text{for } z \neq z_1, z_2 \\ 0 & \text{for } z = z_1 \ \text{ or } \ z = z_2 \end{cases}.$$

For any $x \in \mathcal{H}$ the function $m(z)(z-T)^{-1}x$ is continuous on $\Gamma_1(J_0)$, (by the computation performed above, using the growth condition on the resolvent of T). Hence the integral $y = \int_{\Gamma_1(J_0)} m(z)(z-T)^{-1}x\,dz$ exists by Theorem 2.1. We now show that, for each x, this vector y is in $\mathcal{N}$. Let w be any point in $\rho(T)$ and outside $\Gamma_1(J_0)$. Then, since

$$(w-T)^{-1}(z-T)^{-1} = ((z-T)^{-1}-(w-T)^{-1})(w-z)^{-1},$$

$$(w-T)^{-1}y = \int_{\Gamma_1(J_0)} \frac{1}{w-z}m(z)(z-T)^{-1}x\,dz - \int_{\Gamma_1(J_0)} \frac{1}{w-z}m(z)(w-T)^{-1}x\,dz.$$

The second integral is 0, by Cauchy's theorem (Theorem 2.2). The first integral is an analytic function of w for w outside $\Gamma_1(J_0)$, and it follows that $(w-T)^{-1}y$ has an analytic continuation to the exterior of $\Gamma_1(J_0)$, and thus also to the complement of $\bar{J}_0$. Hence $y \in \mathcal{N}$.

Now $\mathcal{N} \neq \{0\}$ if at least one such y is not 0. Let λ_0 be any point in J_0. Then $\lambda_0 \in \Pi(T)$, and thus for each $\varepsilon > 0$ there is a unit vector x_ε such that $(\lambda_0 - T)x_\varepsilon = h_\varepsilon$ with $\|h_\varepsilon\| < \varepsilon$. Then

$$(z-T)^{-1}x_\varepsilon = \frac{1}{\lambda_0 - z}(z-T)^{-1}h_\varepsilon + \frac{1}{z-\lambda_0}x_\varepsilon,$$

and

$$\int_{\Gamma_1(J_0)} m(z)(z-T)^{-1}x_\varepsilon\,dz = \int_{\Gamma_1(J_0)} \frac{m(z)}{\lambda_0 - z}(z-T)^{-1}h_\varepsilon\,dz + 2\pi i m(\lambda_0)x_\varepsilon$$

by the Cauchy integral formula. The function $m(z)(z-T)^{-1}$ is a continuous operator-valued function on $\Gamma_1(J_0)$, and $(\lambda_0 - z)^{-1}$ is continuous on $\Gamma_1(J_0)$. Hence

$$\left\| \int_{\Gamma_1(J_0)} \frac{m(z)}{\lambda_0 - z}(z-T)^{-1}h_\varepsilon\,dz \right\|$$

approaches 0 as ε approaches 0. Since $\|m(\lambda_0)x_\varepsilon\| = |m(\lambda_0)| \neq 0$, it follows that the vector $\int_{\Gamma_1(J_0)} m(z)(z-T)^{-1}x_\varepsilon\,dz$ is non-zero for sufficiently small ε. $\square$

Theorem 6.4. *If T is an invertible operator such that $\sigma(T)$ contains more than one point, and if there exists a constant M and a positive integer k such that $\|T^n\| \leq M|n|^k$ for $n = \pm 1, \pm 2, \ldots$, then T has a hyperinvariant subspace.*

Proof. First note that $\lim_{n \to \infty} \|T^n\|^{1/n} \leq \lim_{n \to \infty} M^{1/n}(n^{1/n})^k = 1$, and also $\lim_{n \to \infty} \|(T^{-1})^n\|^{1/n} \leq 1$. Hence $\sigma(T)$ and $\sigma(T^{-1})$ are both contained in the unit disk, and it follows that $\sigma(T)$ is contained in the unit circle. If $\sigma(T)$ is disconnected, then it follows from Corollary 2.11 that T has hyperinvariant subspaces. We can assume, then, that $\sigma(T)$ contains an exposed arc of the unit circle. The existence of hyperinvariant subspaces in this case will follow from Theorem 6.3 once the relevant growth condition on the resolvent of T is established.

Suppose that $1 < |z| < 2$. Then

$$\|(z-T)^{-1}\| = \frac{1}{|z|} \left\| 1 + \frac{1}{z} T + \frac{1}{z^2} T^2 + \cdots \right\| \leqq \frac{M}{|z|} \sum_{n=0}^{\infty} \frac{n^k}{|z|^n}.$$

Let $t = 1/|z|$. Then

$$\|(z-T)^{-1}\| \leqq M t \left[1 + \sum_{n=1}^{\infty} n^k t^n \right]$$

$$\leqq M t \left[1 + \sum_{n=1}^{\infty} (n+1) \dots (n+k) t^n \right].$$

Now $\sum_{n=1}^{\infty} (n+1) \dots (n+k) t^n$ is the kth derivative of the function $F(t) = \sum_{n=1}^{\infty} t^{n+k} = t^{k+1}/(1-t)$; hence $\sum_{n=1}^{\infty} (n+1) \dots (n+k) t^n$ has the form $p(t)/(1-t)^{k+1}$ for some polynomial p. Therefore there exists a constant K such that, for $1 < |z| < 2$,

$$\|(z-T)^{-1}\| \leqq K \frac{1}{\left(1 - \dfrac{1}{|z|} \right)^{k+1}}$$

$$\leqq K \frac{2^{k+1}}{(|z|-1)^{k+1}}.$$

Since $|z - z_0| \leqq c(|z|-1)$ for $|z_0| = 1$ and z on a non-tangent line segment with endpoint z_0, (c depending on the segment), it follows that $(z-T)^{-1}$ satisfies the growth condition required by Theorem 6.3 on lines L outside the unit circle.

If $\frac{1}{2} < |z| < 1$, then $(z-T)^{-1} = (1/z) T^{-1} (T^{-1} - (1/z))^{-1}$, and $\|(z-T)^{-1}\| \leqq 2 \|T^{-1}\| \|((1/z) - T^{-1})^{-1}\|$. Since the hypotheses on T and T^{-1} are the same, it follows from the above computation that $((1/z) - T^{-1})^{-1}$ satisfies the necessary growth condition, and thus so does $(z-T)^{-1}$. Theorem 6.3 now gives the result. $\square$

The main existence theorem that we shall prove involves the consideration of certain compact perturbations of normal operators. If T is a compact operator, then T^*T is a positive compact operator, and hence so is its unique positive square root, $(T^*T)^{\frac{1}{2}}$. Note that T^*T and $(T^*T)^{\frac{1}{2}}$ are diagonable, by Theorem 1.4.

Definition. Let T be a compact operator. Arrange the eigenvalues of $(T^*T)^{\frac{1}{2}}$, repeated according to multiplicity, in decreasing order, and let $\mu_n(T)$ denote the nth eigenvalue of $(T^*T)^{\frac{1}{2}}$. For each p, $1 \leq p < \infty$, define

$$|T|_p = \left[\sum_{n=1}^{\infty} (\mu_n(T))^p \right]^{\frac{1}{p}}, \quad \text{and} \quad \mathscr{C}_p = \{T : |T|_p < \infty\}.$$

The $\mathscr{C}_p$ classes are interesting classes of operators because many properties of finite-rank operators have analogues for operators in $\mathscr{C}_p$. The basic properties of $\mathscr{C}_p$ which we require include the following.

Lemma 6.5. If $T \in \mathscr{C}_p$, then
(i) there exists a sequence $\{T_n\}$ of finite-rank operators such that

$$\{\|T - T_n\|\} \to 0 \quad \text{and} \quad \{|T - T_n|_p\} \to 0.$$

(ii) for each bounded operator A the operators AT and TA are in $\mathscr{C}_p$, and

$$|AT|_p \leq \|A\| \, |T|_p, \quad |TA|_p \leq \|A\| \, |T|_p.$$

Proof. We refer to Dunford-Schwartz [2] for the proofs: the proof of (i) is on page 1095, and the proof of (ii) is on page 1093. ☐

Note that $\mathscr{C}_{p_1} \subset \mathscr{C}_{p_2}$ if $p_1 \leq p_2$. Thus every $\mathscr{C}_p$ is contained in some $\mathscr{C}_k$ where k is an integer ≥ 2. A generalized determinant function can be defined on $\mathscr{C}_k$.

Definition. If $T \in \mathscr{C}_k$, with k an integer greater than 1, and if $\{\lambda_1, \lambda_2, \ldots\}$ is an enumeration of the non-zero eigenvalues of T, repeated according to multiplicity, then we define

$$\delta_k(T) = \prod_{i=1}^{\infty} \left[(1 + \lambda_i) \exp\left(-\lambda_i + \frac{\lambda_i^2}{2} + \cdots + \frac{(-1)^{k-1} \lambda_i^{k-1}}{k-1} \right) \right].$$

(If $\Pi_0(T) \subset \{0\}$, define $\delta_k(T) = 1$.)

Lemma 6.6. If $T \in \mathscr{C}_k$, with k an integer greater than 1, then
(i) $\delta_k(T)$ is an absolutely convergent infinite product.
(ii) $\delta_k(T)$ is a continuous function on $\mathscr{C}_k$ (in the topology of $\mathscr{C}_k$).
(iii) there exists a constant Γ_k, depending only on k, such that

$$|\delta_k(T)| \leq \exp(\Gamma_k |T|_k^k).$$

(iv) there exists a constant M_k, depending only on k, such that

$$\|\delta_k(T)(1 + T)^{-1}\| \leq \exp(M_k |T|_k^k)$$

whenever $-1 \notin \sigma(T)$.

(v) *if T is an operator on a finite-dimensional space, then*

$$\delta_k(T) = \det((1+T)) \exp\left(-tr(T) + \cdots + \frac{(-1)^{k-1}}{k-1} tr(T^k)\right),$$

where "det" and "tr" denote "determinant" and "trace" respectively.

Proof. We again merely refer to Dunford-Schwartz [2]: (i), (ii) and (iii) are on page 1106, (iv) is on page 1112, and (v) is on page 1110. □

Lemma 6.7. *If* $B \in \mathscr{C}_k$ *and* A *is any bounded operator, then* $\delta_k(B(z-A)^{-1})$ *is an analytic function of* z *on* $\rho(A)$.

Proof. Fix $z_0 \in \rho(A)$ and $\mathscr{D}$, an open disk centred at z_0 such that $\overline{\mathscr{D}}$ is contained in $\rho(A)$. We construct a sequence of analytic functions converging uniformly to $\delta_k(B(z-A)^{-1})$ on $\mathscr{D}$. For each integer n choose, by Lemma 6.5 (i), a finite-rank operator B_n such that $|B_n - B|_k < 1/n$ and $\|B_n - B\| < 1/n$. Let $\mathscr{M}_n$ denote the range of B_n. For any $z \in \mathscr{D}$, $\mathscr{M}_n$ is invariant under $B_n(z-A)^{-1}$, and

$$\delta_k(B_n(z-A)^{-1}) = \delta_k((B_n(z-A)^{-1})|\mathscr{M}_n),$$

since the range of $B_n(z-A)^{-1}$ is $\mathscr{M}_n$ and thus $B_n(z-A)^{-1}$ and $B_n(z-A)^{-1}|\mathscr{M}_n$ have the same non-zero eigenvalues. Now

$$\det(1 + B_n(z-A)^{-1}|\mathscr{M}_n) \quad \text{and} \quad tr[(1 + B_n(z-A)^{-1}|\mathscr{M}_n)^r], \quad r = 1, \ldots, k,$$

are analytic functions of z on $\rho(A)$, (the components of the matrices of $B_n(z-A)^{-1}|\mathscr{M}_n$ relative to a given basis are analytic functions of z), and it follows that $\delta_k(B_n(z-A)^{-1})$ is analytic on $\rho(A)$.

Also

$$|B_n(z-A)^{-1} - B(z-A)^{-1}|_k = |(B_n - B)(z-A)^{-1}|_k$$

$$\leq |B_n - B|_k \|(z-A)^{-1}\|,$$

by Lemma 6.6 (ii).

Thus if $N = \sup_{z \in \mathscr{D}} \|(z-A)^{-1}\|$, $|B_n(z-A)^{-1} - B(z-A)^{-1}|_k \leq N|B_n - B|_k$. Lemma 6.6 (ii) implies that $\{\delta_k(B_n(z-A)^{-1})\}$ converges pointwise to $\delta_k(B(z-A)^{-1})$, and Lemma 6.6 (iii) implies that $\{\delta_k(B_n(z-A)^{-1})\}$ is uniformly bounded in $\overline{\mathscr{D}}$. Hence $\delta_k(B_n(z-A)^{-1})$ converges uniformly on $\overline{\mathscr{D}}$ to $\delta_k(B(z-A)^{-1})$ by Vitali's theorem, (Titchmarsh [1], page 168). □

Some inequalities from complex analysis will be useful in deriving growth conditions on resolvents of $\mathscr{C}_p$-perturbations of normal operators.

Lemma 6.8 *(Borel-Carathéodory Inequality). If f is a complex-valued function analytic on $\{z : |z| \leq r\}$, then for $|z| < r$,*

$$|f(z)| \leq \frac{2|z|}{r - |z|} \sup\{\operatorname{Re} f(w) : |w| = |z|\} + \frac{r + |z|}{r - |z|}|f(0)|.$$

Proof. This follows readily from the Schwarz lemma; see Titchmarsh [1], page 174. □

Lemma 6.9. *If $p \geq 1$ and f is a function analytic and satisfying $\operatorname{Re} f(z) \leq (1 - |z|)^{-p}$ in $\{z : |z| < 1\}$, then there exists a constant K such that $|f(z)| \leq K(1 - |z|)^{-p-1}$ for $|z| < 1$.*

Proof. For each $r \in (\frac{1}{2}, 1)$, the Borel-Carathéodory inequality (Lemma 6.8) applied to the function $f(z)$ in the disk $\{z : |z| \leq r\}$ yields

$$|f(z)| \leq \frac{r + |z|}{r - |z|}[(1 - r)^{-p} + |f(0)|]$$

for $|z| \leq r$. For $|z| = 2r - 1$ this gives

$$|f(z)| \leq 2^{p+2}[(1 - |z|)^{-p-1} + |f(0)|(1 - |z|)^{-1}].$$

If $|z| < 1$ and $r = (|z| + 1)/2$, then $r \in [\frac{1}{2}, 1)$ and $|z| = 2r - 1$, and thus this inequality holds whenever $|z| < 1$. It follows that $K = 2^{p+2}(1 + |f(0)|)$ satisfies the conditions of the lemma. □

Lemma 6.10. *Let $\mathscr{D}$ be a bounded simply connected region in the complex plane with boundary C. Suppose that C is a smooth Jordan arc. Let $z_0 \in C$ and L be any closed line segment with z_0 as an endpoint which lies entirely in $\mathscr{D}$ except for the point z_0 and which is not tangent to C. Then if f is any function analytic and satisfying $\operatorname{Re} f(z) \leq K(d(z, C))^{-p}$ in $\mathscr{D}$, where $p \geq 1$, $K > 0$, and $d(z, C)$ denotes the distance from z to C, it follows that there exists a constant M such that $|f(z)| \leq M|z - z_0|^{-p-1}$ for $z \in \mathscr{D} \cap L$.*

Proof. First it is geometrically obvious that there is a circle tangent to C at z_0, whose interior is contained in $\mathscr{D}$, which meets C only at z_0; (a proof of this fact is easily constructed using elementary calculus). Let C_0 be such a circle. For z interior to C_0, $\operatorname{Re} f(z) \leq K(d(z, C))^{-p} \leq K(d(z, C_0))^{-p}$. By Lemma 6.9, adapted to the interior of C_0, there exists a constant N such that $|f(z)| \leq N(d(z, C_0))^{-p-1}$ for z interior to C_0. Therefore there exists an M such that $|f(z)| \leq M|z - z_0|^{-p-1}$ for z on L sufficiently close to z_0, (since L is not tangent to C_0). □

Lemma 6.11. *Let $T = A + B$, where $B \in \mathscr{C}_k$ for some integer $k > 1$ and A is normal. Suppose that $\Pi_0(T) = \emptyset$ and that $\sigma(A)$ contains an exposed arc J. Let $z_0 \in J$ and let L be any closed bounded line segment*

with z_0 as an endpoint which is not tangent to J and is such that
$L \cap \sigma(A) = \{z_0\}$. *Then there is a constant K such that* $\|(z-T)^{-1}\|$
$\leq \exp(K|z-z_0|^{-k-1})$ *for all* $z \in L \setminus \{z_0\}$.

Proof. First note that, by Weyl's theorem (Theorem 0.10), $\Pi_0(T) = \emptyset$
implies $\sigma(T) \subset \sigma(A)$. Let $\mathscr{D}$ be an open disk such that $\mathscr{D} \cap \sigma(A) = J$;
then J divides $\mathscr{D}$ into two simply connected regions with simple closed
Jordan curves as boundaries. Given a line L as in the statement of
the lemma, let $\mathscr{D}'$ be a disk tangent to J at z_0 and contained in the sub-
region of $\mathscr{D}$ which meets L, and let C' be the boundary of $\mathscr{D}'$.

Now define the function δ on $\rho(A)$ by $\delta(z) = \delta_k(-B(z-A)^{-1})$.
Then δ is analytic on $\rho(A)$, by Lemma 6.7. Note that, for $z \in \rho(A)$,
$(z-T)^{-1} = (z-A-B)^{-1} = (z-A)^{-1}(1-B(z-A)^{-1})^{-1}$; (this is trivially
verified by multiplying both sides on the left by $(z-T)$). In particular
$-1 \notin \sigma(-B(z-A)^{-1})$, and it follows from the absolute convergence of
δ_k, (Lemma 6.6 (i)), that $\delta(z) \neq 0$. Hence

$$(z-T)^{-1} = (\delta(z))^{-1}(z-A)^{-1}\delta(z)(1-B(z-A)^{-1})^{-1}.$$

We estimate $\|(z-T)^{-1}\|$ by estimating each term in this product.

By Lemma 6.6 (iii)

$$|\delta(z)| \leq \exp(K_1|B(z-A)^{-1}|_k^k)$$

for some constant K_1, and by Lemma 6.5 (ii)

$$|\delta(z)| \leq \exp(K_1|B|_k^k\|(z-A)^{-1}\|^k).$$

Since A is normal, $\|(z-A)^{-1}\| = 1/d(z, \sigma(A))$ (Proposition 6.4). Thus for
$z \in \mathscr{D}'$ we have

$$|\delta(z)| \leq \exp\left(\frac{K_2}{d(z,C')^k}\right)$$

for some constant K_2.

Since $\delta(z)$ is non-vanishing in the simply connected region $\mathscr{D}'$, it
has an analytic logarithm; i. e., there is an analytic function $\alpha(z)$ on $\mathscr{D}'$
such that $\exp(\alpha(z)) = \delta(z)$. Then $\exp(\mathrm{Re}\,\alpha(z)) = |\delta(z)|$, and the above
inequality gives $\mathrm{Re}\,\alpha(z) \leq K_2/d(z,C')^k$. We can now apply Lemma 6.10
and conclude that there is a constant K_3 such that $|\alpha(z)| \leq K_3|z-z_0|^{-k-1}$
for $z \in \mathscr{D}' \cap L$. Hence $\mathrm{Re}\,\alpha(z) \geq -K_3|z-z_0|^{-k-1}$, which yields

$$|(\delta(z))^{-1}| \leq \exp(K_3|z-z_0|^{-k-1}).$$

Now, for $z \in \mathscr{D}' \cap L$,

$$\|(z-T)^{-1}\| = \|(\delta(z))^{-1} \cdot (z-A)^{-1}\delta(z)(1-B(z-A)^{-1})^{-1}\|$$

$$\leq \exp(K_3|z-z_0|^{-k-1})\|(z-A)^{-1}\|\,\|\delta(z)(1-B(z-A)^{-1})^{-1}\|$$

$$\leq \exp(K_3|z-z_0|^{-k-1})\frac{1}{d(z,\sigma(A))}\exp(K_4|B(z-A)^{-1}|_k^k)$$

(by Proposition 6.4 and Lemma 6.6 (iv))

$$\leqq \exp(K_3 |z-z_0|^{-k-1}) \frac{1}{d(z,\sigma(A))} \exp\{K_4 |B|_k^k \|(z-A)^{-1}\|^k\}$$

$$= \exp(K_3 |z-z_0|^{-k-1}) \frac{1}{d(z,\sigma(A))} \exp\left\{\frac{K_4 |B|_k^k}{d(z,\sigma(A))^k}\right\}.$$

Since L is not tangent to J, $|z-z_0|/d(z,\sigma(A))$ is bounded for $z \in L$. It follows that there exists a constant K such that $\|(z-T)^{-1}\| \leqq \exp(K |z-z_0|^{-k-1})$ for $z \in L \cap \mathscr{D}'$.   □

Theorem 6.12. *If* $T = A + B$, *with* $B \in \mathscr{C}_p$ *for some* $p \geqq 1$ *and* A *normal, and if* $\sigma(A)$ *contains an exposed arc, then* T *has a non-trivial hyperinvariant subspace.*

Proof. Let J denote the exposed arc. By Weyl's theorem (Theorem 0.10), $\sigma(A) \subset \sigma(T) \cup \Pi_0(A)$. Since A is a normal operator on a separable space, $\Pi_0(A)$ is countable, and thus a dense subset of J is contained in $\sigma(T)$. Hence $J \subset \sigma(T)$. In particular, T is not a multiple of the identity. Thus if $\Pi_0(T) \neq \emptyset$, T has non-trivial eigenspaces and we are done.

Suppose that $\Pi_0(T) = \emptyset$. Then, by Weyl's theorem, $\sigma(T) \subset \sigma(A)$, and it follows that J is an exposed arc of $\sigma(T)$. Now, by Lemma 6.11, the resolvent of T satisfies the growth condition required by Theorem 6.3, with k the least integer greater than p. The result follows.   □

Corollary 6.13. *If* $T = A + B$, *where* A *is normal,* $B \in \mathscr{C}_p$ *for some* $p \geqq 1$ *and* $\sigma(A)$ *is contained in a smooth Jordan arc, and if* $\sigma(T)$ *contains more than one point, then* T *has a non-trivial hyperinvariant subspace.*

Proof. Let J be a smooth arc such that $\sigma(A) \subset J$. As in the above proof we can assume $\Pi_0(T) = \emptyset$ and get $\sigma(T) \subset J$. If $\sigma(T)$ is disconnected the result follows from the Riesz decomposition theorem, (Corollary 2.11). Hence we can assume that $\sigma(T) = J'$ where J' is a non-trivial subarc of J. Using Weyl's theorem again shows that, in this case, the result follows from Theorem 6.12.   □

Corollary 6.14. *If* $T = A + B$, *where* A *is normal,* $B \in \mathscr{C}_p$ *for some* $p \geqq 1$, *and* $\sigma(A)$ *is contained in a smooth Jordan arc, then* T *has a non-trivial invariant subspace.*

Proof. By the previous corollary we can assume that $\sigma(T)$ contains only one point; translating by an appropriate scalar we can assume that $\sigma(T) = \{0\}$. In this case we show that T is compact.

By Weyl's theorem $\sigma(A) \subset \{0\} \cup \Pi_0(A)$. If A is not compact, then either A has a non-zero eigenvalue λ of infinite multiplicity or the eigenvalues of A have an accumulation point at some $\lambda \neq 0$. In either case there exists a $\lambda \neq 0$ and an orthonormal set $\{x_n\}$ such that $\{\lambda_n\} \to \lambda$ and $A x_n = \lambda_n x_n$. Since B is compact, $\{B x_n\} \to 0$. It follows that $\{\|(A + B - \lambda)x_n\|\} \to 0$, and $\lambda \in \Pi(T)$. But this contradicts $\sigma(T) = \{0\}$. Hence A is compact, and so is $T = A + B$. Thus T has an invariant subspace by Corollary 5.7. $\quad\square$

Corollary 6.15. *If* $T - T^* \in \mathscr{C}_p$ *for some* $p \geq 1$ *then* T *has a non-trivial invariant subspace.*

Proof. Since $T = \frac{1}{2}(T + T^*) + \frac{1}{2}(T - T^*)$, and $\frac{1}{2}(T + T^*)$ is a Hermitian operator, (whose spectrum is thus contained in $\mathbb{R}$), the result follows from Corollary 6.14.

Corollary 6.16. *If* $\sigma(T)$ *contains more than one point, and if* $1 - T^* T \in \mathscr{C}_p$ *for some* $p \geq 1$, *then* T *has a non-trivial hyperinvariant subspace.*

Proof. If T or T^* has non-empty point spectrum then the result is trivially true. Assume that $\Pi_0(T) = \Pi_0(T^*) = \emptyset$. Then the partial isometry occurring in the polar decomposition (Proposition 1.1) of T is unitary; i. e., $T = U P$ where U is unitary and P is positive. Then $P^2 = T^* T$. Since $1 - P^2$ is a compact Hermitian operator, the operator P is diagonable. Suppose that $P e_n = p_n e_n$ for some orthonormal basis $\{e_n\}$. Then $\sum |1 - p_n^2|^p < \infty$. But

$$\sum |1 - p_n^2|^p = \sum |1 - p_n|^p |1 + p_n|^p \geq \sum |1 - p_n|^p.$$

Hence $1 - P \in \mathscr{C}_p$, and $T = U(1 - (1 - P)) = U - U(1 - P)$. Since $\sigma(U)$ is contained in the unit circle and $-U(1 - P) \in \mathscr{C}_p$, the corollary follows from Corollary 6.13. $\quad\square$

The next corollary is a broad generalization of Corollary 3.31.

Corollary 6.17. *If* $1 - T^* T \in \mathscr{C}_p$ *for some* $p \geq 1$, *then* T *has a non-trivial invariant subspace.*

Proof. This result follows from Corollary 6.14 in the same way that Corollary 6.16 followed from Corollary 6.13. $\quad\square$

We can now prove a factorization theorem for the partial-isometry-valued analytic functions studied in Chapter 3.

Corollary 6.18. *If* $\mathscr{K}$ *is a finite-dimensional space and* V *is an analytic* $\mathscr{B}(\mathscr{K})$-*valued function, (i. e.,* V *is in the class* $\mathscr{F}_0$ *defined in Section 3.3), such that* $V(z)$ *is a. e. a partial isometry with initial space* $\mathscr{N}$ *and the dimension of* $\mathscr{H}^2(\mathscr{K}) \ominus \hat{V} \mathscr{H}^2(\mathscr{K})$ *is greater than 1, then there exist non-constant analytic partial-isometry-valued functions* W_1 *and* W_2 *such that* $V(z) = W_1(z) W_2(z)$ *for almost all* z.

Proof. The space $\mathcal{M} = \hat{V}\,\mathcal{H}^2(\mathcal{K})$ is an invariant subspace of the unilateral shift S on $\mathcal{H}^2(\mathcal{K})$. By Theorem 3.32 it suffices to show that S has an invariant subspace $\mathcal{M}_1$ which contains $\mathcal{M}$ and is neither $\mathcal{M}$ nor $\mathcal{H}$. Equivalently, we must show that S^* has a non-trivial invariant subspace $\mathcal{L}$ which is properly contained in $\mathcal{M}^\perp$. Decompose S^* with respect to the decomposition $\mathcal{M}^\perp \oplus \mathcal{M}$ of $\mathcal{H}^2(\mathcal{K})$:

$$S^* = \begin{pmatrix} A & B \\ 0 & C \end{pmatrix}.$$

We show that A satisfies the hypothesis of Corollary 6.17. Note that

$$1 - (S^*)^* S^* = \begin{pmatrix} 1 - A^* A & -A^* B \\ -B^* A & 1 - (B^* B + C^* C) \end{pmatrix}.$$

Now $1 - (S^*)^* S^*$ is the projection onto the first coordinate space of $\mathcal{H}^2(\mathcal{K})$, and hence its rank is the dimension of $\mathcal{K}$. Since $1 - (S^*)^* S^*$ has finite rank, so does its compression $1 - A^* A$. Thus $1 - A^* A \in \mathcal{C}_p$ for all p. By Corollary 6.17, A has a non-trivial invariant subspace. It follows that $Lat(S^* | \mathcal{M}^\perp)$ is non-trivial.   □

6.3 Quasi-similarity and Invariant Subspaces

If $\mathcal{M} \in Lat\,B$ and $A = SBS^{-1}$ for some invertible operator S, then $S\mathcal{M} \in Lat\,A$. If $\mathcal{M}$ is a hyperinvariant subspace for B, then it is trivially seen that $S\mathcal{M}$ is hyperinvariant for A. Thus an operator which is similar to an operator with a non-trivial hyperinvariant subspace has a non-trivial hyperinvariant subspace itself.

We now consider a more general relation between operators than similarity.

Definition. The operators A and B are *quasi-similar* if there exist operators S and T which are one-to-one and have dense range such that $AS = SB$ and $TA = BT$.

Similar operators are, of course, quasi-similar: if $A = SBS^{-1}$, then $AS = SB$ and $S^{-1}A = BS^{-1}$.

Theorem 6.19. *If A and B are quasi-similar, and if B has a nontrivial hyperinvariant subspace, then A has a non-trivial hyperinvariant subspace.*

Proof. Let S and T be the operators implementing the quasi-similarity: $AS = SB$ and $TA = BT$. Let $\mathcal{M}$ be a non-trivial hyperinvariant subspace of B, and let $\mathfrak{A} = \{C \in \mathcal{B}(\mathcal{H}) : AC = CA\}$. Define the subspace $\mathcal{N}$ as the closure of $\{CSx : x \in \mathcal{M}, C \in \mathfrak{A}\}$. Now if $C_1, C_2 \in \mathfrak{A}$ and $y = C_2 Sx$ with $x \in \mathcal{M}$, then $C_1 y = (C_1 C_2) Sx$, and, since $C_1 C_2 \in \mathfrak{A}$, it follows that $\{CSx : x \in \mathcal{M}, C \in \mathfrak{A}\}$ is a hyperinvariant linear manifold for A. Hence $\mathcal{N}$ is a hyperinvariant subspace for A.

Clearly $\mathcal{N} \neq \{0\}$. To show that $\mathcal{N} \neq \mathcal{H}$ note that TCS commutes with B for every $C \in \mathfrak{A}$:

$$(TCS)B = TCAS = TACS = B(TCS).$$

Thus, since $\mathcal{M}$ is hyperinvariant for B, $\mathcal{M} \in Lat\, TCS$. It follows that $\overline{T\mathcal{N}}$ is contained in $\mathcal{M}$. Thus $\mathcal{N} \neq \mathcal{H}$ since T has dense range. ☐

This result implies (by Corollary 1.17) that every operator quasi-similar to a non-scalar normal operator has non-trivial hyperinvariant subspaces.

Definition. The operator A is *power-bounded* if there exists a constant M such that $\|A^n\| \leq M$ for every positive integer n.

Theorem 6.20. *If A is power-bounded, and if the only vector x such that $\{A^n x\}_{n=1}^{\infty}$ or $\{A^{*n} x\}_{n=1}^{\infty}$ converges to 0 is $x=0$, then A is quasi-similar to a unitary operator.*

Proof. Let $\mathcal{L}$ be a Banach limit on ℓ^{∞}; i.e., $\mathcal{L}$ is a bounded linear functional on ℓ^{∞} such that $\mathcal{L}(\{x_1, x_2, x_3, \ldots\}) \geq 0$ whenever $x_i \geq 0$ for all i, $\mathcal{L}(\{1, 1, 1, \ldots\}) = 1$, and $\mathcal{L}(\{x_1, x_2, x_3, \ldots\}) = \mathcal{L}(\{x_2, x_3, \ldots\})$ for all $\{x_1, x_2, x_3, \ldots\} \in \ell^{\infty}$; (see Banach [1] for a proof that Banach limits exist). Define a bilinear form $[\cdot, \cdot]$ by $[x, y] = \mathcal{L}(\{(A^n x, A^n y)\})$ for $x, y \in \mathcal{H}$; (since A is power-bounded, $\{(A^n x, A^n y)\} \in \ell^{\infty}$). Note that $[x, x] = \mathcal{L}(\{\|A^n x\|^2\}) \leq M^2 \|x\|^2$; (the positivity of $\mathcal{L}$ implies that $\mathcal{L}(\{x_1, x_2, \ldots\}) \leq \mathcal{L}(\{y_1, y_2, \ldots\})$ whenever $x_i \leq y_i$ for all i). Hence $[\cdot, \cdot]$ is a bounded bilinear form, and there exists a bounded operator H such that $[x, y] = (Hx, y)$ for all x and y. Since $[x, x]$ is positive, the operator H is positive. Let P denote the positive square root of H, (Corollary 1.8).

We claim that H is one-to-one. For if $x \neq 0$ and $Hx = 0$, then $0 = (Hx, x) = \mathcal{L}\{\|A^n x\|^2\}$; this implies that $\inf \|A^n x\| = 0$. In this case there would be a subsequence $\{A^{n_i} x\} \to 0$, and, since A is power-bounded, it would follow that $\{A^n x\} \to 0$, contradicting the hypothesis. Thus H is one-to-one, and it follows that P is one-to-one also. Let $\mathcal{D}$ denote the range of P; then $\mathcal{D}$ is a dense linear manifold since P is a one-to-one Hermitian operator.

Define the operator $\tilde{U}$ on $\mathcal{D}$ by $\tilde{U}Px = PAx$ for $x \in \mathcal{H}$. Then

$$\begin{aligned}
\|\tilde{U}Px\|^2 &= \|PAx\|^2 = (PAx, PAx) = (HAx, Ax) \\
&= \mathcal{L}(\{(A^{n+1} x, A^{n+1} x)\}) = \mathcal{L}(\{(A^n x, A^n x)\}) \\
&= (Hx, x) = \|Px\|^2.
\end{aligned}$$

Thus $\tilde{U}$ is isometric on $\mathcal{D}$, and it can be extended by continuity to an isometry U on $\mathcal{H}$. Then $UP = PA$. Since the nullspace of A^* is $\{0\}$, A has dense range, and thus so does U. It follows that U is unitary.

We must produce an operator Q such that $QU = AQ$. To do this, first apply the above to A^* to obtain a unitary operator V and a one-to-one positive operator R such that $VR = RA^*$. Then $RV^* = AR$. Now the operator PR is one-to-one and has dense range; hence the partial isometry W occurring in its polar decomposition (Proposition 1.1) is unitary. Thus $PR = WK$, with W unitary and K positive. We claim that $A(RW^{-1}) = (RW^{-1})U$.

Note that $PRV^* = PAR = UPR$. Thus $WKV^* = UWK$, or $KV^* = W^{-1}UWK$. It follows that $V^*(VKV^{-1}) = (W^{-1}UW)K$, and the uniqueness of the polar decomposition implies that $V^* = W^{-1}UW$. The equation $RV^* = AR$ gives $RW^{-1}UW = AR$, or $(RW^{-1})U = A(RW^{-1})$. This completes the proof that A and U are quasi-similar. ☐

The two preceding results lead easily to the following existence theorem.

Theorem 6.21. *If A is power-bounded, and if neither of the sequences $\{A^n\}$ and $\{(A^*)^n\}$ converges strongly to 0, then either A has non-trivial hyperinvariant subspaces or A is a multiple of the identity.*

Proof. Let $\mathcal{M} = \{x : \{A^n x\} \to 0\}$ and $\mathcal{N} = \{x : \{(A^*)^n x\} \to 0\}$. If $\mathcal{M}$ and $\mathcal{N}$ are both $\{0\}$, then A is quasi-similar to a unitary operator U, by Theorem 6.20. If U is a multiple of the identity so is A; if U is not a multiple of the identity, then U has non-trivial hyperinvariant subspaces, (Corollary 1.17), and it follows from Theorem 6.19 that so does A.

We are left with the case where $\mathcal{M}$ and $\mathcal{N}$ are not both $\{0\}$. Note that $\mathcal{M}$ and $\mathcal{N}$ are linear manifolds. If $\{x_k\} \subset \mathcal{M}$ and $\{x_k\} \to x$, then

$$\|A^n x\| \leq \|A^n x - A^n x_k\| + \|A^n x_k\|$$
$$\leq M \|x - x_k\| + \|A^n x_k\|.$$

Hence $\{A^n x\} \to 0$ and $x \in \mathcal{M}$. Thus $\mathcal{M}$ is closed, and the same proof shows that $\mathcal{N}$ is closed. If $AC = CA$ then $\|A^n Cx\| = \|CA^n x\| \leq \|C\| \|A^n x\|$, and it follows that $\mathcal{M}$ is hyperinvariant for A. If $\mathcal{M} \neq \{0\}$, then, since $\mathcal{M} \neq \mathcal{H}$ by assumption, we are done. If $\mathcal{N} \neq \{0\}$, then $\mathcal{N}$ is hyperinvariant for A^*, and therefore $\mathcal{N}^\perp$ is hyperinvariant for A. In all cases A has a non-trivial hyperinvariant subspace. ☐

6.4 Hyperinvariant Subspaces

Most of the above existence theorems actually produce hyperinvariant subspaces; (an exception is the case of compact operators, where existence of hyperinvariant subspaces is given in Corollary 8.25

below). We now consider several results where we assume the existence of invariant subspaces and, under additional hypotheses, derive the existence of hyperinvariant subspaces.

Theorem 6.22. *If A is quasi-similar to an operator in the upper triangular form*

$$\begin{pmatrix} A_{11} & * & \cdots & & * \\ 0 & * & \cdots & & * \\ \vdots & & & & \\ 0 & & \cdots & 0 & A_{nn} \end{pmatrix},$$

where the spectra of A_{11} and A_{nn} are disjoint, then A has a non-trivial hyperinvariant subspace.

Proof. By Theorem 6.19 we can assume that A is an operator in the above form. Let $\mathfrak{A} = \{B \in \mathscr{B}(\mathscr{H}) : AB = BA\}$, and let x be any vector of the form $x_1 \oplus 0 \oplus 0 \oplus \cdots \oplus 0$ with $x_1 \neq 0$. If $\mathscr{M}$ is the closure of $\{Bx : B \in \mathfrak{A}\}$, then $\mathscr{M}$ is clearly a hyperinvariant subspace for A. Since $\mathscr{M}$ is obviously not $\{0\}$ we need only show that $\mathscr{M} \neq \mathscr{H}$. If

$$B = \begin{pmatrix} * & & * \cdots * \\ \vdots & & \vdots \\ B_{n1} & & * \cdots * \end{pmatrix}$$

is any operator in $\mathfrak{A}$, then the fact that the entry in position $(n, 1)$ of AB is equal to the entry in position $(n, 1)$ of BA gives $A_{nn} B_{n1} = B_{n1} A_{11}$. Since the spectra of A_{11} and A_{nn} are disjoint, Rosenblum's corollary (Corollary 0.13) implies that $B_{n1} = 0$. Let y be any non-zero vector of the form $0 \oplus 0 \oplus \cdots \oplus 0 \oplus y_n$. Since $B_{n1} = 0$ it follows that $(Bx, y) = 0$ for every $B \in \mathfrak{A}$. Thus $y \in \mathscr{M}^\perp$, and $\mathscr{M} \neq \mathscr{H}$. $\square$

Corollary 6.23. *If A is not a multiple of the identity and is quasisimilar to an operator in the upper triangular form*

$$\begin{pmatrix} A_{11} & * & \cdots & & * \\ 0 & * & \cdots & & * \\ \vdots & & & & \\ 0 & & 0 & \cdots & 0 & A_{nn} \end{pmatrix}$$

with A_{11} and A_{nn} normal, then A has a non-trivial hyperinvariant subspace.

Proof. By Theorem 6.19 we can assume that A has the above form. If $\sigma(A_{11})$ consists of only one point, then, by the spectral theorem, A_{11} is a multiple of the identity. In this case A has an eigenspace. This eigenspace is not $\mathscr{H}$, since A is not a multiple of the identity, and an eigenspace of an operator is obviously hyperinvariant.

If $\sigma(A_{11})$ consists of more than one point the spectral theorem implies that we can write $A_{11}=A_{11}^0\oplus A_{11}^1$ and $A_{nn}=A_{nn}^0\oplus A_{nn}^1$ where $\sigma(A_{11}^0)\cap\sigma(A_{nn}^1)=\emptyset$. Then A is unitarily equivalent to an operator of the form

$$\begin{pmatrix} A_{11}^0 & 0 & & * & \cdots & * \\ 0 & A_{11}^1 & & * & \cdots & * \\ 0 & 0 & & * & \cdots & * \\ \vdots & & & & & \\ 0 & 0 & \cdots & A_{nn}^0 & & * \\ 0 & 0 & \cdots & 0 & & A_{nn}^1 \end{pmatrix}$$

and Theorem 6.22 applies. ☐

The above corollary will be used in Chapter 7 to show that certain operators ("n-normal" operators) have hyperinvariant subspaces.

The following is a somewhat different kind of existence theorem for hyperinvariant subspaces.

Definition. A segment $[\mathcal{M},\mathcal{N}]$ of a lattice $\mathcal{L}$ (defined in Section 4.1) is a *section* of $\mathcal{L}$ if $\mathcal{K}\in\mathcal{L}$ implies that one of the following holds: $\mathcal{K}<\mathcal{M}$, $\mathcal{K}\in[\mathcal{M},\mathcal{N}]$, or $\mathcal{K}>\mathcal{N}$.

Theorem 6.24. *If $[\mathcal{M},\mathcal{N}]$ is a countable section of $Lat\,A$, then every subspace in $[\mathcal{M},\mathcal{N}]$ is hyperinvariant for A.*

Proof. Let $\mathcal{K}\in[\mathcal{M},\mathcal{N}]$ and suppose that $AB=BA$. Let $\mathcal{S}=\{\lambda\in\mathbb{C}:|\lambda|>\|B\|\}$. For each $\lambda\in\mathcal{S}$, $(B-\lambda)\mathcal{K}$ is a subspace, since $B-\lambda$ is invertible, and $(B-\lambda)\mathcal{K}$ is obviously in $Lat\,A$. A basic fact which we require is that $Lat\,B=Lat(B-\lambda)^{-1}$ for $\lambda\in\mathcal{S}$; (Corollary 2.15). We consider two cases. If $(B-\lambda)\mathcal{K}\in[\mathcal{M},\mathcal{N}]$ for every $\lambda\in\mathcal{S}$, then, since $[\mathcal{M},\mathcal{N}]$ is countable and $\mathcal{S}$ is uncountable, there exist distinct λ_1 and λ_2 in $\mathcal{S}$ such that $(B-\lambda_1)\mathcal{K}=(B-\lambda_2)\mathcal{K}$. Then $\mathcal{K}=(B-\lambda_1)^{-1}(B-\lambda_2)\mathcal{K}$. Now

$$(B-\lambda_1)^{-1}(B-\lambda_2)=(B-\lambda_2+\lambda_1-\lambda_1)(B-\lambda_1)^{-1}$$
$$=(B-\lambda_1)(B-\lambda_1)^{-1}+(\lambda_1-\lambda_2)(B-\lambda_1)^{-1}$$
$$=1+(\lambda_1-\lambda_2)(B-\lambda_1)^{-1}.$$

Thus $\mathcal{K}\in Lat(B-\lambda_1)^{-1}$ and $\mathcal{K}\in Lat\,B$. In the other case $(B-\lambda)\mathcal{K}\notin[\mathcal{M},\mathcal{N}]$ for some $\lambda\in\mathcal{S}$. Then either $(B-\lambda)\mathcal{K}\subset\mathcal{K}$ or $(B-\lambda)\mathcal{K}\supset\mathcal{K}$, since $[\mathcal{M},\mathcal{N}]$ is a section. In the first situation $\mathcal{K}\in Lat\,B$. In the second situation $\mathcal{K}\supset(B-\lambda)^{-1}\mathcal{K}$, and $\mathcal{K}\in Lat(B-\lambda)^{-1}=Lat\,B$. In all cases, then, $\mathcal{K}\in Lat\,B$. ☐

Corollary 6.25. *If $Lat\,A$ is countable, then every invariant subspace of A is hyperinvariant.*

Proof. Since $Lat A = [\{0\}, \mathcal{H}]$ is a section, the result follows immediately from Theorem 6.24. ☐

Corollary 6.26. *If $\mathcal{M} \in Lat A$ and $\mathcal{M}$ is comparable with every element of $Lat A$, then $\mathcal{M}$ is hyperinvariant.*

Proof. In this case $[\mathcal{M}, \mathcal{M}]$ is a section, and therefore this is a special case of Theorem 6.24. ☐

Corollary 6.27. *If A is unicellular, then every invariant subspace of A is hyperinvariant.*

6.5 Additional Propositions

Proposition 6.1. *If $\{\mathcal{M}_i\}$ is a countable collection of subspaces, then $\{A : \{\mathcal{M}_i\} \subset Lat A\}$ is a subset of $\mathcal{B}(\mathcal{H})$ of the first category.*

Proposition 6.2. *If T satisfies the hypotheses of Theorem 6.3, and if $\mathcal{N}$ is as constructed in the proof of Theorem 6.3, then $\sigma(T|\mathcal{N}) \subset \bar{J}_0$.*

Proposition 6.3. *If T is an operator such that there exist $N, \varepsilon > 0$ with $\varepsilon \|x\| \leq \|T^n x\| \leq N \|x\|$ for all x, then T is similar to an isometry. (Define a new norm $\|\|\cdot\|\|$ on $\mathcal{H}$ by $\|\|x\|\| = \mathcal{L}(\{T^n x\}_{n=0}^{\infty})$, where $\mathcal{L}$ is any Banach limit.) In particular, if T is invertible and $\{\|T^n\| : n = \pm 1, \pm 2, \ldots\}$ is bounded, then T is similar to a unitary operator.*

Proposition 6.4. *If A is normal and $z \in \rho(A)$, then $\|(z-A)^{-1}\| = 1/d(z, \sigma(A))$, where $d(z, \sigma(A))$ is the distance from z to $\sigma(A)$.*

Proposition 6.5. *If $\mathcal{H}$ is finite-dimensional and $A \in \mathcal{B}(\mathcal{H})$, then $\mathcal{M}$ is hyperinvariant for A if and only if there exists a polynomial p such that $\mathcal{M}$ is the nullspace or the range of $p(A)$.*

Proposition 6.6. *If A is quasi-similar to a normal operator, then A is not unicellular.*

Proposition 6.7. *Suppose that $\sigma(T) = \sigma_1 \cup \sigma_2$, where σ_1 and σ_2 are non-empty and there is a positive even integer n such that*

$$\sigma_1 \subset \{(x,y) : x \leq -y^n\} \quad \text{and} \quad \sigma_2 \subset \{(x,y) : x \geq y^n\}.$$

Suppose also that $\|(\lambda - T)^{-1}\| \leq K[d(\lambda, \sigma(T))]^{-m}$ for some positive integer m, some constant K, and all λ in the resolvent of T. If Γ is a rectifiable curve as in the diagram on the next page and the operator A is defined by $A = \int_{\Gamma} \lambda^{mn}(\lambda - T)^{-1} d\lambda$, then

   i) A is a bounded operator,
   ii) $AT = TA$, and
   iii) $\overline{A\mathcal{H}}$ is an invariant subspace of T such that $\sigma(T|\overline{A\mathcal{H}}) \subset \sigma_1$.

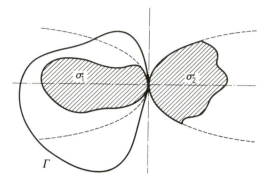

Proposition 6.8. If T is the product of a positive operator and a Hermitian operator and is not a multiple of the identity, then T has a non-trivial hyperinvariant subspace.

Proposition 6.9. If A is a normal operator with spectral measure $\{E_\lambda\}$, then $\mathcal{M}$ is hyperinvariant for A if and only if $\mathcal{M} = E(S)\mathcal{H}$ for some Borel subset S of $\mathbb{C}$.

6.6 Notes and Remarks

Example 6.1 is an unpublished observation of T. Crimmins, and Example 6.2 is due to Shields [1].

The results of Section 6.2 have a long history. Techniques similar to those used in this section were developed by Lorch [2], Godement [1], Wermer [2], Dunford ([1], [2], [3], [4]) and many others, generally for the purpose of obtaining a large number of invariant subspaces analogous (to varying extents) to the family of spectral subspaces of a normal operator. Some of the nicest such results are obtained for spectral operators (see Dunford-Schwartz [3]) and for decomposable operators (see Colojoară-Foiaş [1]). The work of Ljubič and Macaev [1] is also very interesting: they prove a theorem like Theorem 6.3, show that there is no completely general spectral theory by producing an example of an operator A such that $\sigma(A) = [0,1]$ and $\sigma(A|\mathcal{M}) = [0,1]$ for all non-trivial subspaces $\mathcal{M}$ in $Lat\,A$, and derive a number of other results. Other work in this area includes Bartle [1], Leaf [1], Kocan [1] and Wolf [1]—see Colojoară-Foiaş [1] and Dunford-Schwartz [3] for expositions of such results and for further references.

Corollary 6.15 is due to Livsic [1] in the case $p=1$, to Sahanovič in the case $p=2$, and to Gohberg-Krein and Macaev for general p;

(cf. Macaev [1], which contains a stronger result than Corollary 6.15.) The proofs of Theorem 6.3 and Lemma 6.11 presented in the text were suggested by the proof of Schwartz [2] of Corollary 6.15. Kitano [1] generalized the results of Schwartz to obtain a slightly weaker result than Theorem 6.12, and our proofs were motivated by his work although they are closer in spirit to Schwartz's original treatment. The elegant proof at the end of Theorem 6.3 that $\mathcal{N} \neq \{0\}$ was shown to us by J. G. Stampfli—it is much simpler than Schwartz' original proof of this fact. Theorem 6.3 and Lemma 6.11 have recently been used in Nordgren-Radjavi-Rosenthal [2] to show that operators T with the property that all their invariant subspaces are reducing and $1 - T^*T \in \mathscr{C}_p$ must be normal. An interesting extension of Theorem 6.12 has been obtained by C. Apostol [1], and related techniques have recently been used by Stampfli [3] to obtain Proposition 6.7 and other interesting results. Jafarian [1] has used the techniques of this chapter to investigate generalized decomposability of operators.

Theorem 6.4 is due to Wermer [2]; the proof in the text that the resolvent of T satisfies the given growth condition is from Colojoară-Foiaş [1]. If T satisfies the other hypotheses of Theorem 6.4 but $\sigma(T)$ is a singleton, then a theorem of Hille [1] (which generalized an earlier result of Gelfand [2] and is in Hille-Phillips [1] on page 128) implies that $(\lambda - T)^{k+1} = 0$ for some complex λ, and thus T has a hyperinvariant subspace in this case too, (unless T is a multiple of the identity). An interesting related result has been found by Flaschka [1]. Corollary 6.18, (and thus Corollary 6.17 in the case where $\|T\| \leq 1$, $1 - T^*T$ has finite rank and $\{(T^*)^n\} \to 0$), was first proven by Potapov [1], who used a rather complicated theory of multiplicative integrals. A simpler direct approach to this factorization problem is presented in Helson [1]. An interesting study of operators T such that $1 - T^*T \in \mathscr{C}_1$, $\|T\| \leq 1$, and $\sigma(T)$ is not the entire unit disk, ("weak contractions"), is given in Sz.-Nagy-Foiaş ([1], Chapter 8). Another approach to such problems has been developed by de Branges ([1], [2]).

Some work in the spirit of Section 6.2 has been done for operators on locally convex spaces—see Maeda ([1], [2]) and Plafker [1].

The results on operators of class $\mathscr{C}_p$ which we have quoted from Dunford-Schwartz [2] were developed by von Neumann, Schatten, Gohberg, Krein and others. An alternate treatment of most of these results can be found in Gohberg-Krein [1], and references to the original sources are given in Gohberg-Krein [1] and Dunford-Schwartz ([2], p. 1163). An interesting, and relatively elementary, approach to some of these results is given in Ringrose [5].

The concept of quasi-similarity was introduced by Sz.-Nagy and Foiaş, and they obtained Theorem 6.21 in the case $\|T\| \leq 1$ in Sz.-Nagy-

Foiaş [3]. The generalization to the power-bounded case, (and the simpler proof presented in the text), was independently discovered by Sz.-Nagy-Foiaş (cf. [1]) and S. Parrott (unpublished).

Theorem 6.22 and Corollary 6.23 were discovered by Radjavi-Rosenthal [5]. Theorem 6.24 is the generalization by Stampfli [2] of the proof of Corollary 6.26 given by Rosenthal [1]. Corollary 6.26 has been generalized to "inaccessible" members of *Lat A* by Douglas-Pearcy [1], and Douglas-Pearcy-Salinas [1] have recently generalized this result as well as Theorem 6.24.

Proposition 6.1 is a variant of a result of Halmos ([12], p. 900), and the first assertion of Proposition 6.3 is a very slight generalization of the second assertion, which is due to Sz.-Nagy [2]. Proposition 6.5 is an unpublished result of Peter Fillmore's, and Proposition 6.6 is due to Sz.-Nagy-Foiaş [1]. Proposition 6.7 is one of several interesting results of Stampfli [3] which employ techniques related to those of Theorem 6.3 to obtain invariant subspaces for certain operators whose spectra do not contain an exposed arc. Proposition 6.8 was found by Suzuki [1] in the case that both operators are positive; the extension presented here is due to Radjavi-Rosenthal [8].

Proposition 6.9 is from Douglas-Pearcy [1], which also contains a description of the hyperinvariant subspaces of unilateral shifts, (independently found by Sz.-Nagy-Foiaş [1]). More generally, the hyperinvariant subspaces for isometries have been determined by Douglas [2].

A large amount of work has been done on *J*-self-adjoint and *J*-unitary operators; (i.e., operators which behave like Hermitian operators with respect to an indefinite inner product $[x, y] = (Jx, y)$ where J is the direct sum of an identity operator and the negative of an identity operator). This work includes certain results on existence of invariant subspaces—see Pontrjagin [1], Iohvidov-Krein [1], Phillips [1], Krein [1], Helton [1], Masuda [1] and the references given in these papers.

Chapter 7. Certain Results on von Neumann Algebras

In this chapter we consider certain basic results in the theory of von Neumann algebras. This theory is highly developed and quite extensive; we merely consider those aspects of the theory which will be required for the study of more general kinds of operator algebras in Chapters 8 and 9. The reader interested in pursuing the study of von Neumann algebras should consult the books of Schwartz [2], Kaplansky [2], Naimark [1], Sakai [1], Topping [1], and the comprehensive treatises of Dixmier [2], [3].

7.1 Preliminaries

The following notation will be useful in the study of operator algebras. If $\mathscr{H}$ is a Hilbert space and n is a positive integer, then $\mathscr{H}^{(n)}$ denotes the direct sum of n copies of $\mathscr{H}$. If A is an operator on $\mathscr{H}$, then $A^{(n)}$ denotes the direct sum of n copies of A, (regarded as an operator on $\mathscr{H}^{(n)}$), and if $\mathfrak{A}$ is a set of operators on $\mathscr{H}$, then $\mathfrak{A}^{(n)} = \{A^{(n)} : A \in \mathfrak{A}\}$. Clearly $\mathfrak{A}$ is an algebra if and only if $\mathfrak{A}^{(n)}$ is an algebra for each n.

Theorem 7.1. *If $\mathfrak{A}$ is an algebra of operators containing the identity, then the closure of $\mathfrak{A}$ in the strong operator topology is*

$$\{B : Lat\,\mathfrak{A}^{(n)} \subset Lat\,B^{(n)} \text{ for all } n\}.$$

Proof. If B is in the strong closure of $\mathfrak{A}$, then obviously $Lat\,\mathfrak{A} \subset Lat\,B$. Moreover, if the net $\{A_\alpha\}$ converges strongly to B, then $\{A_\alpha^{(n)}\}$ converges strongly to $B^{(n)}$ for each n, and it follows that $Lat\,\mathfrak{A}^{(n)} \subset Lat\,B^{(n)}$ for all n.

Conversely, suppose that B is an operator such that $Lat\,\mathfrak{A}^{(n)} \subset Lat\,B^{(n)}$ for every n. Let $\{x_1, x_2, \ldots, x_k\}$ be any vectors in $\mathscr{H}$ and $\{C : \|Cx_i - Bx_i\| < \varepsilon \text{ for } i = 1, \ldots, k\}$ be any basic strong neighbourhood of B. We must show that $\mathfrak{A}$ contains an operator in this neighbourhood. Let $\mathscr{M}$ be the smallest invariant subspace of $A^{(k)}$ containing the vector $x_1 \oplus \cdots \oplus x_k$. Since $\mathfrak{A}^{(k)}$ is an algebra containing the identity, $\mathscr{M}$ is the closure of $\{Ax_1 \oplus \cdots \oplus Ax_k : A \in \mathfrak{A}\}$. By hypothesis $\mathscr{M} \in Lat\,B^{(k)}$. Therefore $Bx_1 \oplus \cdots \oplus Bx_k$ is in $\mathcal{M}$, and thus there is some $A \in \mathfrak{A}$ such that

$$\|(Ax_1 \oplus \cdots \oplus Ax_k) - (Bx_1 \oplus \cdots \oplus Bx_k)\| < \varepsilon.$$

This A is in the given strong neighbourhood of B. □

Corollary 7.2. *If $\mathfrak{A}$ is an algebra of operators containing the identity, then the closure of $\mathfrak{A}$ in the weak topology coincides with the closure of $\mathfrak{A}$ in the strong topology.*

Proof. Obviously the strong closure of $\mathfrak{A}$ is contained in the weak closure of $\mathfrak{A}$. The converse follows from Theorem 7.1. For suppose that $\{A_\alpha\}$ is a net in $\mathfrak{A}$ and $\{A_\alpha\}$ converges weakly to B. Then, for any n, $\mathcal{M} \in Lat\,\mathfrak{A}^{(n)}$ implies that, for each $x \in \mathcal{M}$ and $y \in \mathcal{M}^\perp$, $(A_\alpha^{(n)} x, y) = 0$. Thus $x \in \mathcal{M}$ and $y \in \mathcal{M}^\perp$ implies $(B^{(n)} x, y) = 0$; i.e., $\mathcal{M} \in Lat\,B^{(n)}$. By Theorem 7.1, then, B is in the strong closure of $\mathfrak{A}$. □

Definition. A *von Neumann algebra* is a weakly closed subalgebra $\mathfrak{A}$ of the algebra of operators on a Hilbert space which contains the identity operator and is self-adjoint, (in the sense that $A \in \mathfrak{A}$ implies $A^* \in \mathfrak{A}$).

Obviously the intersection of any collection of von Neumann algebras of operators on $\mathcal{H}$ is a von Neumann algebra. If $\mathcal{S}$ is any collection of operators on $\mathcal{H}$ we define the *von Neumann algebra generated by $\mathcal{S}$* as the intersection of all von Neumann algebras containing $\mathcal{S}$.

Theorem 7.3. *If $\mathfrak{A}$ is a von Neumann algebra, then $\mathfrak{A}$ is generated by the set of projections that it contains.*

Proof. First note that $A \in \mathfrak{A}$ implies that

$$\mathrm{Re}\,A = \frac{1}{2}(A + A^*) \quad \text{and} \quad \mathrm{Im}\,A = \frac{1}{2i}(A - A^*) \quad \text{are in } \mathfrak{A}.$$

It follows that $\mathfrak{A}$ is generated by its Hermitian operators. If A is a Hermitian element of $\mathfrak{A}$, then the spectral theorem implies (cf. the proof of the second assertion in Example 4.11) that each spectral projection of A is in $\mathfrak{A}$. Since each Hermitian operator is obviously in the von Neumann algebra generated by its spectral projections, it follows that $\mathfrak{A}$ is generated by the set of all spectral projections of Hermitian elements of $\mathfrak{A}$. □

7.2 Commutants

Definition. If $\mathcal{S}$ is a collection of operators on $\mathcal{H}$, then the *commutant of $\mathcal{S}$*, denoted $\mathcal{S}'$, is the set of all operators on $\mathcal{H}$ which commute with every member of $\mathscr{S}$. The *double commutant of* $\mathscr{S}$, denoted $\mathscr{S}''$, is the commutant of $\mathscr{S}'$.

For any $\mathscr{S}$ the set $\mathscr{S}'$ is obviously a weakly closed operator algebra containing the identity. If $\mathscr{S}$ is self-adjoint, then $\mathscr{S}'$ is a von Neumann algebra. Also, for any $\mathscr{S}$, $\mathscr{S} \subset \mathscr{S}''$.

Lemma 7.4. *If $\mathscr{S}$ is any collection of operators on $\mathscr{H}$, then the commutant of $\mathscr{S}^{(n)}$ is the set of all operators on $\mathscr{H}^{(n)}$ of the form*

$$\begin{pmatrix} A_{11} & \cdots & A_{1n} \\ \vdots & & \\ A_{n1} & \cdots & A_{nn} \end{pmatrix}$$

with $A_{ij} \in \mathscr{S}'$ for all (i,j).

Proof. Let $B \in \mathscr{S}$ and

$$A = \begin{pmatrix} A_{11} & \cdots & A_{1n} \\ \vdots & & \\ A_{n1} & \cdots & A_{nn} \end{pmatrix}$$

be any operator on $\mathscr{H}^{(n)}$. Then

$$A B^{(n)} = \begin{pmatrix} A_{11}B & \cdots & A_{1n}B \\ \vdots & & \\ A_{n1}B & \cdots & A_{nn}B \end{pmatrix}$$

and

$$B^{(n)} A = \begin{pmatrix} B A_{11} & \cdots & B A_{1n} \\ \vdots & & \\ B A_{n1} & \cdots & B A_{nn} \end{pmatrix}$$

The result follows immediately. $\quad\square$

The next theorem states one of the most important elementary facts about von Neumann algebras.

Theorem 7.5 *(The Double Commutant Theorem). If $\mathfrak{A}$ is a von Neumann algebra, then $\mathfrak{A} = \mathfrak{A}''$.*

Proof. Since obviously $\mathfrak{A} \subset \mathfrak{A}''$, we need only show that $\mathfrak{A}'' \subset \mathfrak{A}$. Let $B \in \mathfrak{A}''$. By Theorem 7.1 it suffices to show that $Lat\,\mathfrak{A}^{(n)} \subset Lat\,B^{(n)}$ for all n. Since $\mathfrak{A}^{(n)}$ is a self-adjoint algebra, every invariant subspace of $\mathfrak{A}^{(n)}$ is reducing. We therefore need only show that every projection which commutes with $\mathfrak{A}^{(n)}$ commutes with $B^{(n)}$. But this follows directly from Lemma 7.4. $\quad\square$

Corollary 7.6. *If $\mathscr{S}$ is any collection of operators on $\mathscr{H}$, and if $\mathscr{S}^* = \{A^* : A \in \mathscr{S}\}$, then the von Neumann algebra generated by $\mathscr{S}$ is $(\mathscr{S} \cup \mathscr{S}^*)''$.*

Proof. Clearly $(\mathscr{S} \cup \mathscr{S}^*)''$ is a von Neumann algebra containing $\mathscr{S}$. If $\mathfrak{A}$ is any von Neumann algebra containing $\mathscr{S}$, then $\mathfrak{A} \supset \mathscr{S} \cup \mathscr{S}^*$, and thus $\mathfrak{A}' \subset (\mathscr{S} \cup \mathscr{S}^*)'$. It follows that $\mathfrak{A}'' \supset (\mathscr{S} \cup \mathscr{S}^*)''$. But $\mathfrak{A}'' = \mathfrak{A}$ by Theorem 7.5. $\square$

Corollary 7.7. *The operators A and B have the same reducing subspaces if and only if they generate the same von Neumann algebras.*

Proof. Let $\mathfrak{A}_A$ and $\mathfrak{A}_B$ denote the von Neumann algebras generated by A and B respectively. Clearly A and B have the same reducing subspaces if and only if $(\mathfrak{A}_A)'$ and $(\mathfrak{A}_B)'$ contain the same projections. By Theorem 7.3 this is the case if and only if $(\mathfrak{A}_A)' = (\mathfrak{A}_B)'$. Theorem 7.5 shows that this occurs if and only if $\mathfrak{A}_A = \mathfrak{A}_B$. $\square$

7.3 The Algebra $\mathscr{B}(\mathscr{H})$

The collection $\mathscr{B}(\mathscr{H})$ of all operators on a Hilbert space $\mathscr{H}$ is obviously a von Neumann algebra. Certain facts about ideals and generators of this algebra are readily established, and the algebra automorphisms of $\mathscr{B}(\mathscr{H})$ are simply characterized.

Theorem 7.8. *If $\mathscr{I}$ is a two-sided ideal other than $\{0\}$ in $\mathscr{B}(\mathscr{H})$, then $\mathscr{I}$ contains all operators of finite rank.*

Proof. Let A be any operator other than 0 in $\mathscr{I}$, and let x_0 and y_0 be non-zero vectors such that $Ax_0 = y_0$. We first show that, if $x \neq 0$ and y are given, then the operator B such that $Bx = y$ and $B|\{x\}^\perp = 0$ is in $\mathscr{I}$. Given $x \neq 0$ and $y \in \mathscr{H}$, let P denote the operator of rank one such that $Px = x_0$ and $P|\{x\}^\perp = 0$, and let Q be any operator such that $Qy_0 = y$. Now let $B = QAP$. Then $B \in \mathscr{I}$,

$$Bx = QAPx = QAx_0 = Qy_0 = y,$$

and

$$B|\{x\}^\perp = QAP|\{x\}^\perp = 0.$$

Thus for each such x and y, $\mathscr{I}$ contains an operator B as described; i.e., $\mathscr{I}$ contains all operators of rank one. Since every operator of finite rank is a linear combination of operators of rank one, it follows that $\mathscr{I}$ contains all operators of finite rank. $\square$

Theorem 7.9. *Every algebra automorphism of $\mathscr{B}(\mathscr{H})$ is inner; i.e., if $\phi: \mathscr{B}(\mathscr{H}) \to \mathscr{B}(\mathscr{H})$ is a surjective isomorphism, then there exists an invertible operator S such that $\phi(A) = SAS^{-1}$ for all $A \in \mathscr{B}(\mathscr{H})$.*

Proof. Let ϕ be an automorphism of $\mathscr{B}(\mathscr{H})$. Note that ϕ takes idempotents of rank one into idempotents of rank one, for if A is such an idempotent, then $(\phi(A))^2 = \phi(A)$, and, since ϕ is an injective linear map of the one-dimensional vector space $\{ABA : B \in \mathscr{B}(\mathscr{H})\}$ onto $\{\phi(A)B\phi(A) : B \in \mathscr{B}(\mathscr{H})\}$, it follows that this second vector space is one-dimensional.

Now fix any unit vector $e \in \mathscr{H}$, and let A_0 be the idempotent of rank one such that $A_0 e = e$, $A_0|\{e\}^\perp = 0$. By the above remark $\phi(A_0)$ is also an idempotent of rank one. Hence A_0 and $\phi(A_0)$ are similar, (Proposition 7.1). We can assume, with no loss of generality, that this similarity has been composed with ϕ, so that $\phi(A_0) = A_0$.

Define a transformation S on $\mathscr{H}$ by

$$S(Ae) = \phi(A)e \quad \text{for } A \in \mathscr{B}(\mathscr{H}).$$

If $A_1 e = A_2 e$, then $(A_1 - A_2)A_0 e = 0$, and thus $(A_1 - A_2)A_0 = 0$. In this case $\phi(A_1 - A_2)A_0 = 0$, or $(\phi(A_1) - \phi(A_2))e = 0$. Hence S is well-defined, and S is clearly linear. Also, if $\phi(A)e = 0$ then $\phi(A)\phi(A_0) = 0$, and therefore $AA_0 = 0$, or $Ae = 0$. Hence S is injective; it is obviously surjective. Now if $A, B \in \mathscr{B}(\mathscr{H})$, then $SABe = \phi(A)\phi(B)e$, and $SABe = \phi(A)SBe$. Thus $SAx = \phi(A)Sx$ for all $x \in \mathscr{H}$, and it follows that $\phi(A) = SAS^{-1}$.

We need only prove that S is bounded. Assume S is not bounded. Then there is a unit vector x_1 such that $\|Sx_1\| > 1$. If S were bounded on $\{x_1\}^\perp$ (as a transformation from $\{x_1\}^\perp$ into $\mathscr{H}$), then S would be bounded. Hence there is a unit vector x_2 orthogonal to x_1 with $\|Sx_2\| > 2\|Sx_1\|$. Inductively construct an orthonormal sequence $\{x_n\}$ such that $\|Sx_{n+1}\| > (n+1)\|Sx_n\|$ for all n. Define $A \in \mathscr{B}(\mathscr{H})$ by $Ax_n = x_{n+1}$ and $A|\left(\bigvee_n x_n\right)^\perp = 0$. Then $\phi(A)Sx_n = SAx_n = Sx_{n+1}$. Thus $\|\phi(A)Sx_n\| > (n+1)\|Sx_n\|$, which contradicts the fact that $\phi(A)$ is a bounded operator. $\quad\square$

Theorem 7.10. *The set of operators which generate $\mathscr{B}(\mathscr{H})$ as a von Neumann algebra is uniformly dense in $\mathscr{B}(\mathscr{H})$.*

Proof. Given $A \in \mathscr{B}(\mathscr{H})$ and $\varepsilon > 0$, write A in terms of its real and imaginary parts: $A = H + iK$, with H and K Hermitian. By the spectral theorem, (Theorem 1.6), we can assume that $H = M_\phi$ with $\phi \in \mathscr{L}^\infty(X, \mu)$. Since ϕ is a uniform limit of measurable simple functions, it follows that there exists a diagonable Hermitian operator H_1 such that $\|H - H_1\| < \varepsilon/3$. Write H_1 as a diagonal matrix with respect to an orthonormal basis $\{e_n\}_{n=0}^\infty$. By changing the diagonal elements of the matrix slightly we can produce a diagonal matrix H_2 with distinct real diagonal elements such that $\|H_1 - H_2\| < \varepsilon/3$. Then the eigenvectors of H_2 are exactly $\{e_n\}$, and thus, since a Hermitian operator is completely normal,

Theorem 1.25 implies that every invariant subspace of H_2 other than $\{0\}$ has the form $\bigvee_{n \in \mathscr{S}} \{e_n\}$ for some set $\mathscr{S}$ of natural numbers.

Now consider the matrix of K with respect to $\{e_n\}$. This matrix may have some entries (Ke_n, e_m) equal to 0. Obviously there exists a Hermitian operator K_2 such that $(K_2 e_n, e_m)$ is different from 0 for all m and n and such that $\|K - K_2\| < \varepsilon/3$. Then the operator $A_2 = H_2 + iK_2$ is within ε of A. Let $\mathfrak{A}$ denote the von Neumann algebra generated by A_2; we show that $\mathfrak{A} = \mathscr{B}(\mathscr{H})$.

As in the proof of Corollary 7.7, $\mathfrak{A} = \mathscr{B}(\mathscr{H})$ if and only if there is no projection in $\mathfrak{A}'$ other than 0 or 1. This is the case if A_2 is irreducible. If $\mathscr{M}$ is a reducing subspace of A_2, then $\mathscr{M} \in \operatorname{Lat} H_2$ and $\mathscr{M} \in \operatorname{Lat} K_2$. If $\mathscr{M} = \bigvee_{n \in \mathscr{S}} \{e_n\}$ with $\mathscr{S} \neq \emptyset$, then $K_2 \mathscr{M} \subset \mathscr{M}$ implies that $\mathscr{S}$ is the set of all natural numbers. Thus $\mathscr{M} = \{0\}$ or $\mathscr{M} = \mathscr{H}$. $\square$

7.4 Abelian von Neumann Algebras

One class of abelian von Neumann algebras is the class of algebras $\{M_\phi : \phi \in \mathscr{L}^\infty(X, \mu)\}$ for measure spaces (X, μ). We prove a generalization of the spectral theorem: every abelian von Neumann algebra is a subalgebra of an algebra $\{M_\phi : \phi \in \mathscr{L}^\infty(X, \mu)\}$ for some (X, μ). This will follow from the spectral theorem once we prove that every abelian von Neumann algebra is generated by a single operator. The following lemma is the key to this result.

Lemma 7.11. *If $\{P_n\}_{n=0}^\infty$ is a commutative set of projections, then there exists a collection $\{Q_\alpha\}_{\alpha \in [0,1]}$ of projections such that $\alpha < \beta$ implies $Q_\alpha \mathscr{H} \subset Q_\beta \mathscr{H}$ and the von Neumann algebras generated by $\{P_n\}_{n=0}^\infty$ and by $\{Q_\alpha\}_{\alpha \in [0,1]}$ are the same. Moreover $\{Q_\alpha\}$ can be chosen to be left-continuous; i.e., such that $\lim_{\alpha \to \alpha_0^-} Q_\alpha = Q_{\alpha_0}$ (in the strong topology) for $\alpha_0 \in (0,1]$.*

Proof. Given $\{P_n\}$, we construct $\{Q_\alpha\}$ by first defining a chain $\{R_\beta\}$ indexed by a subset of $[0,1]$. Define $R_0 = 0$ and $R_1 = 1$.

Let $\mathscr{S}_1$ denote the set of all n such that $P_n \neq 1$. If $\bigvee_{n \in \mathscr{S}_1} \{P_n \mathscr{H}\} \neq \mathscr{H}$, define $R_\beta = 1$ for $\beta \in (\frac{3}{4}, 1)$. If $\bigvee_{n \in \mathscr{S}_1} \{P_n \mathscr{H}\} = \mathscr{H}$, just proceed. Then, in either case, consider the P_n's in the given order. Let $R_{\frac{1}{2}}$ be the first P_n which is neither 0 nor 1. Let $\mathscr{S}_2$ denote the set of all n such that the range of $P_n R_{\frac{1}{2}}$ is a proper subspace of the range of $R_{\frac{1}{2}}$. If $\bigvee_{n \in \mathscr{S}_2} \{P_n R_{\frac{1}{2}} \mathscr{H}\} \neq R_{\frac{1}{2}} \mathscr{H}$, define a constant interval for R_β: let $R_\beta = R_{\frac{1}{2}}$ for $\beta \in (\frac{1}{4}, \frac{1}{2})$. If $\bigvee_{n \in \mathscr{S}_2} \{P_n R_{\frac{1}{2}} \mathscr{H}\} = R_{\frac{1}{2}} \mathscr{H}$, do not define R_β for any other values of β at this stage.

Now let P be the first P_n which is distinct from $0, 1$, and $R_{\frac{1}{2}}$. If $PR_{\frac{1}{2}}$ is either 0 or $R_{\frac{1}{2}}$, omit $PR_{\frac{1}{2}}$; otherwise define $R_{\beta_0} = PR_{\frac{1}{2}}$ where β_0 is an interior point of the subinterval of $(0, \frac{1}{2})$ on which R_β has not yet been defined; (which subinterval this is depends upon whether or not a constant interval to the left of $\frac{1}{2}$ had been defined). Let $\mathcal{S}_3$ be the set of all n such that $P_n R_{\beta_0} \mathcal{H}$ is a proper subspace of $R_{\beta_0} \mathcal{H}$. If $\bigvee_{n \in \mathcal{S}_3} \{P_n R_{\beta_0} \mathcal{H}\}$ $\neq R_{\beta_0} \mathcal{H}$, define $R_\beta = R_{\beta_0}$ for $\beta \in (\beta_0/2, \beta_0)$; otherwise do not define R_β for other values of β at this stage. Then consider $R_{\frac{1}{2}} + P - PR_{\frac{1}{2}}$. If this projection is equal to one of the R_β which is already defined, omit it. Otherwise define $R_{\frac{5}{8}} = R_{\frac{1}{2}} + P - PR_{\frac{1}{2}}$. As above, determine whether or not a constant interval should be defined to the left of $\frac{5}{8}$; if it should be, define $R_\beta = R_{\frac{5}{8}}$ for $\beta \in (\frac{9}{16}, \frac{5}{8})$.

We inductively continue this construction. At each stage R_β has been defined for β in finitely many intervals, on each of which R_β is constant, (and some of which are single points). Let P be the first P_n which is distinct from the values of R_β already defined. Then for each pair (E_1, E_2) of adjacent already-defined values of R_β, (adjacent in the sense that $E_1 \mathcal{H}$ is a proper subspace of $E_2 \mathcal{H}$ and $E_1 \mathcal{H} \subset R_\beta \mathcal{H} \subset E_2 \mathcal{H}$ implies $R_\beta \mathcal{H} = E_1 \mathcal{H}$ or $R_\beta \mathcal{H} = E_2 \mathcal{H}$), consider the projection $E_1 + E_2 P - E_1 P$. If this projection is distinct from E_1 and E_2, define $R_{\beta_0} = E_1 + E_2 P - E_1 P$ for some β_0 in the interior of the previously unused interval between the values of β which yield E_1 and E_2. Define a constant interval to the left of R_{β_0} if required; i.e., if $\mathcal{P}$ is the collection of those P_n such that $P_n R_{\beta_0} \mathcal{H}$ is a proper subspace of $R_{\beta_0} \mathcal{H}$, and if $\bigvee_{P \in \mathcal{P}} PR_{\beta_0} \mathcal{H} \neq R_{\beta_0} \mathcal{H}$, then define $R_\beta = R_{\beta_0}$ for $\beta \in (\beta_1, \beta_0)$, where β_1 is an interior point of the maximal interval to the left of β_0 containing no β with R_β already defined.

Since the P_n are enumerated, each P_n is eventually considered in this construction, and it is easy to verify that the algebra generated by all the R_β contains $\{P_n\}$. Also, all the R_β which are constructed are in the algebra generated by $\{P_n\}$. Therefore $\{R_\beta\}$ and $\{P_n\}$ generate the same von Neumann algebras.

It is now a simple matter to define $\{Q_\alpha\}$. Let $Q_0 = 0$, and for each $\alpha \in (0, 1]$ define Q_α as the projection onto $\bigvee_{\beta < \alpha} \{R_\beta \mathcal{H}\}$, (considering in the span only those R_β which have been defined, of course). From the definition of Q_α it is easily proven that $\lim_{\alpha \to \alpha_0} Q_\alpha x = Q_{\alpha_0} x$ for each $\alpha_0 \in (0, 1]$ and each $x \in \mathcal{H}$. Hence the von Neumann algebra generated by $[R_\beta]$ includes the von Neumann algebra generated by $\{Q_\alpha\}$. On the other hand, the definition of the constant intervals to the left of the R_β implies that, whenever R_α had been defined, $R_\alpha = Q_\alpha$. Therefore the von Neumann algebra generated by $\{Q_\alpha\}_{\alpha \in [0, 1]}$ is the same as that generated by $\{R_\beta\}$, and thus also the same as the von Neumann algebra generated by $\{P_n\}$. □

Theorem 7.12. *If $\mathfrak{A}$ is an abelian von Neumann algebra, then there is a Hermitian operator A such that $\{A\}'' = \mathfrak{A}$.*

Proof. First of all, $\mathfrak{A}$ is generated by the set $\mathscr{P}$ of projections in $\mathfrak{A}$, by Theorem 7.3. Since the unit ball of $\mathscr{B}(\mathscr{H})$ is separable and metrizable in the weak operator topology, (Proposition 0.2), $\mathscr{P}$ has a countable dense subset $\{P_n\}_{n=0}^\infty$. Then $\{P_n\}_{n=0}^\infty$ generates $\mathfrak{A}$, and thus so does the corresponding family $\{Q_\alpha\}_{\alpha \in [0,1]}$ given by Lemma 7.11. We now define a spectral measure E on $[0,1]$; this is done in exactly the same way as one obtains a real-valued measure from a left-continuous monotone function on $\mathbb{R}$, (cf. Halmos [1]). That is, define $E([a,b)) = Q_b - Q_a$ for intervals $[a,b) \subset [0,1]$, and extend E in the standard fashion to the Borel sets. Then define the Hermitian operator $A = \int \lambda dE_\lambda$. Obviously A generates $\mathfrak{A}$ as a von Neumann algebra. □

Theorem 7.13. *If $\mathfrak{A}$ is an abelian von Neumann algebra, then there exists a finite measure space (X,μ) such that $\mathfrak{A}$ is unitarily equivalent to a subalgebra of the algebra $\mathscr{L}^\infty$ on $\mathscr{L}^2(X,\mu)$.*

Proof. By Theorem 7.12 $\mathfrak{A}$ has a Hermitian generator A. By the spectral theorem, (Theorem 1.6), there is a finite measure space (X,μ) and a $\phi \in \mathscr{L}^\infty(X,\mu)$ such that A is unitarily equivalent to M_ϕ. Obviously the von Neumann algebra generated by M_ϕ is contained in $\mathscr{L}^\infty$. □

Definition. A *maximal abelian self-adjoint algebra*, abbreviated m.a.s.a., is an abelian von Neumann algebra which is not properly contained in any other abelian von Neumann algebra.

Theorem 1.20 implies that each algebra $\mathscr{L}^\infty$ is a m.a.s.a.; the converse of this is an important fact.

Corollary 7.14. *If $\mathfrak{A}$ is a m.a.s.a., then there is a finite measure space (X,μ) such that $\mathfrak{A}$ is unitarily equivalent to $\mathscr{L}^\infty$ on $\mathscr{L}^2(X,\mu)$.*

Proof. By Theorem 7.13 there is a finite measure space (X,μ) such that $\mathfrak{A}$ is unitarily equivalent to a subalgebra of $\mathscr{L}^\infty$. Since $\mathfrak{A}$ is maximal and $\mathscr{L}^\infty$ is an abelian von Neumann algebra, this subalgebra must be $\mathscr{L}^\infty$. □

In Chapters 8 and 9 we will need to consider unbounded linear transformations which commute with bounded operators.

Definition. The linear transformation T defined on the linear manifold $\mathscr{D}$ in $\mathscr{H}$ *commutes with the operator A* if $A\mathscr{D} \subset \mathscr{D}$ and $TAx = ATx$ for $x \in \mathscr{D}$. The transformation T *commutes with a set $\mathfrak{A}$ of bounded operators* if it commutes with every operator in $\mathfrak{A}$.

We require a characterization of the linear transformations which commute with a m. a. s. a.

Lemma 7.15. *Let (X,μ) be a finite measure space. If T is a linear transformation defined on a dense linear manifold $\mathscr{D}$ of $\mathscr{L}^2(X,\mu)$ and T commutes with $\mathscr{L}^\infty$, then there exists a finite-valued (not necessarily bounded) measurable function ϕ on X such that $Tf = \phi f$ for $f \in \mathscr{D}$.*

Proof. For each fixed representative f of an element of $\mathscr{D}$ let $E_f = \{t \in X : f(t) \neq 0\}$ and define a function ϕ_f on E_f by

$$\phi_f(t) = \frac{(Tf)(t)}{f(t)}, \qquad t \in E_f.$$

We first show that the functions in the family $\{\phi_f\}$ agree on the common part of their domains; i.e. if $f, g \in \mathscr{D}$, then $\phi_f = \phi_g$ a.e. on $E_f \cap E_g$. To prove this we write each of the functions f and g as the ratio of two bounded measurable functions: $f = f_1/f_2$ and g_1/g_2, where f_2 and g_2 are non-zero everywhere. (Any measurable function h on X can be so written. Simply let X_1 be that part of X on which the absolute value of h is bounded by 1. Define $h_1 = h$, $h_2 = 1$ on X_1, and $h_1 = 1$, $h_2 = h^{-1}$ on the complement of X_1.)

Now we have $f_1 g_2 g = g_1 f_2 f \in \mathscr{D}$, and, since T commutes with multiplication by $f_1 g_2$ and by $g_1 f_2$, we get

$$(f_1 g_2) Tg = T(f_1 g_2 g) = T(g_1 f_2 f) = (g_1 f_2) Tf \quad \text{a.e.}$$

Then $(f) Tg = (g) Tf$ a.e. Thus $\phi_f = \phi_g$ on $E_f \cap E_g$.

Next we show that there is a measurable function ϕ on X that agrees with ϕ_f on E_f for each f. If every ϕ_f is 0 a.e. define $\phi \equiv 0$. Otherwise, by Zorn's lemma there is a maximal family $\mathscr{F}$ of mutually disjoint measurable subsets of X such that every member of $\mathscr{F}$ has positive measure and is contained in E_f for some f. Note that $\mu\left(E_f \setminus \bigcup_{F \in \mathscr{F}} F\right) = 0$ for every f by maximality. For each $F \in \mathscr{F}$ fix a function $f_F \in \mathscr{D}$ with $F \subset E_{f_F}$. Define ϕ to be equal to ϕ_{f_F} on F for each F and to 0 on the complement of $\bigcup_{F \in \mathscr{F}} F$. Clearly $\phi = \phi_f$ on E_f for every $f \in \mathscr{D}$. Since μ is finite, $\mathscr{F}$ is at most countable, which implies that ϕ is measurable.

To see that T is multiplication by ϕ on $\mathscr{D}$, take any $f \in \mathscr{D}$ and observe that $\phi \cdot f = \phi_f \cdot f = Tf$ on E_f. To get the equation $Tf = \phi \cdot f$ on the complement of E_f let χ represent the characteristic function of this complement. Then, since $\chi \cdot f \equiv 0$, $\chi \cdot Tf = T(\chi \cdot f) = 0$; thus $Tf = 0$ on $X \setminus E_f$ and hence $Tf = \phi \cdot f$ a.e. $\quad\square$

It is worth noting that the above proof does not use the hypothesis that T is linear; T could be taken to be any function satisfying the other conditions of the lemma, and linearity would follow.

Definition. The linear transformation T defined on the linear manifold $\mathcal{D} \subset \mathcal{H}$ is *closable* if the closure of the graph of T is the graph of a linear transformation $\tilde{T}$; i.e., if $x_n \in \mathcal{D}$, $\{x_n\} \to 0$, $\{Tx_n\} \to y$ implies $y = 0$. Then $\tilde{T}$ is the *closure* of T. The transformation T is *closed* if $T = \tilde{T}$; i.e., the graph of T with domain $\mathcal{D}$ is a closed subspace of $\mathcal{H} \oplus \mathcal{H}$.

Theorem 7.16. *If T is a densely defined linear transformation which commutes with a m.a.s.a. $\mathfrak{R}$, then T is closable. Moreover, if $\tilde{T}$ is the closure of T, then, for each strong neighbourhood $\mathcal{U}$ of 1, there exists a projection P in $\mathcal{U} \cap \mathfrak{R}$ such that $P\mathcal{H}$ is contained in the domain of $\tilde{T}$, $P\tilde{T}P \in \mathfrak{R}$, and $P\tilde{T}P$ commutes with PBP for each operator B that commutes with T.*

Proof. By Corollary 7.14 we can assume that $\mathfrak{R}$ is $\mathcal{L}^\infty$ operating on $\mathcal{L}^2(X, \mu)$, with $\mu(X) < \infty$. Then, by Lemma 7.15, T agrees with multiplication by some measurable function ϕ on its domain $\mathcal{D}$. It is easily verified (Proposition 7.2) that multiplication by ϕ is a closed operator on the domain

$$\mathcal{D}_\phi = \{f \in \mathcal{L}^2(X, \mu): \phi f \in \mathcal{L}^2(X, \mu)\}.$$

Hence T is closable.

Now let $\mathcal{U} = \{C: \|Cf_i - f_i\| < \varepsilon, \ i = 1, \dots, n\}$ be any strong neighbourhood of the identity operator on $\mathcal{L}^2(X, \mu)$. For each positive integer k let $\mathcal{S}_k = \{x \in X: |\phi(x)| \leq k\}$. Since $\{\mathcal{S}_k\}_{k=1}^\infty$ is an increasing family of sets whose union is X, for each i there is a k_i such that $\int_{X \setminus \mathcal{S}_{k_i}} |f_i|^2 d\mu < \varepsilon^2$. Let $k_0 = \max\{k_1, \dots, k_n\}$; then, if P denotes multiplication by the characteristic function of $\mathcal{S}_{k_0}$,

$$\|Pf_i - f_i\|^2 = \int |Pf_i - f_i|^2 d\mu$$
$$= \int_{\mathcal{S}_{k_0}} |Pf_i - f_i|^2 d\mu + \int_{X \setminus \mathcal{S}_{k_0}} |Pf_i - f_i|^2 d\mu$$
$$= \int_{X \setminus \mathcal{S}_{k_0}} |f_i|^2 d\mu < \varepsilon^2.$$

Hence $P \in \mathcal{U}$. Obviously $P \in \mathfrak{R}$ and $P\mathcal{H}$ is contained in the domain $\mathcal{D}_\phi$ of $\tilde{T}$. Also $P\tilde{T}P$ is multiplication by the function which agrees with ϕ on $\mathcal{S}_{k_0}$ and is 0 on $X \setminus \mathcal{S}_{k_0}$. Hence $P\tilde{T}P \in \mathfrak{R}$.

Now if B commutes with T, then, as is easily verified, B commutes with $\tilde{T}$. Also

$$(PBP)(P\tilde{T}P) = (PB)(P\tilde{T}P) = (PB)(\tilde{T}P)$$
$$= P\tilde{T}BP = (P\tilde{T}P)(BP)$$
$$= (P\tilde{T}P)(PBP). \quad \square$$

7.5 The Class of *n*-normal Operators

Definition. The operator A is *n-normal* if there exists a m.a.s.a. $\mathfrak{R}$ such that A is in the commutant of $\mathfrak{R}^{(n)}$.

Theorem 7.17. *The operator A is n-normal if and only if it is unitarily equivalent to an operator of the form*

$$\begin{pmatrix} A_{11} & A_{12} & \cdots & A_{1n} \\ A_{21} & A_{22} & \cdots & A_{2n} \\ \vdots & \vdots & & \vdots \\ A_{n1} & A_{n2} & \cdots & A_{nn} \end{pmatrix}$$

on $\mathscr{H}^{(n)}$, with $\{A_{ij}\}$ a collection of commuting normal operators on $\mathscr{H}$.

Proof. If an operator A has the above form, then, since the von Neumann algebra generated by $\{A_{ij}\}$ is abelian, (this requires Fuglede's theorem), there exists at least one m.a.s.a. $\mathfrak{R}$ such that $\{A_{ij}\} \subset \mathfrak{R}$. Clearly $A \in (\mathfrak{R}^{(n)})'$.

Conversely, if $A \in (\mathfrak{R}^{(n)})'$, Lemma 7.4 implies that A has the above form with $A_{ij} \in \mathfrak{R}'$ for all i, j. Then, since $\mathfrak{R}$ is maximal, each A_{ij} is in $\mathfrak{R}$. $\square$

We shall show that each *n*-normal operator has a "triangular" form. The following measure-theoretic lemma is the basis of this result.

Lemma 7.18. *Let n be a positive integer, $\{\phi_1, \ldots, \phi_n\} \subset \mathscr{L}^{\infty}(X, \mu)$ (where $\mu(X) < \infty$), and $P(z, t) = z^n + \phi_1(t)z^{n-1} + \cdots + \phi_n(t)$ be a polynomial with coefficients in $\mathscr{L}^{\infty}(X, \mu)$. Then there exists $\phi \in \mathscr{L}^{\infty}(X, \mu)$ such that $P(\phi(t), t) = 0$ a.e. on X.*

Proof. Let $M = \max\limits_{1 \leq k \leq n} \{\|\phi_k\|_{\infty}\}$. Re-define the functions ϕ_k on sets of measure 0 to ensure $|\phi_k(t)| \leq M$ for all t. For each fixed t define $\phi(t)$ to be the first zero of $P(z, t)$ in the lexicographical order on $\mathbb{C}$; i.e., for all zeros z of $P(z, t)$, $\operatorname{Re}\phi(t) \leq \operatorname{Re}z$, and if $\operatorname{Re}\phi(t) = \operatorname{Re}z$ then $\operatorname{Im}\phi(t) \leq \operatorname{Im}z$. We need only show that $\phi \in \mathscr{L}^{\infty}(X, \mu)$. It is easily seen that ϕ is bounded, and it only remains to be shown that ϕ is measurable.

Let $\mathscr{S} = \{\lambda = (\lambda^{(1)}, \ldots, \lambda^{(n)}) \in \mathbb{C}^n : |\lambda^{(i)}| \leq M$ for all $i\}$. Define the function $\hat{\phi}$ on $\mathscr{S}$ by letting $\hat{\phi}(\lambda)$ be that zero of $Q(z, \lambda) = z^n + \lambda^{(1)}z^{n-1} + \cdots + \lambda^{(n)}$ which is first lexicographically. Then $\phi(t) = \hat{\phi}(\phi_1(t), \ldots, \phi_n(t))$. Let $R(\lambda)$ and $I(\lambda)$ be the real and imaginary parts of $\hat{\phi}(\lambda)$; $\hat{\phi}(\lambda) = R(\lambda) + iI(\lambda)$. We show that $R(\lambda)$ is continuous and $I(\lambda)$ is lower semi-continuous, which implies that ϕ is measurable since it is the composition of $\hat{\phi}$ with the Borel measurable function $t \to (\phi_1(t), \ldots, \phi_n(t))$.

Let $\{\lambda_k\} \in \mathscr{S}$, $\{\lambda_k\} \to \lambda_0$. Then $\{\hat{\phi}(\lambda_k)\}$ is a bounded sequence and, by taking a convergent subsequence, we can assume that $\{\hat{\phi}(\lambda_k)\} \to x + iy$.

Then $Q(x+iy,\lambda_0)=0$. By the definition of $\hat\phi$ we have $R(\lambda_0)\leqq x$; we must prove equality. If C is a sufficiently small circle about $\hat\phi(\lambda_0)$, then the integral

$$\frac{1}{2\pi i}\int_C \frac{\frac{d}{dz}(Q(z,\lambda_0))}{Q(z,\lambda_0)}\,dz$$

gives the multiplicity of the zero $\hat\phi(\lambda_0)$ of $Q(z,\lambda_0)$. Thus

$$\frac{1}{2\pi i}\int_C \frac{\frac{d}{dz}(Q(z,\lambda_k))}{Q(z,\lambda_k)}\,dz\neq 0$$

for sufficiently large k, and it follows that $Q(z,\lambda_k)$ has a zero inside C. For each $\varepsilon>0$, then, $Q(z,\lambda_k)$ has a zero with real part less than $R(\lambda_0)+\varepsilon$ for k sufficiently large. For such k, $R(\lambda_k)<R(\lambda_0)+\varepsilon$ by definition. Therefore $x=\lim R(\lambda_k)\leqq R(\lambda_0)$, and $x=R(\lambda_0)$ as required.

To complete the proof we must show that $I(\lambda)$ is lower semi-continuous. If $\{\lambda_k\}\to\lambda_0$ and $\hat\phi(\lambda_k)\to x+iy$ as above, then we must prove that $I(\lambda_0)\leqq y$. The above paragraph shows that $x=R(\lambda_0)$, and the definition of $I(\lambda_0)$ therefore implies that $I(\lambda_0)\leqq y$. □

Lemma 7.19. *Let ϕ_{ij} be bounded measurable functions on (X,μ) for $i=1,\ldots,k$, $j=1,\ldots,n$ with $k\leqq n$, and assume that the matrix $((\phi_{ij}(t)))$ has rank less than n for all t. Then the system $\left\{\sum_{j=1}^{n}\phi_{ij}\xi_j=0,\ i=1,\ldots,k\right\}$ has a solution $\{\xi_1,\ldots,\xi_n\}$ with $\xi_j\in\mathscr{L}^\infty(X,\mu)$ and $\sum_{j=1}^{n}|\xi_j(t)|^2=1$ a.e.*

Proof. For each square submatrix $M(t)$ of $((\phi_{ij}(t)))$ let S_M denote the set of all t such that $\det M(t)\neq 0$ but $\det N(t)=0$ for every square submatrix N whose size is greater than that of M. Then the sets S_M are measurable; arranging them in some order we obtain measurable sets $T_1,\ldots,T_m$. Let S_0 be the set $\{t:\phi_{ij}(t)=0$ for all $i,j\}$ and let $S_1=T_1$, $S_p=T_p\setminus(T_1\cup\cdots\cup T_{p-1})$. Observe that X is the disjoint union $S_0\cup S_1\cup\cdots\cup S_m$. We shall define the ξ_j on each S_p. If $p\neq 0$, then for some q there exists, by construction, a $q\times q$ submatrix M with $\det M(t)\neq 0$ for $t\in S_p$ and such that $\det N(t)=0$ for $t\in S_p$ whenever N is a submatrix of $\phi_{ij}(t)$ whose size is greater than $q\times q$. Rearrange the indices, if necessary, to assume that M is $((\phi_{ij}))_{i,j=1}^q$. Observe that $q<n$ by assumption. Find the unique solution $\{x_1(t),\ldots,x_q(t)\}$ of the system $\left\{\sum_{j=1}^{q}\phi_{ij}(t)x_j(t)=\phi_{in}(t),\ i=1,\ldots,q\right\}$ for each $t\in S_p$. This determines

$x_1, \ldots, x_q$ as measurable functions. Define $x_{q+1} = \cdots = x_{n-1} \equiv 0$ and $x_n \equiv -1$ on S_p. Let $\xi_j = x_j / \left(\sum_{j=1}^{n} |x_j|^2 \right)^{\frac{1}{2}}$; then the ξ_j are defined on $\bigcup_{p=1}^{m} S_p$. Extend each ξ_j to X by defining $\xi_j \equiv n^{-\frac{1}{2}}$ on S_0. Clearly $\sum_{j=1}^{n} |\xi_j(t)|^2 = 1$ for all t. $\quad \square$

We can now "triangularize" n-normal operators.

Theorem 7.20. *Let $\mathfrak{R}$ be a m.a.s.a. and $A \in (\mathfrak{R}^{(n)})'$. Then there is a unitary operator $U \in (\mathfrak{R}^{(n)})'$ such that*

$$U^{-1} A U = \begin{pmatrix} A_{11} & A_{12} & \cdots & A_{1n} \\ 0 & A_{22} & \cdots & A_{2n} \\ 0 & 0 & & \vdots \\ \vdots & \vdots & & \vdots \\ 0 & 0 & \cdots & 0 & A_{nn} \end{pmatrix}$$

with $A_{ij} \in \mathfrak{R}$ for all i,j.

Proof. By Corollary 7.14 we can assume that $\mathfrak{R} = \mathscr{L}^\infty$ acting on $\mathscr{L}^2(X, \mu)$, where (X, μ) is some finite measure space. We assume, with no loss of generality, that $\mu(X) = 1$. Then, by the proof of Theorem 7.17, A can be expressed in the form

$$\begin{pmatrix} M_{\phi 11} & M_{\phi 12} & \cdots & M_{\phi 1m} \\ \vdots & & & \vdots \\ \vdots & & & \vdots \\ M_{\phi n1} & M_{\phi n2} & \cdots & M_{\phi nn} \end{pmatrix}$$

with $\phi_{ij} \in \mathscr{L}^\infty(X, \mu)$ for all i,j. To obtain the triangular form for A we proceed in the same way as in proving that every matrix on a finite-dimensional space can be triangularized, except that we require the above measure-theoretic lemmas.

Now, by Lemma 7.18, the polynomial equation $\det((\phi_{ij} - z \delta_{ij})) = 0$, (where δ_{ij} is the Kronecker delta), has a solution $\phi \in \mathscr{L}^\infty(X, \mu)$. For each i,j let ψ_{ij} be a fixed bounded measurable function such that $\psi_{ij}(t) = \phi_{ij}(t) - \phi(t) \delta_{ij}$ for almost every $t \in X$. Use Lemma 7.19, with $k = n$, to get $\xi_{11}, \ldots, \xi_{n1}$ with $\sum_{j=1}^{n} |\xi_{j1}|^2 \equiv 1$ such that $\sum_{j=1}^{n} \psi_{ij} \xi_{j1} \equiv 0$ for $i = 1, \ldots, n$. Then use Lemma 7.19 again, this time with $k = 1$, to get $\xi_{12}, \ldots, \xi_{n2} \in \mathscr{L}^\infty(X, \mu)$ with $\sum_{j=1}^{n} |\xi_{j2}|^2 \equiv 1$ and $\sum_{j=1}^{n} \bar{\xi}_{j1} \xi_{j2} \equiv 0$. By Lemma

7.19 with $k=2$ there exist $\{\xi_{13},\dots,\xi_{n3}\}\subset \mathscr{L}^{\infty}(X,\mu)$ such that $\displaystyle\sum_{j=1}^{n}\bar{\xi}_{j1}\xi_{j3}$
$\equiv \displaystyle\sum_{j=1}^{n}\bar{\xi}_{j2}\xi_{j3}\equiv 0$. Continuing this process we obtain a unitary operator

$$U_1 = \begin{pmatrix} M_{\xi_{11}} & M_{\xi_{12}} & \cdots & M_{\xi_{1n}} \\ \vdots & & & \vdots \\ M_{\xi_{n1}} & M_{\xi_{n2}} & \cdots & M_{\xi_{nn}} \end{pmatrix}.$$

Then $U_1^{-1}A U_1$ has the form

$$\begin{pmatrix} M_\phi & C_{12} & \cdots & C_{1n} \\ 0 & C_{22} & \cdots & C_{2n} \\ \vdots & \vdots & & \vdots \\ 0 & C_{n2} & \cdots & C_{nn} \end{pmatrix}$$

for some $C_{ij}\in\mathscr{L}^{\infty}$.

An obvious induction now completes the proof. We apply the above process to

$$\begin{pmatrix} C_{22} & \cdots & C_{2n} \\ \vdots & & \vdots \\ C_{n2} & \cdots & C_{nn} \end{pmatrix},$$

producing a unitary operator V_2 on $\mathscr{H}^{(n-1)}$. Then, if

$$U_2 = 1 \oplus V_2 \quad \text{on } \mathscr{H}^{(n)},$$

$U_2^{-1} U_1^{-1} A U_1 U_2$ has the form

$$\begin{pmatrix} M_\phi & D_{12} & D_{13} & \cdots & & D_{1n} \\ 0 & M_\psi & D_{23} & \cdots & & D_{2n} \\ 0 & 0 & D_{33} & \cdots & & D_{3n} \\ & & D_{43} & \cdots & & D_{4n} \\ \vdots & \vdots & \vdots & & & \vdots \\ 0 & 0 & D_{n3} & \cdots & & D_{nn} \end{pmatrix}.$$

We continue in this manner to obtain the result. □

It is clear from the definition of n-normal operators that they have non-trivial reducing subspaces: if $A\in(\mathfrak{R}^{(n)})'$ and $P\in\mathfrak{R}$, then $P^{(n)}$ is a projection commuting with A. It is not a priori obvious that n-normal operators have hyperinvariant subspaces.

Theorem 7.21. *If A is n-normal and is not a multiple of the identity, then A has a non-trivial hyperinvariant subspace.*

Proof. By Theorem 7.20 we can assume that A is in triangular form; Corollary 6.23 thus gives the result. $\square$

We require an analogue of the Jordan canonical-form theorem for the case of *n*-normal operators; the next lemma contains the essence of the proof.

Lemma 7.22. *Let A be an operator in the commutant of $(\mathscr{L}^{\infty})^{(n)}$ which is partitioned in the upper triangular form*

$$A = \begin{pmatrix} A_{11} & A_{12} & \cdots & A_{1m} \\ 0 & A_{22} & \cdots & A_{2m} \\ \vdots & & & \vdots \\ 0 & & \cdots & 0 \ A_{mm} \end{pmatrix},$$

where each A_{ij} is a matrix with entries in $\mathscr{L}^{\infty}$ and A_{ii} has the upper triangular form

$$A_{ii} = \begin{pmatrix} M_{\phi_i} & * & \cdots & & * \\ 0 & M_{\phi_i} & & & \vdots \\ \vdots & & & & * \\ 0 & & \cdots & 0 & M_{\phi_i} \end{pmatrix}$$

with $\mu\{t : \phi_i(t) = \phi_j(t)\} = 0$ for $i \neq j$. Then for each $\varepsilon > 0$ there exists a measurable set E with $\mu(X \backslash E) < \varepsilon$ such that the restrictions of A and the diagonal operator

$$\begin{pmatrix} A_{11} & 0 & \cdots & & 0 \\ 0 & A_{22} & & & 0 \\ \vdots & & \ddots & & \vdots \\ & & & & 0 \\ 0 & & \cdots & 0 & A_{mm} \end{pmatrix}$$

to the range of $(M_{\chi_E})^{(n)}$ are similar, and the similarity can be implemented by an operator in the restriction of the commutant of $(\mathscr{L}^{\infty})^{(n)}$ to the range of $(M_{\chi_E})^{(n)}$.

Proof. Choose fixed representatives of the equivalence classes in $\mathscr{L}^{\infty}$ for each entry in A. Choose the representatives of the ϕ_i such that

$\phi_i \neq \phi_j$ everywhere for $i \neq j$. We proceed by induction on m; the case $m=1$ is vacuously true. Assume the result is true for $m-1$ and consider an operator A as in the statement of the lemma. Denote the $(m-1) \times (m-1)$ submatrix in the upper left corner by B and regard A as being partitioned in the form

$$\begin{pmatrix} B & C \\ 0 & A_{mm} \end{pmatrix}.$$

We use the techniques of Corollary 0.15. For each fixed $t \in X$ let $B(t)$, $C(t)$ and $A_{mm}(t)$ denote the matrices over $\mathbb{C}$ which are obtained from B, C and A_{mm} by replacing the $\mathscr{L}^\infty$ entries by their respective values at t. The hypothesis obviously implies that $\sigma(B(t)) \cap \sigma(A_{mm}(t)) = \phi$ for each fixed t. By Rosenblum's corollary (Corollary 0.13), for each fixed t there is a unique matrix $F(t)$ such that

$$F(t) A_{mm}(t) - B(t) F(t) = C(t).$$

The entries of $F(t)$ are complex-valued functions of t; matrix multiplication shows that the entries are the unique solutions of linear equations whose coefficients are certain linear combinations of the entries of $A_{mm}(t)$, $B(t)$, and $C(t)$. The uniqueness of the solution implies that it is computable by Cramer's rule, and it follows that the entries of $F(t)$ are rational functions of the entries of $A_{mm}(t)$, $B(t)$, and $C(t)$. Hence the entries of $F(t)$ are (possibly unbounded) measurable functions on X.

Let E_k denote the set of all $t \in X$ such that each entry of $F(t)$ has modulus less than k. Then $\{E_k\}$ is an increasing family of measurable subsets of X whose union is X; since $\mu(X) < \infty$ there is a k such that $\mu(X - E_k) < \varepsilon/2$.

Let P denote multiplication by the characteristic function of E_k. Consider the restriction of A to the range $\mathscr{M}$ of $P^{(n)}$. If F denotes the matrix whose entries are those of $F(t)$ restricted to E_k, then the equation

$$\begin{pmatrix} B & C \\ 0 & A_{mm} \end{pmatrix} \begin{pmatrix} 1 & F \\ 0 & 1 \end{pmatrix} = \begin{pmatrix} 1 & F \\ 0 & 1 \end{pmatrix} \begin{pmatrix} B & 0 \\ 0 & A_{mm} \end{pmatrix}$$

holds on $\mathscr{M}$. Thus the restrictions to $\mathscr{M}$ of

$$\begin{pmatrix} B & C \\ 0 & A_{mm} \end{pmatrix} \quad \text{and} \quad \begin{pmatrix} B & 0 \\ 0 & A_{mm} \end{pmatrix}$$

are similar. The inductive hypothesis implies that there is a subset E of E_k with $\mu(E_k \backslash E) < \varepsilon/2$ such that the restrictions to the range of

$(M_{\chi_E})^{(n)}$ of A and

$$\begin{pmatrix} A_{11} & 0 & \cdots & & 0 \\ 0 & A_{22} & & & \vdots \\ \vdots & & \ddots & & 0 \\ 0 & & \cdots & 0 & A_{mm} \end{pmatrix}$$

are similar. $\square$

Theorem 7.23. *If $\mathfrak{R}$ is a m.a.s.a., $T \in (\mathfrak{R}^{(n)})'$, and $\mathcal{U}$ is any neighbourhood of 1 in the strong operator topology of $\mathcal{B}(\mathcal{H})$, then there exists a projection $P \in \mathfrak{R} \cap \mathcal{U}$ such that $P^{(n)} T = N + Q$, where $N \in (\mathfrak{R}^{(n)})'$, $Q \in (\mathfrak{R}^{(n)})'$, $NQ = QN$, N is similar to a normal operator and $Q^n = 0$.*

Proof. By Corollary 7.14 we can assume that $\mathfrak{R} = \mathcal{L}^\infty$ acting on $\mathcal{L}^2(X, \mu)$, and by Theorem 7.20 we can assume that T has the upper triangular form

$$\begin{pmatrix} M_{\phi_1} & * & \cdots & & * \\ 0 & M_{\phi_2} & & & \vdots \\ \vdots & & & & \\ & & & & * \\ 0 & & \cdots & 0 & M_{\phi_n} \end{pmatrix}$$

We also assume that, for each i, ϕ_i is a fixed everywhere defined measurable function. Consider all partitions $\mathscr{S} = \{S_1, S_2, \ldots, S_k\}$ of the set $\{1, 2, \ldots, n\}$ into non-empty disjoint subsets S_i whose union is $\{1, 2, \ldots, n\}$. For each such $\mathscr{S} = \{S_1, S_2, \ldots, S_k\}$ let

$$E_{\mathscr{S}} = \{t \in X : \phi_i(t) = \phi_j(t) \text{ if and only if } \{i,j\} \subset S_r \text{ for some } r\}.$$

Then X is the disjoint union of the measurable sets $E_{\mathscr{S}}$; thus, if $P_{\mathscr{S}}$ is multiplication by the characteristic function of $E_{\mathscr{S}}$, then $\{P_{\mathscr{S}} : \mu(E_{\mathscr{S}}) > 0\}$ is a collection of mutually orthogonal projections in $\mathfrak{R}$ whose sum is the identity. The range of each $P_{\mathscr{S}}^{(n)}$ reduces T, and therefore it suffices to prove the theorem for the restriction of T to the range of one such $P_{\mathscr{S}}^{(n)}$. In other words, we can assume that $E_{\mathscr{S}} = X$ for some partition $\mathscr{S}$. Then, for each i and j, $\phi_i(t) = \phi_j(t)$ either for all t or for no t.

We now re-triangularize T, using the approach of Theorem 7.20, to get T into the form required for Lemma 7.22. The "characteristic polynomial" of T is $\prod_{i=1}^{n} (\phi_i - z)$. Since ϕ_i is a root, the proof of Theorem 7.20 shows that T is unitarily equivalent to an operator of the form

$$\begin{pmatrix} M_{\phi_1} & C_{12} & \cdots & C_{1n} \\ 0 & C_{22} & \cdots & C_{2n} \\ \vdots & \vdots & & \vdots \\ 0 & C_{n2} & \cdots & C_{nn} \end{pmatrix}$$

in $(\mathfrak{R}^{(n)})'$. If $\phi_1 = \phi_j$ for some $j > 1$, then ϕ_1 must also be a root of the "characteristic polynomial" of

$$\begin{pmatrix} C_{22} & \cdots & C_{2n} \\ \vdots & & \vdots \\ C_{n2} & \cdots & C_{nn} \end{pmatrix},$$

and T is unitarily equivalent to an operator of the form

$$\begin{pmatrix} M_{\phi_1} & D_{12} & D_{13} & \cdots & D_{1n} \\ 0 & M_{\phi_1} & D_{23} & & \\ \vdots & 0 & D_{33} & & \vdots \\ \vdots & \vdots & \vdots & & \\ 0 & 0 & D_{n3} & \cdots & D_{nn} \end{pmatrix}.$$

We repeat this process as many times as ϕ_1 occurs among $\{\phi_1, \phi_2, \ldots, \phi_n\}$, getting T in the form

$$\begin{pmatrix} A_{11} & F_{12} & \cdots & F_{1k} \\ 0 & F_{22} & \cdots & F_{2k} \\ \vdots & \vdots & & \vdots \\ 0 & F_{k2} & \cdots & F_{kk} \end{pmatrix}$$

where A_{11} is as in Lemma 7.22 and where

$$\begin{pmatrix} F_{22} & \cdots & F_{2k} \\ \vdots & & \vdots \\ F_{k2} & \cdots & F_{kk} \end{pmatrix}$$

is in $(\mathfrak{R}^{(r)})'$ for some $r < n$ and has characteristic polynomial

$$\prod_{\phi_j \neq \phi_1} (\phi_j - z).$$

We now pick any $\phi_j \neq \phi_1$ and apply the above process to

$$\begin{pmatrix} F_{22} & \cdots & F_{2k} \\ \vdots & & \vdots \\ F_{k2} & \cdots & F_{kk} \end{pmatrix},$$

producing an upper triangular matrix A_{22} with its diagonal entries all equal to M_{ϕ_j}. We can continue in this manner to obtain the result that T is unitarily equivalent to an operator of the same form as the operator A in Lemma 7.22.

Now Lemma 7.22 leads easily to the Theorem. For let

$$\mathscr{U} = \{C \in \mathscr{B}(\mathscr{L}^2(X, \mu)): \|(1 - C)f_j\| < \varepsilon \text{ for } j = 1, \ldots, k\}$$

be any basic strong neighbourhood of the identity. For each positive
integer r choose a set E_r satisfying the conclusion of Lemma 7.22 with
$\mu(X \setminus E_r) < 1/r$. Then

$$\int |(1 - \chi_{E_r}) f_j|^2 \, d\mu = \int_{X \setminus E_r} |f_j|^2 \, d\mu,$$

and the absolute continuity of the integral implies that there is an r
such that $\int |(1 - \chi_{E_r}) f_j|^2 \, d\mu < \varepsilon^2$ for $j = 1, \dots, k$. Now let P denote multi-
plication by χ_{E_r}; then $P \in \mathscr{U} \cap \mathfrak{R}$, and the restrictions of T and the
diagonal operator

$$\begin{pmatrix} A_{11} & 0 & \dots & 0 \\ 0 & A_{22} & & \vdots \\ \vdots & & 0 & \\ 0 & & \dots & A_{mm} \end{pmatrix}$$

to the range of $P^{(n)}$ are similar. Let S be the operator of Lemma 7.22
which implements this similarity. Clearly each A_{ii} is of the form $N_i + Q_i$
with N_i normal, $Q_i^n = 0$ and $N_i Q_i = Q_i N_i$: simply let N_i be the diagonal
of A_{ii} and Q_i be the strictly upper triangular part of A_{ii}. Now let Q_0
and N_0 be the direct sum of the $\{Q_i\}$ and the $\{N_i\}$ respectively; then

$$P^{(n)} T = P^{(n)} S N_0 S^{-1} + P^{(n)} S Q_0 S^{-1}$$

is the required decomposition of $P^{(n)} T$. □

It will be useful to have information about the commutants of
operators of the form $N + Q$ as above.

Theorem 7.24. *If* $T = N + Q$, *where* N *is similar to a normal operator,*
Q *is quasinilpotent and* $NQ = QN$, *then every operator which commutes*
with T *also commutes with* N *and with* Q.

Proof. Suppose that $S^{-1} N S$ is normal; then

$$S^{-1} T S = S^{-1} N S + S^{-1} Q S.$$

and $S^{-1} Q S$ is quasinilpotent and commutes with $S^{-1} N S$. If $BT = TB$,
then $S^{-1} B S$ commutes with $S^{-1} T S$. Hence it suffices to show that
every operator which commutes with $S^{-1} T S$ also commutes with
$S^{-1} N S$ and $S^{-1} Q S$. Let $\{E_\lambda\}$ be the spectral measure of $S^{-1} N S$.
Fuglede's theorem (Theorem 1.16) implies that $\{E_\lambda\}$ commutes with
$S^{-1} Q S$ and it follows that $E(\mathscr{S})$ commutes with $S^{-1} T S$ for each Borel
set $\mathscr{S} \subset \mathbb{C}$. Note that $(S^{-1} Q S) | (E(\mathscr{S}) \mathscr{H})$ is quasinilpotent for each $\mathscr{S}$,
and thus

$$\sigma((S^{-1} T S) | E(\mathscr{S}) \mathscr{H}) \subset \bar{\mathscr{S}}$$

by Theorem 1.13 and Proposition 0.8.

Now the only relations between the operator A and $\{E_\lambda\}$ which were required in the proof of Fuglede's theorem are the above one and the fact that $E(\mathscr{S})$ commutes with A for each Borel set $\mathscr{S}$. Therefore the proof of Theorem 1.16 shows that every operator C which commutes with $S^{-1}TS$ also commutes with $\{E_\lambda\}$. It follows that every such C commutes with $S^{-1}NS$ and thus also with $S^{-1}QS$. ☐

7.6 Additional Propositions

Proposition 7.1. Any two idempotents of the same finite rank are similar.

Proposition 7.2. If ϕ is a measurable function on the finite measure space (X,μ), and if

$$\mathscr{D} = \{f \in \mathscr{L}^2(X,\mu) : \phi f \in \mathscr{L}^2(X,\mu)\},$$

then multiplication by ϕ is a closed operator with domain $\mathscr{D}$.

Proposition 7.3. If $\mathfrak{A}$ is a von Neumann algebra and $A \in \mathfrak{A}$, then the partial isometry and the positive operator occurring in the polar decomposition (Proposition 1.1) of A are in $\mathfrak{A}$.

Proposition 7.4. If $\mathfrak{A}$ is a von Neumann algebra, then there exists an $A \in \mathfrak{A}$ with $A \neq 0$ and $A^2 = 0$ if and only if $\mathfrak{A}$ is not abelian. This holds also if $\mathfrak{A}$ is merely assumed to be a uniformly closed (not weakly closed) self-adjoint subalgebra of $\mathscr{B}(\mathscr{H})$.

Proposition 7.5. The set of irreducible operators is a G_δ, (a countable intersection of open sets).

Proposition 7.6. If $\mathfrak{A}_n$ is a von Neumann algebra generated by a single operator for $n = 1, 2, 3, \ldots$, then $\mathfrak{A} = \sum_{n=1}^{\infty} \oplus \mathfrak{A}_n$ is generated by a single operator.

Proposition 7.7. Let $\mathfrak{A} = (\mathfrak{R}^{(n)})'$, where $\mathfrak{R}$ is a m.a.s.a. Then $\mathfrak{A}$ is generated by a single operator.

Proposition 7.8. There exist three projections P_1, P_2, P_3 such that the von Neumann algebra generated by $\{P_1, P_2, P_3\}$ is $\mathscr{B}(\mathscr{H})$.

7.7 Notes and Remarks

We have presented only a fragment of the theory of von Neumann algebras. This theory, initiated by von Neumann and Murray and developed by a number of other mathematicians, is a beautiful illustration of the interplay between analysis and algebra. Excellent expositions of aspects of this subject are given in Sakai [1], Schwartz [2], Naimark [1], Topping [1], and Kaplansky [2], and an encyclopaedic account of the theory is presented in Dixmier ([1], [2]).

Theorem 7.5, the first theorem of this subject, was discovered by von Neumann [1].

Theorem 7.9 is due to Eidelheit [1], who proved it in the more general case where $\mathscr{H}$ is a Banach space. Rickart has obtained much more general results, (cf. Rickart [1]). A version of the theorem for C^*-algebras can be found in Gardner [1]. Theorem 7.10 is due to Halmos [15]; the proof in the text is from Radjavi-Rosenthal [3]. Theorem 7.12 is the classical result of von Neumann [1]; a much more elegant, (though perhaps less instructive), proof is given in Rickart [1]. Theorem 7.13 and Corollary 7.14 are due to Segal [1], while Lemma 7.15 and Theorem 7.16 are from Arveson [1].

The class of n-normal operators was first studied by Brown [1]. The results on triangular and Jordan forms for n-normal operators, (Lemmas 7.18, 7.19, 7.22 and Theorem 7.23), are due to Foguel [1], and the proofs presented are modifications of his. Theorem 7.20 was not explicitly stated by Foguel; it was stated and proved, using techniques based on regarding the m.a.s.a. as a space of continuous functions on a Stonian space, in Deckard-Pearcy [3]. Such techniques were subsequently used by Hoover [1] to give an alternate proof of Theorem 7.23. Theorem 7.21 is due to Hoover [1], who derived it as a corollary of Theorem 7.23. The proof of Theorem 7.21 given here is from Radjavi-Rosenthal [5]. Theorem 7.24 is due to Dunford [1].

Proposition 7.4 is an unpublished result of·Kadison's, and Proposition 7.5 is from Halmos [15]. Proposition 7.6 is in Pearcy [1]; (another proof is in Rosenthal [7] and there are undoubtedly many other proofs known). Proposition 7.7 is a special case of a result in Pearcy [1], while Proposition 7.8 is due to Davis [1].

Chapter 8. Transitive Operator Algebras

Definition. A subalgebra $\mathfrak{A}$ of $\mathscr{B}(\mathscr{H})$ is *transitive* if it is weakly closed, contains the identity operator, and has the property that $Lat\,\mathfrak{A} = \{\{0\}, \mathscr{H}\}$.

One transitive operator algebra is $\mathscr{B}(\mathscr{H})$; the question of whether or not there exist any others is a well-known unsolved problem.

Definition. The *transitive algebra problem* is the problem: if $\mathfrak{A}$ is a transitive operator algebra on $\mathscr{H}$, must $\mathfrak{A}$ be equal to $\mathscr{B}(\mathscr{H})$?

Note that an affirmative answer to the transitive algebra problem would imply that every operator which is not a multiple of the identity has a non-trivial hyperinvariant subspace. For if $\mathscr{B}(\mathscr{H})$ were the only transitive operator algebra, then, for any operator $A \in \mathscr{B}(\mathscr{H})$, let $\mathfrak{A}$ denote the commutant of A. If A is not a multiple of the identity, then $\mathfrak{A} \neq \mathscr{B}(\mathscr{H})$, and the affirmative answer to the transitive algebra problem would imply that $\mathfrak{A}$ had a non-trivial invariant subspace, and thus that A had a hyperinvariant subspace.

Thus a positive answer to the transitive algebra problem would be a very powerful result which would include, in particular, all the results of Chapter 6. On the other hand, in spite of a great deal of interest in this problem, no transitive algebras other than $\mathscr{B}(\mathscr{H})$ have yet been discovered. In this chapter we present some special cases of the transitive algebra problem that have been solved. Most of the results are of the form: if $\mathfrak{A}$ is a transitive operator algebra and $\mathfrak{A}$ satisfies some additional hypothesis, then $\mathfrak{A} = \mathscr{B}(\mathscr{H})$. Some of these results lead to corollaries on the existence of hyperinvariant subspaces and on generators of $\mathscr{B}(\mathscr{H})$ as a weakly closed algebra.

We begin by considering operator algebras which have no invariant linear manifolds; this situation can be completely characterized.

8.1 Strictly Transitive Algebras

The algebraic analogue of the transitive algebra problem has an affirmative answer. At the outset we assume that $\mathscr{V}$ is any vector space over any field, and $\mathscr{L}(\mathscr{V})$ is the algebra of all linear transformations on $\mathscr{V}$. We consider subalgebras of $\mathscr{L}(\mathscr{V})$ which contain the identity.

Definition. A subalgebra $\mathfrak{A}$ of $\mathscr{L}(\mathscr{V})$ is *strictly transitive* if the only linear manifolds in $\mathscr{V}$ which are invariant under all the transformations in $\mathfrak{A}$ are $\{0\}$ and $\mathscr{V}$.

If $\mathfrak{A}$ is a subalgebra of $\mathscr{L}(\mathscr{V})$ and $x \in \mathscr{V}$, then $\{Ax : A \in \mathfrak{A}\}$ is an invariant linear manifold of $\mathfrak{A}$. It follows that $\mathfrak{A}$ is strictly transitive if and only if for each pair (x, y) of vectors in $\mathscr{V}$ such that $x \neq 0$ there exists a transformation $A \in \mathfrak{A}$ such that $Ax = y$.

Definition. Let n be a positive integer. A subalgebra $\mathfrak{A}$ of $\mathscr{L}(\mathscr{V})$ is *n-fold strictly transitive* if for each set $\{x_i\}_{i=1}^n$ of n linearly independent vectors in $\mathscr{V}$ and each set $\{y_i\}_{i=1}^n$ of vectors in $\mathscr{V}$ there exists an $A \in \mathfrak{A}$ such that $Ax_i = y_i$ for $i = 1, \ldots, n$.

Note that if $\mathfrak{A}$ is n-fold strictly transitive, then it is also m-fold strictly transitive for each $m < n$.

There is an equivalent definition of n-fold strict transitivity which helps to clarify the concept. If $\mathscr{S}$ is any subset of $\mathscr{L}(\mathscr{V})$, let $Lat_1 \mathscr{S}$ denote the collection of all linear manifolds invariant under $\mathscr{S}$. As in Section 7.1, let $\mathscr{V}^{(n)}$ denote the direct sum of n copies of $\mathscr{V}$, $A^{(n)}$ denote the direct sum of n copies of A for each linear transformation A, and $\mathscr{S}^{(n)} = \{A^{(n)} : A \in \mathscr{S}\}$ for subsets $\mathscr{S}$ of $\mathscr{L}(\mathscr{V})$.

Theorem 8.1. *The subalgebra $\mathfrak{A}$ of $\mathscr{L}(\mathscr{V})$ is n-fold strictly transitive if and only if $Lat_1 \mathfrak{A}^{(n)} = Lat_1(\mathscr{L}(\mathscr{V}))^{(n)}$.*

Proof. If $\mathfrak{A}$ is n-fold strictly transitive and $B \in \mathscr{L}(\mathscr{V})$, we must show that $Lat_1 \mathfrak{A}^{(n)} \subset Lat_1 B^{(n)}$. For this it suffices, since every invariant linear manifold is a span of cyclic ones, to show that every cyclic member of $Lat_1 \mathfrak{A}^{(n)}$ is a member of $Lat_1 B^{(n)}$; i.e., for each $x \in \mathscr{V}^{(n)}$ we must show that $\mathscr{M} = \{A^{(n)}x : A \in \mathfrak{A}\}$ is invariant under $B^{(n)}$. If $x = x_1 \oplus \cdots \oplus x_n$, then we must prove that $B^{(n)}x \in \mathscr{M}$ or, equivalently, that there exists $A \in \mathfrak{A}$ such that $Bx_1 \oplus \cdots \oplus Bx_n = Ax_1 \oplus \cdots \oplus Ax_n$. Since the case where $x = 0$ is trivial, by re-indexing we can assume that, for some k, $1 \leq k \leq n$, $\{x_1, \ldots, x_k\}$ is linearly independent and x_i is a linear combination of $\{x_1, \ldots, x_k\}$ for $i > k$. Since $\mathfrak{A}$ is n-fold strictly transitive, there exists $A \in \mathfrak{A}$ such that $Ax_i = Bx_i$ for $i = 1, \ldots, k$. For $i > k$, x_i is a linear combination of $\{x_1, \ldots, x_k\}$, and the linearity of A and B implies that $Ax_i = Bx_i$ for $i = k + 1, \ldots, n$ also.

The other half of the proof is even easier. Suppose that $Lat_1 \mathfrak{A}^{(n)} = Lat_1(\mathscr{L}(\mathscr{V}))^{(n)}$. Then if $\{x_1, \ldots, x_n\}$ is a linearly independent subset of $\mathscr{V}$ and $\{y_1, \ldots, y_n\}$ is any subset of $\mathscr{V}$, let $B \in \mathscr{L}(\mathscr{V})$ be any transformation such that $Bx_i = y_i$ for all i. The fact that $B^{(n)}(x_1 \oplus \cdots \oplus x_n)$ is in the cyclic invariant linear manifold of $\mathfrak{A}^{(n)}$ generated by $x_1 \oplus \cdots \oplus x_n$ shows that there is an $A \in \mathfrak{A}$ with $A x_i = B x_i = y_i$ for all i. $\square$

This theorem together with Theorem 7.1 shows that if $\mathscr{H}$ is a Hilbert space and $\mathfrak{A}$ is a subalgebra of $\mathscr{B}(\mathscr{H})$ such that $Lat_1 \mathfrak{A}^{(n)} = Lat_1(\mathscr{B}(\mathscr{H}))^{(n)}$ for all n, then $\mathfrak{A}$ is dense in $\mathscr{B}(\mathscr{H})$ in the strong topology; (note that, for a Hilbert space $\mathscr{H}$, $Lat_1(\mathscr{L}(\mathscr{H}))^{(n)} = Lat_1(\mathscr{B}(\mathscr{H}))^{(n)} = Lat(\mathscr{B}(\mathscr{H}))^{(n)}$ for each n). This suggests the following definition.

Definition. The subalgebra $\mathfrak{A}$ of $\mathscr{L}(\mathscr{V})$ is *strictly dense* if $\mathfrak{A}$ is n-fold strictly transitive for every n.

It is a remarkable fact that this definition is very redundant.

Theorem 8.2. *If $\mathfrak{A}$ is a 2-fold strictly transitive algebra of linear transformations on a vector space $\mathscr{V}$, then $\mathfrak{A}$ is strictly dense.*

Proof. To show that $\mathfrak{A}$ is n-fold strictly transitive for each n we proceed by induction on n. Assume that $\mathfrak{A}$ is n-fold strictly transitive and let $\{x_1, \ldots, x_{n+1}\}$ be a linearly independent subset of $\mathscr{V}$. First note that it suffices to show that for each j there exists an $A_j \in \mathfrak{A}$ such that $A_j x_j \neq 0$ and $A_j x_k = 0$ for $k \neq j$. For if such A_j's are given and if $\{y_1, \ldots, y_{n+1}\}$ is any subset of $\mathscr{V}$, then the fact that $\mathfrak{A}$ is strictly transitive implies that for each j there is some $B_j \in \mathfrak{A}$ with $B_j A_j x_j = y_j$. Then $\left(\sum_{k=1}^{n+1} B_k A_k \right)(x_j) = y_j$ for each j.

To prove that such A_j's exist it suffices to show that such an A_{n+1} exists, (simply by permuting the indices). Suppose, then, that no such A_{n+1} exists; i.e., assume that $A \in \mathfrak{A}$ and $A x_k = 0$ for $k = 1, \ldots, n$ implies that $A x_{n+1} = 0$. We will show that this contradicts the strict transitivity of $\mathfrak{A}$.

Since $\mathfrak{A}$ is 2-fold transitive it follows from Theorem 8.1 that $\mathscr{M} \in Lat_1 \mathfrak{A}^{(2)}$ implies that $\mathscr{M}$ is one of the spaces $\{0\}$, $\mathscr{V}^{(2)}$, $\{0\} \oplus \mathscr{V}$, or $\{x \oplus \lambda x : x \in \mathscr{V}\}$ for some scalar λ. Now the induction hypothesis implies that, for each $j \leq n$, there is some $C_j \in \mathfrak{A}$ with $C_j x_j \neq 0$ and $C_j x_k = 0$ for $k \in \{1, 2, \ldots, n\} \setminus \{j\}$. Then $A \in \mathfrak{A}$ and $A C_j x_j = 0$ implies $A C_j x_{n+1} = 0$. For $j \leq n$ the linear manifold $\{A C_j x_j \oplus A C_j x_{n+1} : A \in \mathfrak{A}\}$ is in $Lat_1 \mathfrak{A}^{(2)}$, and therefore there exists a scalar α_j such that $A C_j x_{n+1} = \alpha_j A C_j x_j$ for $A \in \mathfrak{A}$. The induction hypothesis also implies that

$$\mathscr{V}^{(n)} = \{A x_1 \oplus \cdots \oplus A x_n : A \in \mathfrak{A}\},$$

and the assumption that $A x_{n+1} = 0$ whenever $A x_i = 0$ for $i \leq n$ shows that a mapping T from $\mathscr{V}^{(n)}$ to $\mathscr{V}$ is unambiguously defined by

$$T(A x_1 \oplus \cdots \oplus A x_n) = A x_{n+1} \quad \text{for} \quad A \in \mathfrak{A}.$$

A trivial computation proves that T is linear. Then, for $A \in \mathfrak{A}$,

$$A^{(n)} C_j^{(n)}(x_1 \oplus \cdots \oplus x_n) = 0 \oplus \cdots \oplus 0 \oplus A C_j x_j \oplus 0 \oplus \cdots \oplus 0$$

and
$$T(A^{(n)} C_j^{(n)}(x_1 \oplus \cdots \oplus x_n)) = A C_j x_{n+1} = \alpha_j A C_j x_j.$$

Since $C_j x_j \neq 0$, $\{A C_j x_j : A \in \mathfrak{A}\} = \mathscr{V}$ by the strict transitivity of $\mathfrak{A}$; the linearity of T gives $T(y_1 \oplus \cdots \oplus y_n) = \sum_{j=1}^{n} \alpha_j y_j$ for all $(y_1 \oplus \cdots \oplus y_n) \in \mathscr{V}^{(n)}$. Then, for $A \in \mathfrak{A}$,

$$A\left(x_{n+1} - \sum_{j=1}^{n} \alpha_j x_j\right) = A x_{n+1} - T(A x_1 \oplus \cdots \oplus A x_n)$$
$$= A x_{n+1} - A x_{n+1} = 0.$$

Hence $\{x : A x = 0 \text{ for all } A \in \mathfrak{A}\}$ is a non-trivial invariant linear manifold for $\mathfrak{A}$, contradicting the strict transitivity of $\mathfrak{A}$. $\quad\square$

We now return to the study of operators on Hilbert space, although we continue to consider linear manifolds rather than subspaces. The next lemma will enable us to prove that 1-fold strict transitivity implies strict density in this case.

Lemma 8.3. *If $\mathfrak{A}$ is a uniformly closed, strictly transitive subalgebra of $\mathscr{B}(\mathscr{H})$ and T is a linear transformation (not assumed bounded) taking $\mathscr{H}$ into $\mathscr{H}$ which commutes with every $A \in \mathfrak{A}$, then T is a multiple of the identity operator.*

Proof. Fix any vector $x_0 \neq 0$ in $\mathscr{H}$. Let ϕ denote the map from $\mathfrak{A}$ to $\mathscr{H}$ defined by $\phi(A) = A x_0$, and let $\mathscr{I}$ be the kernel of ϕ. Then ϕ induces a map $\hat{\phi}$ from $\mathfrak{A}/\mathscr{I}$ to $\mathscr{H}$; $\hat{\phi}$ is obviously injective, surjective (since $\mathfrak{A}$ is strictly transitive), and linear. Moreover, since $\|A x_0\| \leq \|A\| \cdot \|x_0\|$, $\hat{\phi}$ is bounded. The closed graph theorem implies that $\hat{\phi}^{-1}$ is also bounded.

Fix $x \in \mathscr{H}$, and choose $A \in \mathfrak{A}$ such that $A x_0 = x$. Then
$$\|T x\| = \|T A x_0\| = \|A T x_0\| \leq \|A\| \cdot \|T x_0\|.$$

Therefore, since this is true for every such $A \in \mathfrak{A}$,
$$\|T x\| \leq \|\hat{\phi}^{-1}\| \cdot \|T x_0\| \cdot \|x\|.$$

Hence T is bounded.

Let $\lambda_0 \in \sigma(T)$; we claim that $T = \lambda_0$. If not, then the nullspace of $T - \lambda_0$ is not $\mathscr{H}$. Since $T - \lambda_0 \in \mathfrak{A}'$ the nullspace of $T - \lambda_0$ is in $Lat\,\mathfrak{A}$, and it follows that $T - \lambda_0$ is one-to-one. In addition the range of $T - \lambda_0$ is in $Lat_1\,\mathfrak{A}$, and thus is $\mathscr{H}$. Hence $\lambda_0 \notin \sigma(T)$, which is a contradiction. $\quad\square$

Theorem 8.4. *If $\mathfrak{A}$ is a uniformly closed, strictly transitive subalgebra of $\mathscr{B}(\mathscr{H})$, then $\mathfrak{A}$ is strictly dense.*

Proof. By Theorem 8.2 we need only prove that $\mathfrak{A}$ is 2-fold strictly transitive. Let $\{x_1, x_2\}$ be a linearly independent set of vectors. As in the proof of Theorem 8.2, it suffices to show that there is some $A \in \mathfrak{A}$ with $A x_1 = 0$, $A x_2 \neq 0$. If this were not so, a linear transformation T could be defined by $T A x_1 = A x_2$ for $A \in \mathfrak{A}$. The strict transitivity of $\mathfrak{A}$ implies that T is everywhere defined. Moreover, if A and B are in $\mathfrak{A}$, then $T A B x_1 = A B x_2 = A T B x_1$, and $T A = A T$ for $A \in \mathfrak{A}$. Lemma 8.3 states that $T = \alpha$ for some $\alpha \in \mathbb{C}$; then $A(x_2 - \alpha x_1) = 0$ for all $A \in \mathfrak{A}$, which contradicts the strict transitivity of $\mathfrak{A}$. $\quad\square$

Corollary 8.5. *If $\mathfrak{A}$ is a weakly closed strictly transitive subalgebra of $\mathscr{B}(\mathscr{H})$, then $\mathfrak{A} = \mathscr{B}(\mathscr{H})$.*

Proof. This follows immediately from Theorems 8.4, 8.1 and 7.1. $\quad\square$

Corollary 8.6 *(Burnside's Theorem).* *If $\mathscr{H}$ is finite-dimensional, then the only transitive subalgebra of $\mathscr{B}(\mathscr{H})$ is $\mathscr{B}(\mathscr{H})$.*

Proof. This follows from Corollary 8.5 and the fact that every sub-algebra of $\mathscr{B}(\mathscr{H})$ is weakly closed; (the linear manifolds and sub-spaces of a finite-dimensional space coincide, of course). $\quad\square$

8.2 Partial Solutions of the Transitive Algebra Problem

As Corollary 8.5 states, the transitive algebra problem has an affirmative answer if transitivity is strengthened to strict transitivity. On the other hand, the problem is trivially answered negatively if the assumption that $\mathfrak{A}$ is weakly closed is relaxed: if $\mathfrak{A} = \{\lambda + K : \lambda \in \mathbb{C}, K \text{ compact}\}$, then $Lat \, \mathfrak{A} = \{\{0\}, \mathscr{H}\}$. The problem is not affected if the assumption that $\mathfrak{A}$ has an identity is dropped, (Proposition 8.2), but we leave that as part of the definition for simplicity.

Theorem 8.7. *The only transitive von Neumann algebra is $\mathscr{B}(\mathscr{H})$.*

Proof. If $\mathfrak{A}$ is a transitive von Neumann algebra, then $\mathfrak{A}' = \{\lambda : \lambda \in \mathbb{C}\}$, (by Theorem 7.3). Hence the double commutant theorem (Theorem 7.5) gives the result. $\quad\square$

Operator algebras which are not self-adjoint are much more difficult to deal with. Our basic approach to the transitive algebra problem is via the following obvious corollary of Theorem 7.1: if $\mathfrak{A}$ is a weakly closed operator algebra and if $Lat \, \mathfrak{A}^{(n)} = Lat (\mathscr{B}(\mathscr{H}))^{(n)}$ for all n, then $\mathfrak{A} = \mathscr{B}(\mathscr{H})$. It will therefore be important to study $Lat \, \mathfrak{A}^{(n)}$.

Definition. If $\mathfrak{A}$ is a subalgebra of $\mathscr{B}(\mathscr{H})$ and $\mathscr{M} \in Lat \, \mathfrak{A}^{(n)}$, then $\mathscr{M}$ is an *invariant graph subspace* for $\mathfrak{A}^{(n)}$ if there exist linear trans-formations $T_1, \ldots, T_{n-1}$ with a common domain $\mathscr{D}$, ($\mathscr{D}$ a linear mani-fold different from $\{0\}$ in $\mathscr{H}$), such that

$$\mathscr{M} = \{x \oplus T_1 x \oplus \cdots \oplus T_{n-1} x : x \in \mathscr{D}\}.$$

A linear transformation T is a *graph transformation for* $\mathfrak{A}$ if, for some n, T occurs as one of the T_i's in an invariant graph subspace for $\mathfrak{A}^{(n)}$.

Note that if $\mathcal{M} = \{x \oplus T_1 x \oplus \cdots \oplus T_{n-1} x : x \in \mathcal{D}\}$ then $\mathcal{M}$ is an invariant graph subspace for $\mathfrak{A}^{(n)}$ if and only if each T_i commutes (as in Section 7.4) with every $A \in \mathfrak{A}$ and $\mathcal{D}$ is an invariant linear manifold of $\mathfrak{A}$; this follows directly from the fact that

$$A^{(n)} \mathcal{M} = \{A x \oplus A T_1 x \oplus \cdots \oplus A T_{n-1} x : x \in \mathcal{D}\} \quad \text{for} \quad A \in \mathfrak{A}.$$

If $\mathfrak{A}$ is transitive, then each domain $\mathcal{D}$ of a graph transformation for $\mathfrak{A}$ is dense in $\mathcal{H}$, since $\overline{\mathcal{D}}$ is an invariant subspace for $\mathfrak{A}$.

Also note that every graph transformation for $\mathcal{B}(\mathcal{H})$ is a multiple of the identity operator. The following converse of this is the basis of all the subsequent results on transitive algebras.

Lemma 8.8 *(Arveson's Lemma). If* $\mathfrak{A}$ *is a transitive subalgebra of* $\mathcal{B}(\mathcal{H})$, *and if the only graph transformations for* $\mathfrak{A}$ *are the multiples of the identity, then* $\mathfrak{A} = \mathcal{B}(\mathcal{H})$.

Proof. It follows immediately from Theorem 7.1 that it suffices to show that $Lat\, \mathfrak{A}^{(n)} \subset Lat(\mathcal{B}(\mathcal{H}))^{(n)}$ for all n. The hypothesis that $\mathfrak{A}$ is transitive states that this holds for $n=1$, and the assumption that the only graph transformations for $\mathfrak{A}$ are multiples of the identity states that $\mathfrak{A}^{(n)}$ and $(\mathcal{B}(\mathcal{H}))^{(n)}$ have the same invariant graph subspaces for all n.

Suppose that $Lat\, \mathfrak{A}^{(k)} \subset Lat(\mathcal{B}(\mathcal{H}))^{(k)}$ for $k \leq n$. To prove that $Lat\, \mathfrak{A}^{(n+1)} \subset Lat(\mathcal{B}(\mathcal{H}))^{(n+1)}$ it suffices to show that every cyclic invariant subspace of $\mathfrak{A}^{(n+1)}$, (i.e., subspace of the form $\overline{\{A^{(n+1)} x : A \in \mathfrak{A}\}}$ for some $x \in \mathcal{H}^{(n+1)}$), is in $Lat(\mathcal{B}(\mathcal{H}))^{(n+1)}$, since every invariant subspace is the span of the cyclic invariant subspaces that it contains. Let $x = x_1 \oplus \cdots \oplus x_n \oplus x_{n+1} \in \mathcal{H}^{(n+1)}$ and let $\mathcal{M}$ be the closure of

$$\{A x_1 \oplus \cdots \oplus A x_n \oplus A x_{n+1} : A \in \mathfrak{A}\}.$$

Let $B \in \mathcal{B}(\mathcal{H})$; we must show that $\mathcal{M} \in Lat\, B^{(n+1)}$.

We consider two distinct cases.

Case (i): The set $\{x_1, \ldots, x_n, x_{n+1}\}$ is a linearly dependent set. In this case, by permuting the components we can assume that $x_{n+1} = \sum_{i=1}^{n} k_i x_i$ for some scalars $\{k_i\}$. Let $\mathcal{N}$ be the closure (in $\mathcal{H}^{(n)}$) of $\{A x_1 \oplus \cdots \oplus A x_n : A \in \mathfrak{A}\}$. By the inductive hypothesis $\mathcal{N} \in Lat\, B^{(n)}$. Clearly

$$\mathcal{M} = \left\{ y_1 \oplus \cdots \oplus y_n \oplus \sum_{i=1}^{n} k_i y_i : (y_1 \oplus \cdots \oplus y_n) \in \mathcal{N} \right\}.$$

If $\left(y \oplus \cdots \oplus y_n \oplus \sum_{i=1}^{n} k_i y_i \right) \in \mathcal{M}$, then

$$B^{(n+1)} \left(y_1 \oplus \cdots \oplus y_n \oplus \sum_{i=1}^{n} k_i y_i \right) = B y_1 \oplus \cdots \oplus B y_n \oplus \sum_{i=1}^{n} k_i B y_i$$

is in $\mathcal{M}$ since $(B y_1 \oplus \cdots \oplus B y_n)$ is in $\mathcal{N}$. Thus $\mathcal{M} \in Lat \, B^{(n+1)}$.

Case (ii): The set $\{x_1, \ldots, x_n, x_{n+1}\}$ is linearly independent. In the proof that $\mathcal{M} \in Lat \, B^{(n+1)}$ in this case we will need the following fact: if $k \leq n$ and $\{y_1, \ldots, y_k\}$ is a linearly independent set, then the closure of $\{A y_1 \oplus \cdots \oplus A y_k : A \in \mathfrak{A}\}$ is $\mathcal{H}^{(k)}$. This follows directly from the inductive hypothesis, for $Lat \, \mathfrak{A}^{(k)} = Lat \, \mathcal{B}(\mathcal{H})^{(k)}$, and obviously the only invariant subspace of $\mathcal{B}(\mathcal{H})^{(k)}$ which contains a vector with linearly independent components is $\mathcal{H}^{(k)}$. We will show that $\mathcal{M} = \mathcal{H}^{(n+1)}$ in this case.

Our basic hypothesis about $\mathfrak{A}$ implies that every vector in an invariant graph subspace of $\mathcal{H}^{(n+1)}$ has linearly dependent components; hence $\mathcal{M}$ is certainly not a graph subspace. We claim that $\mathcal{M}$ contains a non-zero vector of the form $0 \oplus y_1 \oplus \cdots \oplus y_n$. If $\mathcal{M}$ did not contain such a vector, then it would be the case that the vectors in $\mathcal{M}$ are determined by their first components. Let $\mathcal{D}$ denote the linear manifold in $\mathcal{H}$ consisting of all first components of vectors in $\mathcal{M}$. If $T_i z$ denotes the $(i+1)^{st}$ component of the vector in $\mathcal{M}$ whose first component is z, then

$$\mathcal{M} = \{z \oplus T_1 z \oplus \cdots \oplus T_n z : z \in \mathcal{D}\}.$$

Each T_i is obviously a linear transformation, (since $\mathcal{M}$ is a subspace), and thus $\mathcal{M}$ is a graph subspace. This is a contradiction and we conclude that $\mathcal{M}$ does contain a vector $0 \oplus y_1 \oplus \cdots \oplus y_n$ different from 0.

By permuting the components we can assume that, for some k, $1 \leq k \leq n$, $\{y_1, \ldots, y_k\}$ is a linearly independent set and y_i is a linear combination of $\{y_1, \ldots, y_k\}$ for each $i > k$. It follows from the induction hypothesis that the invariant subspace of $\mathfrak{A}^{(k)}$ generated by $y_1 \oplus \cdots \oplus y_k$ is $\mathcal{H}^{(k)}$. In particular, there is a sequence $\{A_m\} \in \mathfrak{A}$ such that $\lim_{m \to \infty} (A_m y_1 \oplus \cdots \oplus A_m y_k) = x_2 \oplus 0 \oplus \cdots \oplus 0$. Thus, for some scalars $\{c_1, \ldots, c_{n-k}\}$,

$$\lim_{m \to \infty} (A_m y_1 \oplus \cdots \oplus A_m y_k \oplus A_m y_{k+1} \oplus \cdots \oplus A_m y_n)$$

$$= x_2 \oplus 0 \oplus \cdots \oplus 0 \oplus c_1 x_2 \oplus \cdots \oplus c_{n-k} x_2.$$

Thus

$$\lim_{m \to \infty} A_m^{(n+1)} (0 \oplus y_1 \oplus \cdots \oplus y_n)$$

$$= 0 \oplus x_2 \oplus 0 \oplus \cdots \oplus 0 \oplus c_1 x_2 \oplus \cdots \oplus c_{n-k} x_2$$

is in $\mathcal{M}$. Subtracting this vector from the vector $x_1 \oplus \cdots \oplus x_{n+1}$ shows that

$$x_1 \oplus 0 \oplus x_3 \oplus \cdots \oplus x_{k+1} \oplus (x_{k+2} - c_1 x_2) \oplus \cdots \oplus (x_{n+1} - c_{n-k} x_2)$$

is also in $\mathcal{M}$. The n non-zero components of this vector are linearly independent, and the induction hypothesis applied to $\mathfrak{A}^{(n)}$ on $\mathcal{H}^{(n)}$, (expressed in the form $\mathcal{H} \oplus \{0\} \oplus \mathcal{H}^{(n-1)}$), shows that $\mathcal{M}$ contains all vectors in $\mathcal{H}^{(n+1)}$ whose second components are 0. Also

$$0 \oplus x_2 \oplus 0 \oplus \cdots \oplus 0 = (x_1 \oplus x_2 \oplus \cdots \oplus x_{n+1}) - (x_1 \oplus 0 \oplus x_3 \oplus \cdots \oplus x_{n+1})$$

is in $\mathcal{M}$, and it follows that $0 \oplus z \oplus 0 \oplus \cdots \oplus 0 \in \mathcal{M}$ for all $z \in \mathcal{H}$. Thus $\mathcal{M} = \mathcal{H}^{(n+1)}$. $\square$

Our first application of Arveson's lemma is a strengthening of the theorem that a weakly closed strictly transitive algebra is $\mathcal{B}(\mathcal{H})$, (Corollary 8.5). The next result is, in a sense, half-way between this theorem and an affirmative answer to the transitive algebra problem.

Definition. A linear manifold $\mathcal{D} \subset \mathcal{H}$ is an *operator range* if there exists $A \in \mathcal{B}(\mathcal{H})$ such that $\mathcal{D} = \{A x : x \in \mathcal{H}\}$.

Some of the properties of operator ranges are listed in the additional propositions; (see Propositions 8.4 to 8.6). Obviously every closed subspace is an operator range, (let A be the projection onto the subspace), and it is not hard to show that the collection of operator ranges is a proper subset of the collection of all linear manifolds, (cf. Proposition 8.4).

Theorem 8.9. *If $\mathfrak{A}$ is a weakly closed subalgebra of $\mathcal{B}(\mathcal{H})$ containing the identity operator, and if the only operator ranges which are invariant under $\mathfrak{A}$ are $\{0\}$ and $\mathcal{H}$, then $\mathfrak{A} = \mathcal{B}(\mathcal{H})$.*

Proof. By Lemma 8.8 we need only show that all the graph transformations for $\mathfrak{A}$ are multiples of the identity. Let

$$\mathcal{M} = \{x \oplus T_1 x \oplus \cdots \oplus T_{n-1} x : x \in \mathcal{D}\}$$

be an invariant graph subspace for $\mathfrak{A}^{(n)}$. If P is the projection of $\mathcal{M}$ onto the first coordinate space, (i.e., $P(x \oplus T_1 x \oplus \cdots \oplus T_{n-1} x) = x$ for $x \in \mathcal{D}$), then $P\mathcal{M} = \mathcal{D}$. Since $\mathcal{M}$ is a subspace of $\mathcal{H}^{(n)}$ and $\mathcal{H}$ is isomorphic to $\mathcal{H}^{(n)}$, it follows that $\mathcal{D}$ is an operator range. Now $\mathcal{D}$ is invariant under $\mathfrak{A}$, and we conclude that $\mathcal{D} = \mathcal{H}$. Let T denote the linear transformation from $\mathcal{H}$ to $\mathcal{H}^{(n-1)}$ defined by $Tx = T_1 x \oplus \cdots \oplus T_{n-1} x$ for $x \in \mathcal{H}$. Then the graph of T is $\mathcal{M}$, and the closed graph theorem implies that T is bounded. It follows that each T_i is bounded, and $T_i \in \mathcal{B}(\mathcal{H})$ for all i.

Fix an i and choose $\lambda_i \in \sigma(T_i)$. We claim that $T_i = \lambda_i$; the proof of this is similar to the proof of the corresponding part of Lemma 8.3. First note that $(T_i - \lambda_i) \in \mathfrak{A}'$. It follows that the nullspace and range of

$T_i - \lambda_i$ are invariant under $\mathfrak{A}$. Both of these linear manifolds are operator ranges, and therefore they are either $\{0\}$ or $\mathscr{H}$. If the nullspace is $\mathscr{H}$ or the range is $\{0\}$, then we are done. If the nullspace is $\{0\}$ and the range is $\mathscr{H}$, then $T_i - \lambda_i$ is invertible, which is impossible since $\lambda_i \in \sigma(T_i)$. $\square$

Theorem 8.10. *If $\mathfrak{A}$ is a transitive operator algebra, and if $\mathfrak{A}$ contains a m.a.s.a., then $\mathfrak{A} = \mathscr{B}(\mathscr{H})$.*

Proof. By Lemma 8.8 it suffices to show that the only graph transformations for $\mathfrak{A}$ are the multiples of the identity. Let T be a graph transformation for $\mathfrak{A}$; then T is a densely defined linear transformation which commutes with every operator in $\mathfrak{A}$. In particular, T commutes with every operator in the m.a.s.a. contained in $\mathfrak{A}$. By Theorem 7.16, then, T has a closure $\tilde{T}$, and for each neighborhood $\mathscr{U}$ of the identity operator there is a projection $P \in \mathscr{U}$ such that $P \tilde{T} P$ is a bounded normal operator which commutes with PAP for all $A \in \mathfrak{A}$. It obviously suffices to show that $P \tilde{T} P$ is a multiple of P for each such P. If, for some such P, $P \tilde{T} P$ is not a multiple of P, then $P \tilde{T} | (P \mathscr{H})$ is a bounded normal operator on $P \mathscr{H}$ which is not a multiple of the identity. Thus $P \tilde{T} | (P \mathscr{H})$ has some non-trivial spectral projection E. By Fuglede's theorem (Theorem 1.16) PAP commutes with E for all $A \in \mathfrak{A}$. Choose an $x \neq 0$ in $E \mathscr{H}$ and a $y \neq 0$ in $(1 - E) P \mathscr{H}$. Then, for $A \in \mathfrak{A}$, $(PAPx, y) = 0$. But $(PAPx, y) = (PAx, Py) = (Ax, Py) = (Ax, y)$. Thus the closure of $\{Ax : A \in \mathfrak{A}\}$ is a non-trivial invariant subspace of $\mathfrak{A}$. Hence the fact that $\mathfrak{A}$ is transitive implies $P \tilde{T} P$ is a multiple of P. $\square$

Arveson's lemma can be slightly strengthened; the following lemma will be useful in obtaining other special cases of the transitive algebra problem.

Lemma 8.11. *If $\mathfrak{A}$ is a transitive operator algebra, and if every graph transformation for $\mathfrak{A}$ has an eigenvector, then $\mathfrak{A} = \mathscr{B}(\mathscr{H})$.*

Proof. By Lemma 8.8 it suffices to show that every graph transformation for $\mathfrak{A}$ is a multiple of the identity. Let

$$\mathscr{M} = \{x \oplus T_1 x \oplus \cdots \oplus T_{n-1} x : x \in \mathscr{D}\}$$

be an invariant graph subspace for $\mathfrak{A}^{(n)}$. By hypothesis there exist $\lambda_1 \in \mathbb{C}$ and $x_1 \in \mathscr{D}$ such that $x_1 \neq 0$ and $T_1 x_1 = \lambda_1 x_1$. Let

$$\mathscr{D}_1 = \{x \in \mathscr{D} : T_1 x = \lambda_1 x\}.$$

If $x \in \mathscr{D}_1$ and $A \in \mathfrak{A}$, then $T_1 A x = A T_1 x = \lambda_1 A x$. Hence $\mathscr{D}_1$ is an invariant linear manifold for $\mathfrak{A}$. Let $\mathscr{M}_1 = \{x \oplus \lambda_1 x \oplus T_2 x \oplus \cdots \oplus T_{n-1} x : x \in \mathscr{D}_1\}$. Then $\mathscr{M}_1$ is a closed subspace of $\mathscr{M}$, and $\mathscr{M}_1 \in Lat\, \mathfrak{A}^{(n)}$. Thus by hypothesis $T_2|\mathscr{D}_1$ has an eigenvector. If λ_2 is an eigenvalue of $T_2|\mathscr{D}_1$ let

$$\mathscr{D}_2 = \{x \in \mathscr{D}_1 : T_2 x = \lambda_2 x\},$$

and let

$$\mathscr{M}_2 = \{x \oplus \lambda_1 x \oplus \lambda_2 x \oplus T_3 x \oplus \cdots \oplus T_{n-1} x : x \in \mathscr{D}_2\}.$$

Then $\mathscr{M}_2 \in \operatorname{Lat} \mathfrak{A}^{(n)}$. Continue in this manner and obtain a linear manifold $\mathscr{D}_{n-1}$ and scalars $\{\lambda_1, \ldots, \lambda_{n-1}\}$ such that the subspace

$$\mathscr{M}_{n-1} = \{x \oplus \lambda_1 x \oplus \cdots \oplus \lambda_{n-1} x : x \in \mathscr{D}_{n-1}\}$$

is in $\operatorname{Lat} \mathfrak{A}^{(n)}$, is contained in $\mathscr{M}$, and is not $\{0\}$. Then $\mathscr{D}_{n-1}$ is an invariant linear manifold of $\mathfrak{A}$, and thus is dense in $\mathscr{H}$. Since $\mathscr{M}_{n-1}$ is closed, $\mathscr{D}_{n-1}$ is closed, and thus $\mathscr{D}_{n-1} = \mathscr{H}$. Since $\mathscr{M}_{n-1} \subset \mathscr{M}$, it follows that $\mathscr{H} = \mathscr{D}_{n-1} = \mathscr{D}$ and, for each i, $T_i x = \lambda_i x$ for all $x \in \mathscr{H}$. $\quad\square$

Theorem 8.12. *A transitive operator algebra which contains a non-zero operator of finite rank is* $\mathscr{B}(\mathscr{H})$.

Proof. We use Lemma 8.11. Let

$$\mathscr{M} = \{x \oplus T_1 x \oplus \cdots \oplus T_{n-1} x : x \in \mathscr{D}\}$$

be an invariant graph subspace for the algebra, and let F be a non-zero operator of finite rank in the algebra.

We first show that the range of F is contained in $\mathscr{D}$. To see this let $y = Fx$. Since $\mathscr{D}$ is dense in $\mathscr{H}$, there exists a sequence $\{x_n\} \subset \mathscr{D}$ such that $\{x_n\} \to x$. Now $\mathscr{D}$ is an invariant linear manifold for the algebra, and thus $Fx_n \in \mathscr{D}$ for all n. Hence $\{Fx_n\}$ is contained in the intersection of $\mathscr{D}$ and the range of F. This intersection is finite-dimensional, hence closed, and it follows that $Fx = \lim_{n \to \infty} Fx_n$ is also in this intersection. In particular, $y = Fx$ is in $\mathscr{D}$.

For each i, $FT_i = T_i F$ implies that the range of F is invariant under T_i. Hence there is a finite-dimensional subspace which is invariant under T_i, and it follows that T_i has an eigenvector. Thus Lemma 8.11 gives the result. $\quad\square$

Theorem 8.13. *If* $\mathfrak{A}$ *is a transitive algebra, and if* $\mathfrak{A}$ *contains an operator* A *such that*

(i) *every eigenspace of* A *is one-dimensional and*

(ii) *for every n, each non-trivial invariant subspace of* $A^{(n)}$ *contains an eigenvector of* $A^{(n)}$,

then $\mathfrak{A} = \mathscr{B}(\mathscr{H})$.

Proof. By Lemma 8.11 we need only show that each graph transformation for $\mathfrak{A}$ has an eigenvector. Let $\mathscr{M} \in \operatorname{Lat} \mathfrak{A}^{(n)}$ and

$$\mathscr{M} = \{x \oplus T_1 x \oplus \cdots \oplus T_{n-1} x : x \in \mathscr{D}\}.$$

By (ii) there is a vector $x_0 \in \mathcal{D}$ such that $x_0 \oplus T_1 x \oplus \cdots \oplus T_{n-1} x_0$ is an eigenvector of $A^{(n)}$. Then clearly x_0 is an eigenvector of A; let $A x_0 = \lambda x_0$. Then, for each i, $T_i A x_0 = A T_i x_0$ implies $\lambda T_i x_0 = A T_i x_0$; i.e., $T_i x_0$ is in the eigenspace of A corresponding to λ. Hence (i) implies $T_i x_0$ is a multiple of x_0; i.e., x_0 is an eigenvector of T_i. $\square$

Corollary 8.14. *The only transitive algebra containing a Donoghue operator is $\mathscr{B}(\mathscr{H})$.*

Proof. This follows directly from the fact that Donoghue operators satisfy conditions (i) and (ii) of Theorem 8.13. The only eigenspace of a Donoghue operator is the one-dimensional space spanned by $\{e_0\}$, and Proposition 4.13 states that Donoghue operators satisfy (ii). $\square$

The next special case of the transitive algebra problem that we consider is the case where the algebra contains a unilateral shift of finite multiplicity. Several preliminary results are required.

Definition. The densely defined linear transformation T will be said to *have compression spectrum* if there exists $\lambda \in \mathbb{C}$ such that the range of $T - \lambda$ is not dense in $\mathscr{H}$; i.e., if $\mathscr{D}$ is the domain of T, then the closure of $\{T x - \lambda x : x \in \mathscr{D}\}$ is not $\mathscr{H}$.

The following is another variant of Arveson's lemma.

Lemma 8.15. *If $\mathfrak{A}$ is a transitive operator algebra such that every graph transformation for $\mathfrak{A}$ has compression spectrum, then $\mathfrak{A} = \mathscr{B}(\mathscr{H})$.*

Proof. Let T be a graph transformation for $\mathfrak{A}$; by Lemma 8.8 we need only show that T is a multiple of the identity. Choose $\lambda \in \mathbb{C}$ such that the range of $T - \lambda$ is not dense; we show that $T = \lambda$. Suppose that there were some x in the domain of T such that $(T - \lambda) x \neq 0$. Then the transitivity of $\mathfrak{A}$ would imply that $\{A(T - \lambda) x : A \in \mathfrak{A}\}$ is dense in $\mathscr{H}$. But the graph transformations for $\mathfrak{A}$ commute with every operator in $\mathfrak{A}$, and therefore $A(T - \lambda) x = (T - \lambda) A x$. Thus $\{(T - \lambda) A x : A \in \mathfrak{A}\}$ would be dense in $\mathscr{H}$, contradicting the assumption that the range of $T - \lambda$ is not dense. We conclude that $T = \lambda$. $\square$

The next two lemmas deal with certain properties of $\mathscr{H}^2(\mathscr{K})$ which will be needed. The notation is as in Chapter 3.

Lemma 8.16. *Suppose that $F \in \mathscr{F}_0$ and $\mathscr{N}_1$ and $\mathscr{N}_2$ are subspaces of $\mathscr{K}$ with $\dim \mathscr{N}_1 > \dim \mathscr{N}_2$ such that $F(z) \mathscr{N}_1 = \mathscr{N}_2$ for almost all z. Then there is a non-zero $f \in \mathscr{H}^2(\mathscr{N}_1)$ such that $\hat{F} f = 0$.*

Proof. Let $m = \dim \mathscr{N}_2$; $m < \infty$. Let $\{x_n\}$ be an orthonormal basis for $\mathscr{N}_1$ and $\{y_n\}_{n=1}^m$ be an orthonormal basis for $\mathscr{N}_2$. Choose a fixed representative F of its equivalence class. Suppose that, for a particular z, $F(z) \mathscr{N}_1 = \mathscr{N}_2$. There must exist m vectors $\{x_{n(1)}, \ldots, x_{n(m)}\}$ from the basis for $\mathcal{N}_1$ such that $\{F(z)x_{n(i)} : i = 1, \ldots, m\}$ is a basis for $\mathcal{N}_2$; (simply choose a maximal collection $\{x_{n(i)}\}$ of x_n's such that $\{F(z)x_{n(i)}\}$ is linearly independent). Then the determinant of the matrix

$$
\begin{pmatrix}
(F(z)x_{n(1)}, y_1) & (F(z)x_{n(1)}, y_2) & \cdots & (F(z)x_{n(1)}, y_m) \\
(F(z)x_{n(2)}, y_1) & (F(z)x_{n(2)}, y_2) & \cdots & (F(z)x_{n(2)}, y_m) \\
\vdots & \vdots & & \vdots \\
(F(z)x_{n(m)}, y_1) & (F(z)x_{n(m)}, y_2) & \cdots & (F(z)x_{n(m)}, y_m)
\end{pmatrix}
$$

is different from 0. Since $\{x_n\}$ has only a countable number of subsets of m elements, and since $F(z)\mathcal{N}_1 = \mathcal{N}_2$ a.e., it follows that there exist vectors $\{x_{n(1)}, \ldots, x_{n(m)}\}$ such that the above matrix has non-zero determinant for z in some set of positive measure. Let $M(z)$ denote such a matrix.

For each z let $\delta(z)$ denote the determinant of $M(z)$. Then $\delta(z) \in \mathcal{H}^\infty$, since each of the functions $z \to (F(z)x_{n(i)}, y_i)$ is in $\mathcal{H}^\infty$, and Corollary 3.10 implies that, for almost every z, $\delta(z) \neq 0$.

Now, for each z, let $N(z)$ denote the classical adjoint of $M(z)$. Then $N(z)M(z) = \delta(z)I$, where I is the $m \times m$ identity matrix. We now regard $M(z)$ and $N(z)$ as the operators from $\bigvee_{i=1}^m \{x_{n(i)}\}$ to $\mathcal{N}_2$ and $\mathcal{N}_2$ to $\bigvee_{i=1}^m \{x_{n(i)}\}$ whose matrices they represent relative to the bases $\{x_{n(i)}\}_{i=1}^m$ and $\{y_i\}_{i=1}^m$.

Choose a vector $x_{n(0)}$ in $\{x_n\}$ with $n(0)$ different from $n(1), \ldots, n(m)$, and define the function f by $f(z) = N(z)F(z)x_{n(0)} - \delta(z)x_{n(0)}$. Then $f \in \mathcal{H}^2(\mathcal{N}_1)$, and f is not 0 since $N(z)$ takes $F(z)x_{n(0)}$ into $\bigvee_{i=1}^m \{x_{n(i)}\}$ and $x_{n(0)} \notin \bigvee_{i=1}^m \{x_{n(i)}\}$. Also

$$
\begin{aligned}
F(z)f(z) &= (F(z)N(z))F(z)x_n - F(z)\delta(z)x_n \\
&= (M(z)N(z))F(z)x_n - F(z)\delta(z)x_n \\
&= \delta(z)F(z)x_n - \delta(z)F(z)x_n = 0. \quad \square
\end{aligned}
$$

Lemma 8.17. *If* $F \in \mathcal{F}_0$ *with the range of* $\hat{F}$ *dense in* $\mathcal{H}^2(\mathcal{K})$, *and if* $F \in \mathcal{B}(\mathcal{K})$ *is defined by*

$$
(Fx, y) = \int (F(z)x, y)\,d\mu
$$

for $x, y \in \mathcal{K}$, *then the range of* F *is dense in* $\mathcal{K}$.

Proof. We first show that $f \in \mathcal{H}^2(\mathcal{K})$ and $x \in \mathcal{K}$ implies

$$\int (F(z)f(z), x) d\mu = \int (Ff(z), x) d\mu.$$

This is the case since, if we write $f = \sum_{n=0}^{\infty} x_n e_n$ (Fourier expansion), then

$$\int (F(z)f(z), x) d\mu = \sum_{n=0}^{\infty} \int (F(z)x_n, x) z^n d\mu.$$

Since the function $z \to (F(z)x_n, x)$ is in $\mathcal{H}^2$, $\int (F(z)x_n, x) z^n d\mu = 0$ for $n > 0$. Thus

$$\int (F(z)f(z), x) d\mu = \int (F(z)x_0, x) d\mu = (Fx_0, x).$$

Also

$$\int (Ff(z), x) d\mu = \sum_{n=0}^{\infty} \int (Fx_n, x) z^n d\mu = (Fx_0, x).$$

Now if x is orthogonal to the range of F, then the above equation shows that $\hat{F}f$ is orthogonal to $x e_0$ for all $f \in \mathcal{H}^2(\mathcal{K})$. The result follows. □

Theorem 8.18. *The only transitive operator algebra which contains a unilateral shift of finite multiplicity is* $\mathcal{B}(\mathcal{H})$.

Proof. Let $\mathfrak{A}$ be a transitive algebra containing a shift and let $\mathcal{M} = \{x \oplus T_1 x \oplus \cdots \oplus T_{n-1} x : x \in \mathcal{D}\}$ be an invariant graph subspace for $\mathfrak{A}^{(n)}$. By Lemma 8.15 it is sufficient to prove that each T_i has compression spectrum.

Let S be a unilateral shift of finite multiplicity k which is contained in $\mathfrak{A}$; we represent $\mathcal{H}$ as $\mathcal{H}^2(\mathcal{K})$ so that S is multiplication by the function e_1. Then $\mathcal{H}^{(n)} = (\mathcal{H}^2(\mathcal{K}))^{(n)}$. There is an obvious identification of $(\mathcal{H}^2(\mathcal{K}))^{(n)}$ with $\mathcal{H}^2(\mathcal{K}^{(n)})$, so that $S^{(n)}$ is multiplication by e_1 on $\mathcal{H}^2(\mathcal{K}^{(n)})$.

Now, since $\mathcal{M}$ is invariant under the shift $S^{(n)}$, by Corollary 3.26 there exist a subspace $\mathcal{N}$ of $\mathcal{K}^{(n)}$ and a function $V \in \mathcal{F}_0$ such that $V(z)$ is a partial isometry with initial space $\mathcal{N}$ for almost all z and such that $\mathcal{M} = \hat{V} \mathcal{H}^2(\mathcal{N})$. For each j between 1 and n, let P_j denote the projection of $\mathcal{K}^{(n)}$ onto its jth coordinate space $\mathcal{K}_j$. Define the function $F_j \in \mathcal{F}_0$ by $F_j(z) = P_j V(z)$ for all z. Then

$$\mathcal{M} = \{\hat{F}_1 f \oplus \hat{F}_2 f \oplus \cdots \oplus \hat{F}_n f : f \in \mathcal{H}^2(\mathcal{N})\}.$$

Note that $\hat{F}_j f = T_{j-1} \hat{F}_1 f$ for $j = 2, \ldots, n$.

We now prove that $\dim \mathcal{N} = k$. The range of $\hat{F}_1$ is $\mathcal{D}$, which is dense in $\mathcal{H}^2(\mathcal{K})$. Therefore for each $x \in \mathcal{K}_1$ there is a sequence $\{f_n\} \subset \mathcal{H}^2(\mathcal{N})$ such that $\{\hat{F}_1 f_n\} \to x e_0$, or

$$\lim_{n \to \infty} \int \| F_1(z) f_n(z) - x \|^2 d\mu = 0.$$

Thus some subsequence of $\{F_1(z)f_n(z)\}$ converges to x almost everywhere. That is, each vector x in $\mathscr{K}_1$ is in the closure of the range of $F_1(z)$ for almost all z. Since the range of $F_1(z)$ is contained in $\mathscr{K}_1$, which is finite-dimensional, the range of $F_1(z)$ is closed and hence coincides with $\mathscr{K}_1$ for almost all z. Since the operator $F_1(z)$ is 0 on $\mathscr{N}^\perp$ this shows that $\dim \mathscr{N} \geqq \dim \mathscr{K}_1 = k$.

To see that $\dim \mathscr{N}$ cannot be greater than k, first observe that $\hat{F}_1 f = 0$ for $f \in \mathscr{H}^2(\mathscr{N})$ implies $\hat{F}_j f = T_j \hat{F}_1 f = 0$ for all j, which shows that $\hat{V} f = 0$ and thus $f = 0$. Now if $\dim \mathscr{N} > k$, then Lemma 8.16 would imply that there is some non-zero $f \in \mathscr{H}^2(\mathscr{N})$ such that $\hat{F}_1 f = 0$, which would contradict this fact.

Since $\dim \mathscr{N} = \dim \mathscr{K}$ we can assume that $\mathscr{N} = \mathscr{K}$; (simply replace $\hat{V}$ by $\hat{V}\hat{W}$ where $\hat{W}$ is a constant partial isometry on $\mathscr{K}^{(n)}$ with initial space $\mathscr{K}$ and final space $\mathscr{N}$). Now identify $\mathscr{K}_j$ with $\mathscr{K}$, in the obvious way, for all j, and define the operator F_j on $\mathscr{K}$ to be the integral of F_j as in Lemma 8.17. Since the range of $\hat{F}_1$ is dense in $\mathscr{H}^2(\mathscr{K})$, Lemma 8.17 shows that the range of F_1 is dense in $\mathscr{K}$. Since $\mathscr{K}$ is finite-dimensional, F_1 is invertible. For $j = 1, \ldots, n-1$ let λ_j be an eigenvalue of $F_{j+1} F_1^{-1}$. Then the range of $F_{j+1} - \lambda_j F_1$ is not dense, and, by Lemma 8.17, the range of $\hat{F}_{j+1} - \lambda_j \hat{F}_1$ is not dense in $\mathscr{H}^2(\mathscr{K})$. The domain $\mathscr{D}$ of the graph transformations is the range of F_1, and $T_j = F_{j+1} \hat{F}^{-1}$ on $\mathscr{D}$. Thus the range of $T_j - \lambda_j$ consists of all vectors of the form

$$(\hat{F}_{j+1} \hat{F}_1^{-1} - \lambda_j) \hat{F}_1 f = (\hat{F}_{j+1} - \lambda_j \hat{F}_1) f,$$

and it follows that T_j has compression spectrum. ☐

Corollary 8.19. *If A is not a multiple of the identity, and if A has the form*

$$A = \begin{pmatrix} A_{11} & A_{12} & \cdots & A_{1n} \\ A_{21} & A_{22} & \cdots & A_{2n} \\ \vdots & & & \vdots \\ A_{n1} & A_{n2} & \cdots & A_{nn} \end{pmatrix}$$

where each A_{ij} is an analytic Toeplitz operator, then A has a non-trivial hyperinvariant subspace.

Proof. Let $\mathfrak{A}$ be the commutant of $\{A\}$; we must prove that $\mathfrak{A}$ is not transitive. If S is the unilateral shift of multiplicity 1, then $S^{(n)}$, a unilateral shift of multiplicity n, is in $\mathfrak{A}$. Thus if $\mathfrak{A}$ were transitive, Theorem 8.18 would imply that $\mathfrak{A} = \mathscr{B}(\mathscr{H})$. But $\mathfrak{A} \neq \mathscr{B}(\mathscr{H})$ since A is not a multiple of the identity. ☐

We shall generalize Theorem 8.10 to the case where $\mathfrak{A}$ contains $\mathfrak{R}^{(n)}$ for some m.a.s.a. $\mathfrak{R}$ and positive integer n. For this we require a lemma about $Lat\,\mathfrak{R}^{(n)}$.

Definition. The projection P on $\mathcal{H}^{(n)}$ is *special* if $P = P_1 \oplus P_2 \oplus \cdots \oplus P_n$ where each P_i is either the operator 0 or the operator 1 on $\mathcal{H}$. What is "special" about special projections is their relation to the decomposition $\mathcal{H} \oplus \cdots \oplus \mathcal{H}$ of $\mathcal{H}^{(n)}$.

We shall often confuse $P\mathcal{H}^{(n)}$ with the corresponding $\mathcal{H}^{(k)}$, where k denotes the number of those P_i's in the expression for P which are equal to 1.

Lemma 8.20. *If $\mathfrak{R}$ is a m.a.s.a., n a positive integer, $\mathcal{M} \in \operatorname{Lat} \mathfrak{R}^{(n)}$, and $\mathcal{U}$ a strong neighbourhood of the identity on $\mathcal{H}$, then there exists a projection $Q \in \mathcal{U} \cap \mathfrak{R}$ such that $PQ^{(n)}\mathcal{M}$ is closed for every special projection P on $\mathcal{H}^{(n)}$.*

Proof. By Corollary 7.14 we can assume that $\mathfrak{R}$ is $\mathscr{L}^{\infty}$ on the space $\mathscr{L}^2(X, \mu)$ for some finite measure space (X, μ). We proceed by induction on n. The result is trivially true for $n = 1$ because the only special projections in this case are 0 and 1; $Q = 1$ will thus serve for any $\mathcal{U}$.

We now consider the case $n = 2$. There is really no necessity for considering this case separately since the general inductive step below applies here too. We present this case first to show the essence of the proof without too many technical details. Let $\mathcal{M} \in \operatorname{Lat}(\mathscr{L}^{\infty})^{(2)}$, and let $\mathcal{U} = \{C : \|Cf_i - f_i\| < \varepsilon, \ i = 1, \ldots, k\}$ be a neighbourhood of the identity. Let $\mathcal{N}$ be the second coordinate space of $\mathcal{M}$; i.e., $\mathcal{N} = \{0 \oplus f : (0 \oplus f) \in \mathcal{M}\}$. Then $\mathcal{N}$ and $\mathcal{M} \ominus \mathcal{N}$ are in $\operatorname{Lat}(\mathscr{L}^{\infty})^{(2)}$. If $(0 \oplus f) \in (\mathcal{M} \ominus \mathcal{N})$ then $f = 0$. Hence $\mathcal{M} \ominus \mathcal{N} = \{g \oplus Tg : g \in \mathcal{D}\}$ for some invariant linear manifold $\mathcal{D}$ of $\mathscr{L}^{\infty}$ and some linear transformation T. Now $\overline{\mathcal{D}}$ is an invariant subspace of $\mathscr{L}^{\infty}$, and therefore there is a measurable set E such that $\overline{\mathcal{D}} = \{f \in \mathscr{L}^2(X, \mu) : f = 0 \text{ a.e. on } X \backslash E\}$; (cf. Example 4.11). Let P_E denote the projection of $\mathscr{L}^2(X, \mu)$ onto $\overline{\mathcal{D}}$; i.e., P_E is multiplication by the characteristic function of E, and, in particular, $P_E \in \mathscr{L}^{\infty}$. If $(g \oplus Tg) \in (\mathcal{M} \ominus \mathcal{N})$, then $(1 - P_E)^{(2)}(g \oplus Tg) \in (\mathcal{M} \ominus \mathcal{N})$, and thus $0 \oplus (1 - P_E)Tg \in (\mathcal{M} \ominus \mathcal{N})$. It follows that $(1 - P_E)Tg = 0$. Thus the range of T is contained in $\overline{\mathcal{D}}$. Now $\overline{\mathcal{D}}$ has an obvious identification with $\mathscr{L}^2(E, \mu)$, and T is a densely defined linear transformation on $\mathscr{L}^2(E, \mu)$ which commutes with $\{M_\psi : \psi \in \mathscr{L}^{\infty}(E, \mu)\}$.

By Lemma 7.15 there exists a measurable function ϕ on (E, μ) such that $Tf = \phi f$ for $f \in \mathcal{D}$. As in the proof of Theorem 7.16, the fact that $\bigcup_{m=1}^{\infty} \{x : |\phi(x)| \le m\} = E$ implies that there exists a measurable subset F of E such that ϕ is bounded on F and $\int_{E \backslash F} |f_i|^2 \, d\mu < \varepsilon^2/4$ for $i = 1, \ldots, k$.

Similarly, the fact that $E = \left(\bigcup_{m=1}^{\infty} \{x : |\phi(x)| \ge 1/m\}\right) \cup \{x : \phi(x) = 0\}$ implies that, for some m, there exists a measurable subset G of E such that

$\int\limits_{E\setminus G} |f_i|^2 \, d\mu < \varepsilon^2/4$ for $i=1,\dots,k$ and such that $\phi(x)=0$ or $|\phi(x)|\geq 1/m$
for all $x\in G$. ∎

Now let Q be multiplication by the characteristic function of $X\setminus(F\cup G)$. Then $Q\in\mathscr{L}^\infty$, and we claim that Q satisfies the conditions of the lemma. By construction $Q\in\mathscr{U}\cap\mathfrak{R}$. We must show that $PQ^{(2)}\mathscr{M}$ is closed for each special projection P. There are only two non-trivial special projections in this case, the projection P_1 onto the first coordinate space and the projection P_2 onto the second coordinate space. Since $Q\in\mathscr{L}^\infty$, $Q^{(2)}\mathscr{K}\in Lat(\mathscr{L}^\infty)^{(2)}$ whenever $\mathscr{K}\in Lat(\mathscr{L}^\infty)^{(2)}$. Also $\mathscr{M}=(\mathscr{M}\ominus\mathscr{N})\oplus\mathscr{N}$ implies $Q^{(2)}\mathscr{M}=Q^{(2)}(\mathscr{M}\ominus\mathscr{N})\oplus Q^{(2)}\mathscr{N}$. Now $P_1 Q^{(2)}\mathscr{M}=P_1 Q^{(2)}(\mathscr{M}\ominus\mathscr{N})=\{f\in\mathscr{L}^2(X,\mu): f=0 \text{ a.e. on } (X\setminus E)\cup F\cup G\}$, and $P_1 Q^{(2)}\mathscr{M}$ is closed.

Note that $P_2(\mathscr{M}\ominus\mathscr{N})$ is orthogonal to $P_2\mathscr{N}$, and thus $P_2 Q^{(2)}(\mathscr{M}\ominus\mathscr{N})$ is orthogonal to $P_2 Q^{(2)}\mathscr{N}$. Therefore

$$P_2 Q^{(2)}\mathscr{M} = P_2 Q^{(2)}(\mathscr{M}\ominus\mathscr{N})\oplus P_2 Q^{(2)}\mathscr{N}.$$

Clearly $P_2 Q^{(2)}\mathscr{N}=\{f:f=Qf \text{ and } 0\oplus f\in\mathscr{M}\}$ is closed, and we need only consider $P_2 Q^{(2)}(\mathscr{M}\ominus\mathscr{N})$. By construction $P_2 Q^{(2)}(\mathscr{M}\ominus\mathscr{N})=\{\phi g\in\mathscr{L}^2(E,\mu):g=0 \text{ a.e. on } F\cup G\}$. Since $x\in(E\setminus(F\cup G))$ implies $|\phi(x)|\notin(0,(1/m))$ the function Ψ defined by

$$\psi(x)=\begin{cases} \dfrac{1}{\phi(x)} & \text{for } x\in E\setminus(F\cup G\cup\{x:\phi(x)=0\}) \\ 0 & \text{for } x\in(F\cup G\cup\{x:\phi(x)=0\}) \end{cases}$$

is in $\mathscr{L}^\infty(E,\mu)$. Thus if the sequence $\{\phi g_m\}$ in $P_2 Q^{(2)}(\mathscr{M}\ominus\mathscr{N})$ converges to $h\in\mathscr{L}^2(E,\mu)$, it follows that $\{\psi\phi g_m\}$ converges to ψh, and $h=\phi(\psi h)$ is in $P_2 Q^{(2)}(\mathscr{M}\ominus\mathscr{N})$. This completes the proof in the case $n=2$.

The proof of the general inductive step is similar to the case $n=2$ but somewhat more involved. Assume the lemma known in the case $n-1$ and consider any invariant subspace $\mathscr{M}$ of $(\mathscr{L}^\infty)^{(n)}$. To simplify notation we shall say that a projection Q with certain stated properties is "arbitrarily close to 1" if for every strong neighbourhood $\mathscr{U}$ of 1 there exists a Q in $\mathscr{U}$ with the stated properties.

Note that if Q_1 and Q_2 are arbitrarily close to 1, and if $\mathscr{U}=\{C:\|Cf_i-f_i\|<\varepsilon, i=1,\dots,k\}$, then choosing a particular Q_1 such that $\|Q_1 f_i-f_i\|<\varepsilon/2$ and a particular Q_2 such that $\|Q_2 Q_1 f_i-Q_1 f_i\|<\varepsilon/2$ shows that $Q_2 Q_1$ is arbitrarily close to 1. Thus the product of any finite number of operators which are arbitrarily close to 1 is also arbitrarily close to 1. Note also that if P is a special projection and $\mathscr{K}$ is in $Lat(\mathscr{L}^\infty)^{(n)}$ with $P\mathscr{K}$ closed, then $PQ^{(n)}\mathscr{K}$ is closed whenever Q is a projection in $\mathscr{L}^\infty$; (this follows directly from the fact that $PQ^{(n)}$

$= Q^{(n)} P$ is a projection). Also, for each fixed n, there is a finite number (2^n) of special projections on $\mathscr{H}^{(n)}$. These remarks show that it suffices to prove that for each given special projection P there is a $Q \in \mathscr{L}^\infty$ arbitrarily close to 1 such that $P Q^{(n)} \mathscr{M}$ is closed.

Fix a special projection P. If P is the identity on $\mathscr{H}^{(n)}$ take $Q = 1$; otherwise $P = P_1 \oplus \cdots \oplus P_n$ with at least one $P_i = 0$. By re-indexing if necessary we can assume that $P_1 = 0$. Let R denote the projection of $\mathscr{H}^{(n)}$ onto the orthocomplement of its first component; i.e.,

$$R(g_1 \oplus g_2 \oplus \cdots \oplus g_n) = 0 \oplus g_2 \oplus \cdots \oplus g_n.$$

Since $P_1 = 0$, $P \mathscr{M} = P R \mathscr{M}$. The proof would be completed if we knew that there exists a $Q_0 \in \mathscr{L}^\infty$ arbitrarily close to 1 such that $R Q_0^{(n)} \mathscr{M}$ is closed. This suffices because, using the obvious identification of $R Q_0^{(n)} \mathscr{M}$ with an invariant subspace of $(\mathscr{L}^\infty)^{(n-1)}$, the induction hypothesis implies that $P R Q^{(n)} Q_0^{(n)} \mathscr{M}$ is closed for some $Q \in \mathscr{L}^\infty$ arbitrarily close to 1, and $P R Q^{(n)} Q_0^{(n)} \mathscr{M} = P(Q Q_0)^{(n)} \mathscr{M}$. The rest of the proof, therefore, is devoted to proving that $R Q_0^{(n)} \mathscr{M}$ is closed for some $Q_0 \in \mathscr{L}^\infty$ arbitrarily close to 1.

Let $\mathscr{N} = R \mathscr{H}^{(n)} \cap \mathscr{M} = \{(g_1 \oplus g_2 \oplus \cdots \oplus g_n) \in \mathscr{M} : g_1 = 0\}$. Then $\mathscr{M} \ominus \mathscr{N} \in \mathrm{Lat}(\mathscr{L}^\infty)^{(n)}$, and $R(\mathscr{M} \ominus \mathscr{N})$ is orthogonal to $R \mathscr{N}$. Since $R Q^{(n)} \mathscr{N}$ is obviously closed for each $Q \in \mathscr{L}^\infty$, it suffices to show that $R Q_0^{(n)}(\mathscr{M} \ominus \mathscr{N})$ is closed for some suitable Q_0.

Let

$$\mathscr{K} = [\mathscr{M} \ominus \mathscr{N}] \ominus \{(g_1 \oplus g_2 \oplus \cdots \oplus g_n) \in \mathscr{M} \ominus \mathscr{N} : g_i = 0 \text{ for all } i > 1\}.$$

Then $R Q^{(n)}(\mathscr{M} \ominus \mathscr{N}) = R Q^{(n)} \mathscr{K}$ for all $Q \in \mathscr{L}^\infty$, and therefore we need only show that $R Q_0^{(n)} \mathscr{K}$ is closed for some $Q_0 \in \mathscr{L}^\infty$ arbitrarily close to 1. The special form of $\mathscr{K}$ will enable us to do this.

Since the first component of each vector in $\mathscr{K}$ determines all the other components, there is a linear manifold $\mathscr{D}$ invariant under $\mathscr{L}^\infty$ and linear transformations T_i on $\mathscr{D}$ such that

$$\mathscr{K} = \{f \oplus T_1 f \oplus \cdots \oplus T_{n-1} f : f \in \mathscr{D}\}.$$

As in the proof of the case $n = 2$,

$$\bar{\mathscr{D}} = \{f \in \mathscr{L}^2(X, \mu) : f = 0 \text{ a.e. on } X \backslash E\}$$

for some measurable set E. If P_E denotes the projection onto $\bar{\mathscr{D}}$, then $(1 - P_E)^{(n)} \mathscr{K} \subset \mathscr{K}$, and it follows that the range of T_i is contained in $\bar{\mathscr{D}}$ for all i. Thus each T_i is multiplication by some measurable function ϕ_i on $\mathscr{L}^2(E, \mu)$. As in the case $n = 2$ there exist integers p and m and a measurable subset F of E such that multiplication by the characteristic function of $X \backslash F$ is arbitrarily close to 1 and such that $x \in (X \backslash F)$ implies $|\phi_i(x)| \leq p$ and $|\phi_i(x)| \notin (0, 1/m)$ for $i = 1, \ldots, n-1$.

Let Q_0 be multiplication by the characteristic function of $X\setminus F$; we claim that $R Q_0^{(n)} \mathcal{K}$ is closed. Now

$$R Q_0^{(n)} \mathcal{K} = \{\phi_1 f \oplus \cdots \oplus \phi_{n-1} f : f = 0 \text{ a.e. on } F \cup (X\setminus E)\}.$$

Note that the fact that $g \oplus 0 \oplus 0 \oplus \cdots \oplus 0 \in \mathcal{K}$ only for $g = 0$ implies

$$\mu\left(\bigcap_{i=1}^{n-1} \{x \in E : \phi_i(x) = 0\}\right) = 0. \text{ For each } i \text{ let } E_i = \{x \in E\setminus F : \phi_i(x) \neq 0\}, \text{ so}$$

that $\mu\left((E\setminus F)\setminus \bigcup_{i=1}^{n-1} E_i\right) = 0.$ Suppose that the sequence $\{\phi_1 f_k \oplus \cdots \oplus \phi_{n-1} f_k\}$

in $R Q_0^{(n)} \mathcal{K}$ converges to $g_1 \oplus \cdots \oplus g_{n-1}$; we must prove that $g_1 \oplus \cdots \oplus g_{n-1}$ is in $R Q_0^{(n)} \mathcal{K}$. Recall that for each i, $|\phi_i(x)| \geq 1/m$ for $x \in E_i$; it follows that $f_k | E_i$ converges to $(g_i | E_i)/(\phi_i | E_i)$ in $\mathcal{L}^2(E_i, \mu)$.

Since $\mu\left((E\setminus F)\setminus \bigcup_{i=1}^{n-1} E_i\right) = 0$, this implies that $\{f_k\}$ converges to some f

in $\mathcal{L}^2(E\setminus F, \mu)$. Extend f by defining $f \equiv 0$ on $F \cup (X\setminus E)$. Then $\{\phi_i f_k\} \to \phi_i f$ for each i. Hence $g_i = \phi_i f$ for all i, and $g_1 \oplus \cdots \oplus g_{n-1}$ is in $R Q_0^{(n)} \mathcal{K}$. This proves the lemma. $\square$

This lemma will be needed in proving a result about certain operator algebras containing m.a.s.a.'s (Theorem 9.15) as well as in proving the following generalization of Theorem 8.10.

Theorem 8.21. *If $\mathfrak{A}$ is a transitive operator algebra on $\mathcal{H}$ which contains $\mathfrak{R}^{(k)}$ for some m.a.s.a. $\mathfrak{R}$ and some positive integer k, then $\mathfrak{A} = \mathcal{B}(\mathcal{H})$.*

Proof. By Lemma 8.8 we need only show that every graph transformation for $\mathfrak{A}$ is a multiple of the identity. Let $\mathcal{M}$ be an invariant graph subspace for $\mathfrak{A}^{(n)}$:

$$\mathcal{M} = \{x \oplus T_1 x \oplus \cdots \oplus T_{n-1} x : x \in \mathcal{D}\}.$$

Now let $\mathcal{K}$ be the space on which $\mathfrak{R}$ acts; then identify $\mathcal{H}$ with $\mathcal{K}^{(k)}$. The space $\mathcal{H}^{(n)}$, in which $\mathcal{M}$ lies, is then identified with $(\mathcal{K}^{(k)})^{(n)} = \mathcal{K}^{(kn)}$. Since $\mathfrak{R}^{(k)} \subset \mathfrak{A}$, $R^{(kn)}$ is contained in $\mathfrak{A}^{(n)}$, and $\mathcal{M} \in Lat\, \mathfrak{R}^{(kn)}$. By Lemma 8.20, for each strong neighbourhood $\mathcal{U}_0$ of the identity on $\mathcal{K}$ there exists a $Q_0 \in \mathcal{U}_0 \cap \mathfrak{R}$ such that $Q_0^{(kn)} \mathcal{M}$ has the property that $P Q_0^{(kn)} \mathcal{M}$ is closed for every special projection P on $\mathcal{K}^{(kn)}$. In particular, $P(Q_0^{(k)})^{(n)} \mathcal{M}$ is closed for every special projection P on $\mathcal{H}^{(n)}$. Let $Q = Q_0^{(k)}$ for such a Q_0. If we show that, for each such Q, $Q\mathcal{D} \subset \mathcal{D}$ and $Q T_i Q$ is a multiple of Q, then it will follow that T_i is a multiple of 1, since T_i commutes with Q.

Fix such a Q. Then $Q^{(n)} \mathcal{M} = \{Qx \oplus Q T_1 x \oplus \cdots \oplus Q T_{n-1} x : x \in \mathcal{D}\}$. Since $Q \in \mathfrak{A}$, $Q^{(n)} \mathcal{M} \subset \mathcal{M}$ and it follows that $Q\mathcal{D} \subset \mathcal{D}$ and $Q T_i = T_i Q = Q T_i Q$. Also, since $P Q^{(n)} \mathcal{M}$ is closed for $P = 1 \oplus 0 \oplus \cdots \oplus 0$, $Q\mathcal{D}$ is a closed subspace of $\mathcal{H}$ and

$$Q^{(n)} \mathcal{M} = \{x \oplus Q T_1 x \oplus \cdots \oplus Q T_{n-1} x : x \in Q\mathcal{D}\}.$$

For each i, the graph $\{x \oplus Q T_i x : x \in Q \mathcal{D}\}$ of $Q T_i$ is closed, and the closed graph theorem implies that $Q T_i$ is a bounded operator on $Q \mathcal{D}$.

Since $\mathcal{M} \in Lat\, \mathfrak{A}^{(n)}$, $Q^{(n)} \mathcal{M}$ is obviously in $Lat(Q^{(n)} \mathfrak{A}^{(n)} | Q^{(n)} \mathcal{H}^{(n)})$ $= Lat(Q \mathfrak{A} | Q \mathcal{H})^{(n)}$. Thus, for every i, $Q A Q$ commutes with $Q T_i Q$ for each $A \in \mathfrak{A}$. In particular, $Q T_i Q$ commutes with $Q R^{(k)} Q = (Q_0 R)^{(k)}$ for each $R \in \mathfrak{R}$. Now $\{Q_0 R : R \in \mathfrak{R}\}$ is a m.a.s.a. on $Q_0 \mathcal{K}$, and thus $Q T_i Q | (Q \mathcal{H})$ is a k-normal operator. If $Q T_i Q | (Q \mathcal{H})$ is not a multiple of the identity on $Q \mathcal{H}$, then, by Theorem 7.21, $Q T_i Q | (Q \mathcal{H})$ has a non-trivial hyperinvariant subspace $\mathcal{N}$. Then $\mathcal{N} \in Lat(Q A Q | Q \mathcal{H})$ for all $A \in \mathfrak{A}$. Let x and y be non-zero vectors in $Q \mathcal{H}$, with $x \in \mathcal{N}$, $y \in Q \mathcal{H} \ominus \mathcal{N}$. Then, for any $A \in \mathfrak{A}$, $(Ax, y) = (A Q x, Q y) = (Q A Q x, y) = 0$. Hence the closure of $\{Ax : A \in \mathfrak{A}\}$ is a non-trivial invariant subspace of $\mathfrak{A}$, contradicting transitivity. It follows that $Q T_i Q | (Q \mathcal{H})$ is a multiple of the identity operator on $Q \mathcal{H}$, or $Q T_i Q$ is a multiple of Q. Since this is true for Q's arbitrarily close to 1, it follows that each T_i is a multiple of the identity. $\square$

The following remarkable lemma leads easily to the result that a transitive algebra containing a non-zero compact operator is $\mathcal{B}(\mathcal{H})$.

Lemma 8.22 (*Lomonosov's Lemma*). *If $\mathfrak{A}$ is a subalgebra of $\mathcal{B}(\mathcal{H})$ (not necessarily closed in any topology) with $Lat\, \mathfrak{A} = \{\{0\}, \mathcal{H}\}$, and if $K \neq 0$ is any compact operator on $\mathcal{H}$, then there is an $A \in \mathfrak{A}$ such that $1 \in \Pi_0(A K)$.*

Proof. The idea of the proof is to construct a function ψ of the form $\psi(x) = \sum_{i=1}^{n} \gamma_i(x) A_i K x$ with $A_i \in \mathfrak{A}$ which maps a compact convex subset of $\mathcal{H}$ into itself. Then the Schauder fixed point theorem gives a point x such that $\left[\sum_{i=1}^{n} \gamma_i A_i \right] K x = x$ for some $x \neq 0$ and some scalars $\{\gamma_i\}$.

We begin by assuming, without loss of generality, that $\|K\| = 1$. Choose any $x_0 \in \mathcal{H}$ such that $\|K x_0\| > 1$, (so that, in particular, $\|x_0\| > 1$), and let $\mathcal{S} = \{x : \|x - x_0\| \leq 1\}$. For each $A \in \mathfrak{A}$ let $\mathcal{U}(A) = \{y : \|A y - x_0\| < 1\}$. Since every non-zero vector is a cyclic vector for $\mathfrak{A}$, it follows that

$$\bigcup_{A \in \mathfrak{A}} \mathcal{U}(A) = \mathcal{H} \setminus \{0\}.$$

Since K is compact, $\overline{K \mathcal{S}}$ is a compact subset of $\mathcal{H}$. Note that $\|K\| = 1$ and $\|K x_0\| > 1$ implies $0 \notin \overline{K \mathcal{S}}$. Hence $\bigcup_{A \in \mathfrak{A}} \mathcal{U}(A)$ contains $\overline{K \mathcal{S}}$, and, since each $\mathcal{U}(A)$ is open, there exist operators $\{A_1, \ldots, A_n\} \subset \mathfrak{A}$ such that $\overline{K \mathcal{S}} \subset \bigcup_{i=1}^{n} \mathcal{U}(A_i)$.

Now for $y \in \overline{K\mathcal{S}}$ and $i=1,\ldots,n$ define

$$\alpha_i(y) = \max\{0, 1 - \|A_i y - x_0\|\} \,.$$

Then $0 \leqq \alpha_i(y) \leqq 1$, and for each $y \in \overline{K\mathcal{S}}$ there is an i such that $y \in \mathcal{U}(A_i)$, which implies $\alpha_i(y) > 0$. Thus $\sum\limits_{i=1}^{n} \alpha_i(y) > 0$ for all $y \in \overline{K\mathcal{S}}$, and we can define

$$\beta_i(y) = \frac{\alpha_i(y)}{\sum\limits_{j=1}^{n} \alpha_j(y)}$$

for $i=1,\ldots,n$ and $y \in \overline{K\mathcal{S}}$. Each β_i is a continuous function from $\overline{K\mathcal{S}}$ (with the norm topology) into the unit interval. Therefore the function ψ defined by $\psi(x) = \sum\limits_{i=1}^{n} \beta_i(Kx) A_i Kx$ is a continuous function from $\mathcal{S}$ into $\mathcal{H}$.

We want to show that $\psi(\mathcal{S}) \subset \mathcal{S}$. First observe that, for each $x \in \mathcal{S}$, $\sum\limits_{i=1}^{n} \beta_i(Kx) = 1$. Thus $x \in \mathcal{S}$ implies

$$\|\psi(x) - x_0\| = \left\| \sum_{i=1}^{n} \beta_i(Kx) A_i Kx - \sum_{i=1}^{n} \beta_i(Kx) x_0 \right\|$$

$$= \left\| \sum_{i=1}^{n} \beta_i(Kx)[A_i Kx - x_0] \right\|$$

$$\leqq \sum_{i=1}^{n} \beta_i(Kx) \|A_i Kx - x_0\| \,.$$

Now $\alpha_i(Kx) = 0$ whenever $\|A_i Kx - x_0\| \geqq 1$. Therefore

$$\|\psi(x) - x_0\| \leqq \sum_{i=1}^{n} \beta_i(Kx) = 1, \quad \text{and} \quad \psi(x) \in \mathcal{S}.$$

The set $\mathcal{S}$ is closed and convex, but it is not compact. However, for each i the operator $A_i K$ is compact and therefore $\bigcup\limits_{i=1}^{n} \overline{A_i K\mathcal{S}}$ is compact. It follows by Mazur's theorem (Dunford-Schwartz [1], p. 416) that the closed convex hull $\mathcal{C}$ of $\bigcup\limits_{i=1}^{n} \overline{A_i K\mathcal{S}}$ is also compact. Now $\sum\limits_{i=1}^{n} \beta_i(Kx) = 1$ for $x \in \mathcal{S}$ implies $\psi(\mathcal{S}) \subset \mathcal{C}$. Hence $\psi(\mathcal{S} \cap \mathcal{C}) \subset \mathcal{S} \cap \mathcal{C}$. The set $\mathcal{S} \cap \mathcal{C}$ is convex, and, since it is a closed subset of $\mathcal{C}$, it is also compact. Note that $\psi(\mathcal{S}) \subset \mathcal{S} \cap \mathcal{C}$, so $\mathcal{S} \cap \mathcal{C} \neq \emptyset$.

The Schauder fixed point theorem (cf. Dunford-Schwartz [1], p. 456) states that a continuous function mapping a compact convex subset of a Banach space into itself has a fixed point. Thus $\psi(x) = x$ for some $x \in \mathscr{S}$. Define the operator A by $A = \sum_{i=1}^{n} \beta_i(Kx) A_i$. Then $A \in \mathfrak{A}$, $AKx = x$, and, since $x \neq 0$ $(0 \notin \mathscr{S})$, $1 \in \Pi_0(AK)$.   □

Lomonosov's lemma easily yields remarkable improvements of the existence theorems of Chapter 5 as well as of Theorem 8.12 and Corollary 8.14.

Theorem 8.23. *A transitive algebra which contains a non-zero compact operator is $\mathscr{B}(\mathscr{H})$.*

Proof. Let $\mathfrak{A}$ be a transitive algebra and K be a non-zero compact operator in $\mathfrak{A}$. Choose, by Lomonosov's lemma, an operator $A \in \mathfrak{A}$ such that $1 \in \Pi_0(AK)$. Then AK is a compact operator in $\mathfrak{A}$. Let $\mathscr{S}$ be an open set such that $\mathscr{S} \cap \sigma(AK) = \{1\}$, and let f be the characteristic function of $\mathscr{S}$. Define $f(AK)$ by the Riesz functional calculus; then $f(AK)$ is a (not necessarily self-adjoint) projection onto an invariant subspace $\mathscr{M}$ of AK such that $\sigma((AK)|\mathscr{M}) = \{1\}$, (cf. Theorem 2.10). Since $(AK)|\mathscr{M}$ is compact and $0 \notin \sigma((AK)|\mathscr{M})$ it follows that $\mathscr{M}$ is finite-dimensional, and $f(AK)$ is a finite-rank operator.

Now $f(AK)$ is in the weakly closed algebra generated by 1 and AK. For $\eta(\sigma((AK)^{(n)})) = \sigma((AK)^{(n)})$ for all n (since $(AK)^{(n)}$ is compact), hence $\mathrm{Lat}\,(AK)^{(n)} \subset \mathrm{Lat}(f(AK))^{(n)}$ (by Corollary 2.13), and Theorem 7.1 implies $f(AK)$ is in the weakly closed algebra generated by 1 and AK; (cf. Proposition 8.10).

Thus $\mathfrak{A}$ contains the finite-rank operator $f(AK)$, and the result follows from Theorem 8.12.   □

Corollary 8.24. *Every operator which commutes with a non-zero compact operator and is not a multiple of the identity has a non-trivial hyperinvariant subspace.*

Proof. Let $\mathfrak{A}$ be the commutant of the given operator. Then $\mathfrak{A}$ is not $\mathscr{B}(\mathscr{H})$ since the operator is not a multiple of the identity. Also, $\mathfrak{A}$ contains a compact operator. Hence, by Theorem 8.23, $\mathfrak{A}$ has a non-trivial invariant subspace.   □

Corollary 8.25 (*Lomonosov's Theorem*). *Every non-zero compact operator has a non-trivial hyperinvariant subspace.*

Proof. There is a very easy proof of this special case of Corollary 8.24. Let K be a compact operator; if $\Pi_0(K) \neq \emptyset$ the result is trivial, so assume that $\sigma(K) = \{0\}$. Let $\mathfrak{A} = \{K\}'$, and suppose that $\mathfrak{A}$ is transitive. Find operators $\{A_1, \ldots, A_n\} \subset \mathfrak{A}$ as in the second paragraph of the proof of Lemma 8.22 (page 156), and let $c = \max \{\|A_i\|\}$. Then $A_{i_1} Kx_0 \in \mathscr{S}$

for some i_1, $A_{i_2} K A_{i_1} K x_0 \in \mathscr{S}$ for some i_2, etc. Continuing this procedure k times gives

$$A_{i_k} K \ldots A_{i_2} K A_{i_1} K x_0 \in \mathscr{S} .$$

Since K commutes with each A_i,

$$(c^{-1} A_{i_k}) \ldots (c^{-1} A_{i_2})(c^{-1} A_{i_1})(c K)^k x_0 \in \mathscr{S} ,$$

which contradicts the fact that $(c K)^k \Rightarrow 0$ as $k \to \infty$. □

Definition. If $\mathscr{H}$ is a separable Hilbert space, then the *tensor product* of $\mathscr{H}$ with itself, denoted $\mathscr{H} \otimes \mathscr{H}$, is the space $\sum_{n=1}^{\infty} \oplus \mathscr{H}_n$ with $\mathscr{H}_n = \mathscr{H}$ for all n. If $\mathfrak{A}$ is a weakly closed subalgebra of $\mathscr{B}(\mathscr{H})$, then the *tensor product* of $\mathfrak{A}$ and $\mathscr{B}(\mathscr{H})$, denoted $\mathfrak{A} \otimes \mathscr{B}(\mathscr{H})$, is the set of all operators on $\mathscr{H} \otimes \mathscr{H}$ of the form

$$\begin{pmatrix} A_{11} & A_{12} & A_{13} & \cdots \\ A_{21} & A_{22} & A_{23} & \cdots \\ A_{31} & A_{32} & A_{33} & \cdots \\ \vdots & \vdots & \vdots & \end{pmatrix}$$

such that $A_{ij} \in \mathfrak{A}$ for all i,j; (with respect to the decomposition $\sum_{n=1}^{\infty} \oplus \mathscr{H}_n$ of $\mathscr{H} \otimes \mathscr{H}$). It is easily verified that $\mathfrak{A} \otimes \mathscr{B}(\mathscr{H})$ is a weakly closed algebra whenever $\mathfrak{A}$ is.

The definitions given above are particular instances of much more general concepts of tensor products; (cf. Dixmier [1]).

Theorem 8.26. *If $\mathfrak{A}$ is a transitive operator algebra, then $\mathfrak{A} \otimes \mathscr{B}(\mathscr{H})$ is transitive; if $\mathfrak{A} \otimes \mathscr{B}(\mathscr{H}) = \mathscr{B}(\mathscr{H} \otimes \mathscr{H})$, then $\mathfrak{A} = \mathscr{B}(\mathscr{H})$.*

Proof. The proof is very easy. Suppose that $\mathfrak{A}$ is a transitive operator algebra, and that $\mathscr{M}$ is an invariant subspace of $\mathfrak{A} \otimes \mathscr{B}(\mathscr{H})$ other than $\{0\}$. Let $x = x_1 \oplus x_2 \oplus \cdots \in \mathscr{M}$, and suppose that $x_k \neq 0$. Since the operator $\sum_{i=1}^{\infty} \oplus B_i$ with $B_i = 0$ for $i \neq k$ and $B_k = 1$ is in $\mathfrak{A} \otimes \mathscr{B}(\mathscr{H})$, the vector $0 \oplus \cdots \oplus 0 \oplus x_k \oplus 0 \oplus \cdots$ is in $\mathscr{M}$. Thus so is $0 \oplus \cdots \oplus 0 \oplus A x_k \oplus 0 \oplus \cdots$ for all $A \in \mathfrak{A}$, and the transitivity of $\mathfrak{A}$ implies that $\mathscr{M}$ contains all vectors of the form $0 \oplus \cdots \oplus x \oplus 0 \oplus \cdots$ where $x \in \mathscr{H}$ and x is in the kth coordinate space. That is, $\mathscr{M}$ contains the kth coordinate space. Now for any j consider the operator

$$\begin{pmatrix} A_{11} & A_{12} & A_{13} & \cdots \\ A_{21} & A_{22} & A_{23} & \cdots \\ A_{31} & A_{32} & A_{33} & \cdots \\ \vdots & \vdots & \vdots & \end{pmatrix}$$

where $A_{mn}=0$ unless $m=j$ and $n=k$, and where $A_{jk}=1$. This operator is in $\mathfrak{A}\otimes\mathscr{B}(\mathscr{H})$ and takes the kth coordinate space onto the jth. Hence $\mathscr{M}$ contains all the coordinate spaces, and $\mathscr{M}=\mathscr{H}\otimes\mathscr{H}$.

The second assertion is trivial. $\square$

Note that, for any transitive algebra $\mathfrak{A}$, the algebra $\mathfrak{A}\otimes\mathscr{B}(\mathscr{H})$ contains the unilateral shift of multiplicity $\aleph_0$, since the shift

$$\begin{pmatrix} 0 & 0 & 0 & 0 & \dots \\ 1 & 0 & 0 & 0 & \dots \\ 0 & 1 & 0 & 0 & \dots \\ 0 & 0 & 1 & 0 & \dots \\ 0 & 0 & 0 & 1 & \dots \\ \vdots & \vdots & \vdots & \vdots & \end{pmatrix}$$

is in $\mathfrak{A}\otimes\mathscr{B}(\mathscr{H})$, many projections, (e.g., the operators $\sum_{i=1}^{\infty}\oplus B_i$ with some $B_i=1$ and the remaining $B_i=0$), algebraic operators, and so on. Thus Theorem 8.26 shows that an affirmative answer to the transitive algebra problem under the additional hypothesis that the algebra contain an operator of any of these types would give an affirmative answer in general. Some possible special cases of the transitive algebra problem that are apparently not equivalent to the general case are discussed in Chapter 10.

8.3 Generators of $\mathscr{B}(\mathscr{H})$

In this section we shall say that a subset $\mathscr{S}$ of $\mathscr{B}(\mathscr{H})$ generates $\mathscr{B}(\mathscr{H})$ if the only weakly closed subalgebra of $\mathscr{B}(\mathscr{H})$ which contains $\mathscr{S}$ and 1 is $\mathscr{B}(\mathscr{H})$; (by Proposition 8.2 it is not necessary to state that it contains 1). Some of the partial solutions to the transitive algebra problem given above yield results about generators of $\mathscr{B}(\mathscr{H})$.

Theorem 8.27. *If A is a Donoghue operator and B is any operator with $\Pi_0(B)=\emptyset$, then $\{A,B\}$ generates $\mathscr{B}(\mathscr{H})$.*

Proof. Let $\mathfrak{A}$ be the weakly closed algebra generated by $\{1,A,B\}$. Suppose that $\mathscr{M}$ is a non-trivial invariant subspace of $\mathfrak{A}$. Then $\mathscr{M}\in Lat A$, and Theorem 4.12 shows that $\mathscr{M}$ is finite-dimensional. Now $\mathscr{M}\in Lat B$, and $\Pi_0(B|\mathscr{M})\neq\emptyset$ since $B|\mathscr{M}$ is an operator on a finite-dimensional space. But $\Pi_0(B|\mathscr{M})\subset\Pi_0(B)$, which gives a contradiction. Hence $\mathscr{M}$ is trivial.

Thus $\mathfrak{A}$ is a transitive algebra containing a Donoghue operator, and Corollary 8.14 implies $\mathfrak{A}=\mathscr{B}(\mathscr{H})$. $\square$

The next result about generators requires a preliminary theorem about matrix representations of operators.

Lemma 8.28. *Let $\{e_n\}_{n=1}$ be an orthonormal basis for $\mathscr{H}$. If $A \in \mathscr{B}(\mathscr{H})$ is not a multiple of 1, and if n and m are fixed positive integers, then for each $\varepsilon > 0$ there exists a unitary operator U such that $\|1 - U\| < \varepsilon$ and $(U^* A U e_n, e_m) \neq 0$.*

Proof. Suppose that $(A e_n, e_m) = 0$. If $A e_n$ is not a multiple of e_n define $U_1 = I$. If $A e_n = \lambda e_n$ choose a k such that $A e_k \neq \lambda e_k$. Then there exist arbitrarily small positive numbers δ such that $(1 - \delta^2)^{\frac{1}{2}} e_n + \delta e_k$ is not an eigenvector of A. Choose such a δ with $(1 - (1 - \delta^2)^{\frac{1}{2}})^2 + \delta^2 < \varepsilon^2 / 8$, and let $\delta' = (1 - \delta^2)^{\frac{1}{2}}$. Now define U_1 by

$$U_1 e_n = \delta' e_n + \delta e_k,$$
$$U_1 e_k = -\delta e_n + \delta' e_k, \quad \text{and}$$
$$U_1 e_i = e_i \quad \text{if } i \neq n, k.$$

Note that $\|1 - U_1\| < \varepsilon/2$. If $(A U_1 e_n, U_1 e_m) \neq 0$, then U_1 satisfies the conclusion of the lemma. Suppose that $(A U_1 e_n, U_1 e_m) = 0$. Then there is some $i \neq n$ such that $(A U_1 e_n, U_1 e_i) \neq 0$, since $A U_1 e_n$ is not a multiple of $U_1 e_n$. Then $i \neq m$. We consider two cases.

Case 1: $n \neq m$.

Define

$$U_2 U_1 e_m = \delta' U_1 e_m + \delta U_1 e_i,$$
$$U_2 U_1 e_i = -\delta U_1 e_m + \delta' U_1 e_i, \quad \text{and}$$
$$U_2 U_1 e_p = U_1 e_p \quad \text{for } p \neq i, m.$$

Let $U = U_2 U_1$. Then $\|1 - U\| < \varepsilon$, and

$$(U^* A U e_n, e_m) = (A U_2 U_1 e_n, U_2 U_1 e_m)$$
$$= (A U_1 e_n, \delta' U_1 e_m + \delta U_1 e_i)$$
$$= 0 + \delta(A U_1 e_n, U_1 e_i)$$
$$\neq 0.$$

Case 2: $n = m$.

Let $\omega_1, \omega_2, \omega_3, \omega_4$ denote the fourth roots of unity, and consider the vectors

$$f_j = \delta' U_1 e_n + \omega_j \delta U_1 e_i$$

for $j = 1, 2, 3, 4$. There is a j such that $(A f_j, f_j) \neq 0$, since $4 \delta \delta' (A U_1 e_n, U_1 e_i)$
$= \sum_{j=1}^{4} \omega_j (A f_j, f_j)$. For such a j define

$$U_2 U_1 e_n = f_j,$$

$$U_2 U_1 e_i = -\bar{\omega}_j \delta U_1 e_n + \delta' U_1 e_i, \quad \text{and}$$

$$U_2 U_1 e_p = U_1 e_p \quad \text{for } p \neq n, i.$$

Now let $U = U_2 U_1$. Clearly $\|1 - U\| < \varepsilon$, and

$$(U^* A U e_n, e_n) = (A U_2 U_1 e_n, U_2 U_1 e_n)$$
$$= (A f_j, f_j) \neq 0. \quad \Box$$

Theorem 8.29. *Let $\{A_i\}$ be a countable set of operators, none of which is a multiple of 1. Then there exists an orthonormal basis $\{e_n\}_{n=1}^{\infty}$ for $\mathcal{H}$ such that $(A_i e_m, e_n) \neq 0$ for all i, m, n.*

Proof. Let $\{f_n\}_{n=1}^{\infty}$ be any fixed orthonormal basis for $\mathcal{H}$. For each triple (i, m, n) of natural numbers, let $\mathcal{S}_{i,m,n}$ denote the set of all unitary operators U such that $(U^* A_i U f_m, f_n) = 0$. Regard the set $\mathcal{U}$ of all unitary operators on $\mathcal{H}$ as a topological subspace of $\mathcal{B}(\mathcal{H})$ with the norm topology. Since $\mathcal{U}$ is closed in $\mathcal{B}(\mathcal{H})$, $\mathcal{U}$ is a complete metric space. Obviously $\mathcal{S}_{i,m,n}$ is a closed subset of $\mathcal{U}$ for each (i, m, n).

We show that each $\mathcal{S}_{i,m,n}$ is nowhere dense in $\mathcal{U}$. For suppose $U_0 \in \mathcal{S}_{i,m,n}$. By Lemma 8.28, applied to $\mathcal{H}$ with the orthonormal basis $\{U_0 f_n\}_{n=1}^{\infty}$, for each $\varepsilon > 0$ there exists a $U \in \mathcal{U}$ such that $\|1 - U\| < \varepsilon$ and $(U^* A_i U U_0 f_m, U_0 f_n) \neq 0$. Then

$$0 \neq (U_0^* U^* A_i U U_0 f_m, f_n) = ((U U_0)^* A_i U U_0 f_m, f_n),$$

and it follows that $U U_0 \notin \mathcal{S}_{i,m,n}$. Also $\|U_0 - U U_0\| < \varepsilon$, and thus $\mathcal{S}_{i,m,n}$ is nowhere dense.

Now, by the Baire category theorem, the union of the countable collection of sets $\{\mathcal{S}_{i,m,n}\}$ is not $\mathcal{U}$. Choose $U \in \mathcal{U}$ such that $U \notin \bigcup \mathcal{S}_{i,m,n}$ and let $e_n = U f_n$. Then $(A_i e_m, e_n) = (A_i U f_m, U f_n) = (U^* A_i U f_m, f_n)$ is not 0 for any i, m, n. $\quad \Box$

Theorem 8.29 can be rephrased: if $\{A_i\}$ is any collection of operators none of which is a multiple of 1, then there is an orthonormal basis $\{e_n\}$ such that the matrices of all the A_i with respect to $\{e_n\}$ have no entries equal to 0.

Theorem 8.30. *If A is not a multiple of 1, then there exists a compact Hermitian operator K such that $\{A, K\}$ generates $\mathcal{B}(\mathcal{H})$.*

Proof. Choose, by Theorem 8.29, an orthonormal basis $\{e_n\}_{n=1}^{\infty}$ for $\mathcal{H}$ such that $(A e_m, e_n) \neq 0$ for all m, n. Define the operator K by $K e_n = (1/n) e_n$ for $n = 1, 2, 3, \ldots$. We claim that $\{A, K\}$ generates $\mathcal{B}(\mathcal{H})$.

Let $\mathfrak{A}$ be any weakly closed algebra containing $\{A, K\}$. The weakly closed algebra generated by $\{K\}$ is the m.a.s.a. consisting of all operators B which are diagonal with respect to $\{e_n\}_{n=1}^{\infty}$, and every non-trivial invariant subspace of K has the form $\bigvee_{n \in \mathscr{S}} \{e_n\}$ for subsets $\mathscr{S}$ of $\{1, 2, 3, \ldots\}$; (cf. Example 4.10). The fact that $(A e_m, e_n) \neq 0$ for all m and n implies that none of the subspaces $\bigvee_{n \in \mathscr{S}} \{e_n\}$ are invariant under A. Hence $\mathfrak{A}$ is a transitive algebra containing a m.a.s.a., and Theorem 8.10, (or Proposition 8.3), implies $\mathfrak{A} = \mathscr{B}(\mathscr{H})$. $\quad \square$

A different application of Theorem 8.29 shows that $\mathscr{B}(\mathscr{H})$ is the linear span of the set of irreducible operators.

Theorem 8.31. *Every operator is the sum of two irreducible operators.*

Proof. Let $A \in \mathscr{B}(\mathscr{H})$. If A is a multiple of 1, let C be any irreducible operator, (e. g., the unilateral shift of multiplicity 1). Then the equation $A = (A - C) + C$ expresses A as the sum of two irreducible operators.

Suppose that A is not a multiple of 1. Then at least one of the two operators $\operatorname{Re} A = \frac{1}{2}(A + A^*)$ and $\operatorname{Im} A = (1/2i)(A - A^*)$ is not a multiple of 1. Assume first that neither $\operatorname{Re} A$ nor $\operatorname{Im} A$ is a multiple of 1. Then use Theorem 8.29 as in the proof of Theorem 8.30 to obtain a compact Hermitian operator K such that

$$Lat\, K \cap Lat(\operatorname{Re} A) = Lat\, K \cap Lat(\operatorname{Im} A) = \{\{0\}, \mathscr{H}\}.$$

Then $A = [\operatorname{Re} A - K - iK] + [K + i(\operatorname{Im} A + K)]$. Each reducing subspace of $(\operatorname{Re} A - K) - iK$ is invariant under $(\operatorname{Re} A - K)$ and K, hence also under $\operatorname{Re} A$, and thus is trivial. Similarly, $[K + i(\operatorname{Im} A + K)]$ is irreducible.

The only remaining case is where exactly one of the operators $\operatorname{Re} A$ and $\operatorname{Im} A$ is a multiple of 1. By multiplying by i if necessary we can assume that $\operatorname{Re} A = \alpha$. Use Theorem 8.29 to produce a compact Hermitian operator K such that $Lat\, K \cap Lat(\operatorname{Im} A) = \{\{0\}, \mathscr{H}\}$. Then the operators $(K + \operatorname{Re} A + (i/2) \operatorname{Im} A)$ and $(-K + (i/2) \operatorname{Im} A)$ are both irreducible, and their sum is A. $\quad \square$

8.4 Additional Propositions

Proposition 8.1. If $A \in \mathscr{B}(\mathscr{H})$ and A is not a multiple of 1, then A has a non-trivial hyperinvariant linear manifold.

Proposition 8.2. If $\mathscr{S} \subset \mathscr{B}(\mathscr{H})$ and the only weakly closed subalgebra of $\mathscr{B}(\mathscr{H})$ containing $\mathscr{S} \cup \{1\}$ is $\mathscr{B}(\mathscr{H})$, then the only weakly closed subalgebra of $\mathscr{B}(\mathscr{H})$ containing $\mathscr{S}$ is $\mathscr{B}(\mathscr{H})$. (This can be proved by observing that the smallest weakly closed algebra containing $\mathscr{S}$ is an ideal under the given hypotheses.)

Proposition 8.3. The following special case of Theorem 8.10 has a simple proof which does not require Lemma 8.8: if $\{e_n\}_{n=1}^{\infty}$ is an orthonormal basis for $\mathscr{H}$, and if $\mathfrak{A}$ is a transitive subalgebra of $\mathscr{B}(\mathscr{H})$ which contains all operators whose matrices with respect to $\{e_n\}_{n=1}^{\infty}$ are diagonal, then $\mathfrak{A} = \mathscr{B}(\mathscr{H})$.

Proposition 8.4. If $\mathscr{D}$ is any operator range, then the algebraic dimension of $\mathscr{D}$ is finite or the power of the continuum.

Proposition 8.5. If $A, B \in \mathscr{B}(\mathscr{H})$ then $A = BC$ for some $C \in \mathscr{B}(\mathscr{H})$ if and only if the range of A is contained in the range of B.

Proposition 8.6. If $\mathscr{D}$ is an operator range and $\mathscr{D} \neq \mathscr{H}$, then $\mathscr{D}$ is a set of the first category in $\mathscr{H}$.

Proposition 8.7. If $A|\mathscr{M}$ has a non-trivial hyperinvariant subspace for some reducing subspace $\mathscr{M} \neq \{0\}$, then A has a non-trivial hyperinvariant subspace.

Proposition 8.8. If $\lambda \neq 0$, A is a Donoghue operator, and $\mathfrak{A}$ is a transitive algebra containing $A \oplus (A + \lambda)$, then $\mathfrak{A} = \mathscr{B}(\mathscr{H})$.

Proposition 8.9. If T is a linear transformation taking $\mathscr{H}$ into $\mathscr{H}$, and if the graph of T is an operator range, then T is bounded.

Proposition 8.10. If K is a compact operator, $1 \in \Pi_0(K)$, and P is the Riesz projection of K corresponding to $\{1\}$, then P is in the uniformly closed algebra generated by K and the identity operator.

8.5 Notes and Remarks

The transitive algebra problem was raised by Kadison [1], after the algebraic results of Section 8.1 had been obtained. Theorem 8.2 was discovered by Jacobson [1]; see Jacobson [2] for a more algebraic and more general version of this result. Theorem 8.4 and Lemma 8.3 were independently discovered by Yood [1] and Rickart [2]—their result is slightly more general than Theorem 8.4; (see Rickart [1]). A more direct proof of Burnside's theorem (Corollary 8.6) can be found in Jacobson[3].

The first partial solutions of the transitive algebra problem were given by Arveson [1]. Lemma 8.8 is implicitly contained in Arveson [1]; the formulation and proof in the text are from Radjavi-Rosenthal [2]. Theorem 8.9 is due to Foiaş ([2], [3]). Foiaş [3] contains a number of interesting related results on intransitive operator algebras. Theorem 8.10 was the first partial solution of the transitive algebra problem; it was obtained by Arveson [1]. Lemma 8.11, Theorem 8.12, Theorem 8.13

and Corollary 8.14 are from Nordgren-Radjavi-Rosenthal [1]. An interesting strengthening of Theorem 8.12 is given in Barnes [1]. Theorem 8.18 was obtained by Arveson [1] in the case where the shift has multiplicity 1; the general case, (as well as Lemmas 8.15, 8.16 and 8.17), was discovered by Nordgren [3]. Corollary 8.19 was observed in Rosenthal [6].

The generalization of Theorem 8.10 given in Theorem 8.21 is due to Douglas-Pearcy [3], where results on matrices of continuous functions on Stonian spaces are used rather than Lemma 8.20. Lemma 8.20 is a strengthening of a result in Radjavi-Rosenthal [7].

Lemma 8.22 is implicitly contained in the work of Lomonosov [1] which states the theorem (slightly weaker than Corollary 8.25) that every operator which commutes with a non-zero compact operator has an invariant subspace. W. B. Arveson, Carl Pearcy, Allen Shields (and undoubtedly others) realized that Lemma 8.22 is the essence of Lomonosov's result, and that Theorem 8.23 and Corollaries 8.24 and 8.25 followed from Lemma 8.22. We are grateful to Carl Pearcy and James Rovnyak for describing these results to us just in time for them to be included in this book. Lomonosov's beautiful result and its corollaries contain affirmative answers to several problems that had been unsolved for years: Theorem 5.3 extends to operators with the property that the weakly closed algebra they generate contains a compact operator, two commuting compact operators have a common invariant subspace, and $\mathscr{B}(\mathscr{H})$ is the only transitive algebra containing a compact operator. The easy proof of Corollary 8.25 is due to H. M. Hilden.

Lomonosov's Lemma (Lemma 8.22), Proposition 8.10, and the result of Barnes [1] imply that a uniformly closed algebra of operators on a Hilbert space which has only trivial invariant subspaces and which contains a non-zero compact operator must contain all compact operators. This is a striking generalization of the well-known result (cf. Dixmier [2]) that an irreducible C^*-algebra which contains one non-zero compact operator contains all compact operators.

Theorem 8.26 and the subsequent remarks are due to Nordgren [3]. Theorem 8.27 is from Nordgren-Radjavi-Rosenthal [1]. Lemma 8.28, Theorem 8.29 and Corollary 8.30 were obtained in Radjavi-Rosenthal [4], which also contains analogous results for real Hilbert spaces. Theorem 8.31 is due to Radjavi [1]; it strengthens the result of Fillmore-Topping [1]. Proposition 8.2 can be found in Davis-Radjavi-Rosenthal [1], and was previously discovered by W. B. Arveson.

An elegant exposition of the known results on operator ranges, including Propositions 8.4, 8.5, 8.6 and 8.9, is given in Fillmore-Williams [1]. Proposition 8.4 is due to Dixmier [1], Proposition 8.5 to Douglas [3], Proposition 8.6 to Banach [1], and Proposition 8.9 to Foiaş ([2],

[3]). Proposition 8.7 is from Douglas-Pearcy [3] and Proposition 8.8 from Nordgren-Radjavi-Rosenthal [1].

Ismagilov [1] has shown that the only "symmetric" transitive algebra in a Pontrjagin space is $\mathscr{B}(\mathscr{H})$. Kadison [2] (cf. Dixmier [2]) has obtained an interesting extension of Theorem 8.7: he shows that a C^*-algebra (a uniformly closed self-adjoint subalgebra of $\mathscr{B}(\mathscr{H})$) which has only trivial invariant subspaces is strictly dense.

Chapter 9. Algebras Associated with Invariant Subspaces

In this chapter we consider subalgebras $\mathfrak{A}$ of $\mathcal{B}(\mathcal{H})$ which are not transitive, and establish certain relations between $Lat\,\mathfrak{A}$ and various properties of $\mathfrak{A}$. We begin with the following problem.

Definition. The *reductive algebra problem* is the question: if $\mathfrak{A}$ is a weakly closed subalgebra of $\mathcal{B}(\mathcal{H})$ which contains 1 and has the property that every invariant subspace of $\mathfrak{A}$ is reducing, (i.e., $\mathcal{M} \in Lat\,\mathfrak{A}$ implies $\mathcal{M}^{\perp} \in Lat\,\mathfrak{A}$), must $\mathfrak{A}$ be self-adjoint; (i.e., must $A \in \mathfrak{A}$ imply $A^* \in \mathfrak{A}$)?

This problem, like the transitive algebra problem, is unsolved, but certain partial results are presented in Section 9.1.

In Section 9.2 we discuss sufficient conditions that a subalgebra of $\mathcal{B}(\mathcal{H})$ be the algebra of all operators which leave invariant a given family of subspaces of $\mathcal{H}$. These results often produce a concrete description of the operators whose invariant subspace lattices contain a given collection of subspaces. They also lead to characterizations of certain weakly closed operator algebras.

Section 9.3 is concerned with a class of subalgebras of $\mathcal{B}(\mathcal{H})$ which are generalizations of certain algebras of triangular matrices. "Triangular operator algebras" include some of the algebras of Section 9.2, although their properties are related to their invariant subspaces in a less direct manner.

9.1 Reductive Algebras

Definition. A subalgebra of $\mathcal{B}(\mathcal{H})$ is *reductive* if it is weakly closed, contains 1, and has the property that all of its invariant subspaces are reducing.

Obviously every von Neumann algebra is reductive; the reductive algebra problem is the question of whether or not the class of reductive operator algebras is identical with the class of von Neumann algebras. Note that an affirmative answer to the reductive algebra problem would imply an affirmative answer to the transitive algebra problem. For if

$\mathfrak{A}$ is a transitive algebra, $Lat\,\mathfrak{A} = \{\{0\}, \mathscr{H}\}$ implies that $\mathfrak{A}$ is reductive; if this implied $\mathfrak{A}$ was a von Neumann algebra, then Theorem 8.7 would show that $\mathfrak{A} = \mathscr{B}(\mathscr{H})$. Thus it is not surprising that we shall not solve the reductive algebra problem in general. We do obtain several special cases.

The basic technique for establishing special cases of the reductive algebra problem is use of Lemma 7.1. For reductive algebras, this lemma can be rephrased: if $\mathfrak{A}^{(n)}$ is reductive for every n, then $\mathfrak{A}$ is a von Neumann algebra. In the case of reductive algebras it is easy to reduce the study of $Lat\,\mathfrak{A}^{(n)}$ to the study of invariant graph subspaces, and we shall therefore be concerned with graph transformations for reductive algebras. We begin with some results on commutants.

Lemma 9.1. *If $\mathfrak{A} \subset \mathscr{B}(\mathscr{H})$ and every invariant subspace of $\mathfrak{A}$ is reducing, then every invariant subspace of $\mathfrak{A}|P\mathscr{H}$ is reducing for each projection P in $\mathfrak{A}'$.*

Proof. Let $\mathscr{M} \in Lat(\mathfrak{A}|P\mathscr{H})$. Then $\mathscr{M} \in Lat\,\mathfrak{A}$, and $\mathscr{M}^\perp \in Lat\,\mathfrak{A}$. Thus $P\mathscr{M}^\perp \in Lat(\mathfrak{A}|P\mathscr{H})$, and $P\mathscr{H} \ominus \mathscr{M} = P\mathscr{M}^\perp$. ☐

Lemma 9.2. *If $\mathfrak{A}$ is a reductive algebra, $T \in \mathfrak{A}'$, and $T^2 = 0$, then $T \in (\mathfrak{A}^*)'$.*

Proof. Given $\mathfrak{A}$ and T, let $\mathscr{K}$ denote the nullspace of T. Then $\mathscr{K}$ is invariant, and hence reducing, for $\mathfrak{A}$. Decompose T with respect to the decomposition $\mathscr{K} \oplus \mathscr{K}^\perp$ of $\mathscr{H}: T = \begin{pmatrix} 0 & C \\ 0 & 0 \end{pmatrix}$. Let $A \in \mathfrak{A}$; by hypothesis $AT = TA$, and it must be shown that $A^*T = TA^*$. Now A has the form $\begin{pmatrix} A_1 & 0 \\ 0 & A_2 \end{pmatrix}$ with respect to the above decomposition of $\mathscr{H}$. Since

$$\begin{pmatrix} 0 & C \\ 0 & 0 \end{pmatrix} \begin{pmatrix} A_1 & 0 \\ 0 & A_2 \end{pmatrix} = \begin{pmatrix} A_1 & 0 \\ 0 & A_2 \end{pmatrix} \begin{pmatrix} 0 & C \\ 0 & 0 \end{pmatrix},$$

the equation $CA_2 = A_1 C$ is satisfied.

Let $\mathscr{M}$ denote the subspace $\{Cx \oplus x : x \in \mathscr{K}^\perp\}$ of $\mathscr{H}$. The above equation shows that $\mathscr{M} \in Lat\,\mathfrak{A}$, and it follows that $\mathscr{M} \in Lat\,\mathfrak{A}^*$. Now

$$\begin{pmatrix} A_1^* & 0 \\ 0 & A_2^* \end{pmatrix} \begin{pmatrix} Cx \\ x \end{pmatrix} = \begin{pmatrix} A_1^* Cx \\ A_2^* x \end{pmatrix}$$

which shows that $A_1^* C = C A_2^*$. Therefore

$$\begin{pmatrix} 0 & C \\ 0 & 0 \end{pmatrix} \begin{pmatrix} A_1^* & 0 \\ 0 & A_2^* \end{pmatrix} = \begin{pmatrix} A_1^* & 0 \\ 0 & A_2^* \end{pmatrix} \begin{pmatrix} 0 & C \\ 0 & 0 \end{pmatrix},$$

or $TA^* = A^*T$. ☐

This lemma can be strengthened.

Lemma 9.3. *If $\mathfrak{A}$ is a reductive algebra, $T \in \mathfrak{A}'$, and T is algebraic, then $T \in (\mathfrak{A}^*)'$.*

Proof. Choose $\lambda_1 \in \Pi_0(T)$, and let $\mathcal{M}_1 = \{x : Tx = \lambda_1 x\}$. Then $\mathcal{M}_1 \in Lat\,\mathfrak{A}$, and thus the projection P_1 onto $\mathcal{M}_1$ is in $\mathfrak{A}' \cap (\mathfrak{A}^*)'$. Choose λ_2 in the point spectrum of $(1 - P_1)T|(\mathcal{M}_1^\perp)$, let $\mathcal{M}_2 = \{x \in \mathcal{M}_1^\perp : Tx = \lambda_2 x\}$, and use P_2 to denote the projection of $\mathcal{H}$ onto $\mathcal{M}_2$. Then $P_2 \in \mathfrak{A}' \cap (\mathfrak{A}^*)'$. Choose λ_3 in the point spectrum of the compression of T to $(\mathcal{M}_1 \oplus \mathcal{M}_2)^\perp$, and let $\mathcal{M}_3$ be the eigenspace of the compression and P_3 the projection of $\mathcal{H}$ onto $\mathcal{M}_3$. Continue in this manner to obtain a decomposition of $\mathcal{H}$ into a direct sum $\mathcal{H} = \sum_{i=1}^{n} \oplus \mathcal{M}_i$ such that, for each i, the projection P_i onto $\mathcal{M}_i$ is in $\mathfrak{A}' \cap (\mathfrak{A}^*)'$ and $P_i T P_i$ is a multiple of P_i. (The fact that this process terminates in a finite number of steps can be shown as follows. Let T have minimal polynomial $p(z)$. Then $p(z) = (z - \lambda_1)q(z)$, and, with respect to the decomposition $\mathcal{H} = \mathcal{M}_1 \oplus \mathcal{M}_1^\perp$, T has the form

$$T = \begin{pmatrix} \lambda_1 & K \\ 0 & S \end{pmatrix}.$$

Now $q(S) = 0$, for if $q(S)y \neq 0$ for some $y \in \mathcal{M}_1^\perp$, then

$$0 = (T - \lambda_1)q(T)(0 \oplus y) = (T - \lambda_1)(u \oplus q(S)y)$$

for some $u \in \mathcal{M}_1$, which contradicts the fact that the nullspace of $T - \lambda_1$ is $\mathcal{M}_1$. Thus the minimal polynomial of S has lower degree than that of T. Since each step in the above process gives a compression with minimal polynomial of lower degree, the process must stop in a finite number of steps.)

Now $T = \sum_{i,j=1}^{n} P_i T P_j$. For $i = j$, $P_i T P_i = \lambda_i P_i$ is in $(\mathfrak{A}^*)'$. For $i \neq j$, $(P_i T P_j)^2 = 0$ and it follows from Lemma 9.2 that $(P_i T P_j) \in (\mathfrak{A}^*)'$. Hence $T \in (\mathfrak{A}^*)'$. □

Lemma 9.4. *If $\mathfrak{A}$ is reductive, every densely defined graph transformation for $\mathfrak{A}$ is bounded, and $\mathfrak{A}' \subset (\mathfrak{A}^*)'$, then $\mathfrak{A}$ is self-adjoint.*

Proof. By Lemma 7.1, it suffices to show that $Lat\,\mathfrak{A}^{(n)} \subset Lat(\mathfrak{A}^*)^{(n)}$ for each positive integer n. Since $\mathfrak{A}$ is reductive this holds for $n = 1$. To prove this inclusion by induction, assume that $Lat\,\mathfrak{A}^{(n)} \subset Lat(\mathfrak{A}^*)^{(n)}$ and let $\mathcal{M} \in Lat\,\mathfrak{A}^{(n+1)}$.

Let $\mathcal{N}$ be the subspace of $\mathcal{M}$ consisting of all vectors whose first component is 0; i.e., $\mathcal{N} = \mathcal{M} \cap (\{0\} \oplus \mathcal{H} \oplus \cdots \oplus \mathcal{H})$. The induction hypothesis implies, using the obvious identification of $\mathcal{N}$ with an element of $Lat\,\mathfrak{A}^{(n)}$, that $\mathcal{N} \in Lat(\mathfrak{A}^*)^{(n+1)}$; hence $(\mathcal{M} \ominus \mathcal{N}) \in Lat\,\mathfrak{A}^{(n+1)}$. The first component of each vector in $\mathcal{M} \ominus \mathcal{N}$ uniquely determines its other components, and it follows that $\mathcal{M} \ominus \mathcal{N}$ is a graph subspace. Thus

$$\mathcal{M} \ominus \mathcal{N} = \{(x \oplus T_1 x \oplus \cdots \oplus T_n x\} : x \in \mathcal{D}\}$$

for some linear manifold $\mathcal{D}$ invariant under $\mathfrak{A}$ and some linear transformations $\{T_i\}$ with common domain $\mathcal{D}$ such that $T_i A = A T_i$ for all $A \in \mathfrak{A}$.

In order to apply the hypothesis we must extend the $\{T_i\}$ to a dense domain. Let P denote the orthogonal projection onto $\bar{\mathcal{D}}$, and let $\tilde{\mathcal{D}} = \{x : Px \in \mathcal{D}\}$. Then $\tilde{\mathcal{D}}$ is dense in $\mathcal{H}$. Extend each T_i to $\tilde{\mathcal{D}}$ by defining $\tilde{T}_i x = T_i P x$ for $x \in \tilde{\mathcal{D}}$. Then

$$\{x \oplus \tilde{T}_1 x \oplus \cdots \oplus \tilde{T}_n x : x \in \tilde{\mathcal{D}}\}$$

is equal to

$$(\mathcal{M} \ominus \mathcal{N}) \oplus \{y \oplus 0 \oplus 0 \oplus \cdots \oplus 0 : y \in \mathcal{D}^\perp\},$$

and it follows that the $\{\tilde{T}_i\}$ are graph transformations for $\mathfrak{A}$. By hypothesis each $\tilde{T}_i$ is bounded, and, since $\tilde{T}_i \in \mathfrak{A}'$, it follows that $\tilde{T}_i \in (\mathfrak{A}^*)'$ for each i. Now each T_i is bounded, $\mathcal{D}$ is closed, and $P \in \mathfrak{A}'$. Hence $T_i A^* x = A^* T_i x$ for $x \in \mathcal{D}$, and $\mathcal{M} \ominus \mathcal{N} \in Lat(\mathfrak{A}^*)^{(n+1)}$. Therefore $\mathcal{M} = (\mathcal{M} \ominus \mathcal{N}) \oplus \mathcal{N}$ is also in $Lat(\mathfrak{A}^*)^{(n+1)}$, and the result follows by induction. $\square$

The preceding lemmas provide an immediate solution to the reductive algebra problem in the finite-dimensional case.

Theorem 9.5. *If $\mathfrak{A}$ is a reductive algebra of operators on a finite-dimensional space, then $\mathfrak{A}$ is self-adjoint.*

Proof. Every densely defined graph transformation is bounded, since all linear transformations on finite-dimensional spaces are bounded. Lemma 9.3 implies that $\mathfrak{A}' \subset (\mathfrak{A}^*)'$, since every operator on a finite-dimensional space is algebraic. Now Lemma 9.4 establishes the result. $\square$

Lemmas 9.2 and 9.4 can also be used to obtain some infinite-dimensional cases of the reductive algebra problem. In fact, we shall show that the hypotheses $\mathfrak{A}' \subset (\mathfrak{A}^*)'$ can be omitted from Lemma 9.4.

Lemma 9.6. *If $\mathfrak{A}$ is a collection of operators such that every closed densely-defined linear transformation commuting with $\mathfrak{A}$ is bounded, then*
   (i) *$\mathfrak{A}$ cannot have an infinite collection of mutually orthogonal non-trivial reducing subspaces, and*
   (ii) *$T \in \mathfrak{A}'$ implies $\sigma(T) = \Pi_0(T) \cup \Gamma(T)$.*

Proof. To establish (i) suppose that $\{\mathcal{N}_j\}_{j=1}^\infty$ is an infinite collection of mutually orthogonal non-trivial reducing subspaces for $\mathfrak{A}$. Let $\mathcal{N}_0$ denote the orthogonal complement of the span of $\{\mathcal{N}_j\}_{j=1}^\infty$, and define

$\mathscr{D}$ as the set of vectors of the form $\sum\limits_{j=0}^{\infty} x_j$ with $x_j \in \mathscr{N}_j$ and $\sum\limits_{j=0}^{\infty} j^2 \|x_j\|^2 < \infty$.

Now $\mathscr{H} = \sum\limits_{j=0}^{\infty} \oplus \mathscr{N}_j$, and we define a closed linear transformation T with domain $\mathscr{D}$ by $T = \sum\limits_{j=0}^{\infty} \oplus j I_j$, where I_j is the identity operator on $\mathscr{N}_j$. Clearly T is a densely-defined closed linear transformation which commutes with $\mathfrak{A}$, and T is certainly not bounded. This contradiction establishes (i).

To prove (ii), suppose that $T \in \mathfrak{A}'$. If $\lambda \notin \Pi_0(T) \cup \Gamma(T)$, then $(T - \lambda)$ is one-to-one and has dense range, and it follows that $(T - \lambda)^{-1}$ is a closed densely-defined linear transformation that commutes with $\mathfrak{A}$. By hypothesis, $(T - \lambda)^{-1}$ is bounded, and thus $\lambda \notin \sigma(T)$. $\quad\square$

Theorem 9.7. *If $\mathfrak{A}$ is a reductive algebra and every densely defined graph transformation for $\mathfrak{A}$ is bounded, then $\mathfrak{A}$ is self-adjoint.*

Proof. We must show that the hypothesis $\mathfrak{A}' \subset (\mathfrak{A}^*)'$ of Lemma 9.4 is redundant. Let $T \in \mathfrak{A}'$; the proof of the fact that $T \in (\mathfrak{A}^*)'$ is similar to the proof of Lemma 9.3. Note first that $\mathfrak{A}$ satisfies the hypothesis of Lemma 9.6, since a closed linear transformation commuting with $\mathfrak{A}$ is a graph transformation for $\mathfrak{A}$.

Choose $\lambda_1 \in \sigma(T)$. By Lemma 9.6 (ii), $\lambda_1 \in \Pi_0(T)$ or $\bar{\lambda}_1 \in \Pi_0(T^*)$. If $\lambda_1 \in \Pi_0(T)$, let $\mathscr{M}_1 = \{x_i : T x = \lambda_1 x\}$. Then $\mathscr{M}_1$ reduces $\mathfrak{A}$. If $\bar{\lambda}_1 \in \Pi_0(T^*)$, let $\mathscr{M}_1 = \{x : T^* x = \bar{\lambda}_1 x\}$. Then $\mathscr{M}_1 \in \operatorname{Lat} \mathfrak{A}^*$, $\mathscr{M}_1^\perp \in \operatorname{Lat} \mathfrak{A}$ and again $\mathscr{M}_1$ reduces $\mathfrak{A}$. In either case, if P_1 is the orthogonal projection onto $\mathscr{M}_1$, then $P_1 \in \mathfrak{A}'$ and $P_1 T P_1$ is a multiple of P_1.

Now $(1 - P_1) T | \mathscr{M}_1^\perp$ commutes with $\mathfrak{A} | \mathscr{M}_1^\perp$. Since any closed linear transformation commuting with $\mathfrak{A} | \mathscr{M}_1^\perp$ can be extended to a closed linear transformation commuting with $\mathfrak{A}$, it follows that $\mathfrak{A} | \mathscr{M}_1^\perp$ satisfies the hypothesis of Lemma 9.6 (ii). Also, by Lemma 9.1, every invariant subspace of $\mathfrak{A} | \mathscr{M}_1^\perp$ is reducing. Choose $\lambda_2 \in \sigma((1 - P) T | \mathscr{M}_1^\perp)$ and obtain an $\mathscr{M}_2 \subset \mathscr{M}_1^\perp$ such that, for P_2 the projection of $\mathscr{H}$ onto $\mathscr{M}_2$, $P_2 T P_2$ is a multiple of P_2 and $P_2 \in \mathfrak{A}'$.

Next consider $(1 - P_1 - P_2) T | (\mathscr{M}_1 \oplus \mathscr{M}_2)^\perp$, and produce a $P_3 \in \mathfrak{A}'$ such that $P_3 T P_3$ is a multiple of P_3. Continue this procedure to obtain a collection $\{P_i\}_{i=1}^n$ of mutually orthogonal projections such that $P_i \in \mathfrak{A}'$, $P_i T P_i$ is a multiple of P_i, and $\sum\limits_{i=1}^n P_i = 1$; (Lemma 9.6 (i) implies that the above procedure will terminate after a finite number of steps, producing such a collection $\{P_i\}_{i=1}^n$).

Now $T = \sum\limits_{i,j=1}^n P_i T P_j$, so in order to prove $T \in (\mathfrak{A}^*)'$ it suffices to show $P_i T P_j \in (\mathfrak{A}^*)'$ for all i, j. For $i = j$, $P_i T P_i$ is a multiple of P_i and thus commutes with $\mathfrak{A}^*$. For $i \neq j$, $P_i T P_j \in \mathfrak{A}'$ and $(P_i T P_j)^2 = 0$ imply, by Lemma 9.1, that $P_i T P_j \in (\mathfrak{A}^*)'$. $\quad \square$

One corollary of Theorem 9.7 is a generalization of Theorem 8.9.

Corollary 9.8. *If $\mathfrak{A}$ is a reductive algebra such that every operator range invariant under $\mathfrak{A}$ is closed, then $\mathfrak{A}$ is self-adjoint.*

Proof. The result will follow from Theorem 9.7 if we show that every graph transformation for $\mathfrak{A}$ is bounded. If

$$\mathcal{M} = \{x \oplus T_1 x \oplus \cdots \oplus T_{n-1} x : x \in \mathcal{D}\}$$

is an invariant graph subspace for $\mathfrak{A}^{(n)}$, then $\mathcal{D}$ is an invariant operator range of $\mathfrak{A}$ (cf. Theorem 8.9) and is therefore closed. It follows as in the proof of Theorem 8.9 that each T_i is bounded. $\quad \square$

Definition. The subalgebra $\mathfrak{A}$ of $\mathcal{B}(\mathcal{H})$ is *strictly cyclic* if there exists a vector $x_0 \in \mathcal{H}$ such that

$$\{A x_0 : A \in \mathfrak{A}\} = \mathcal{H}.$$

Corollary 9.9. *If $\mathfrak{A}$ is a strictly cyclic reductive algebra, then $\mathfrak{A}$ is self-adjoint.*

Proof. By Theorem 9.7 it suffices to show that every densely-defined graph transformation for $\mathfrak{A}$ is bounded. In fact, we show that the only dense linear manifold invariant under a strictly cyclic algebra is $\mathcal{H}$; the fact that all densely-defined graph transformations are bounded then follows from the closed graph theorem, (cf. the proof of Theorem 8.9).

Let $\{A x_0 : A \in \mathfrak{A}\} = \mathcal{H}$ and let $\mathcal{D}$ be a dense linear manifold invariant under $\mathfrak{A}$. Define the map $\mathcal{S} : \mathfrak{A} \to \mathcal{H}$ by $\mathcal{S}(A) = A x_0$ for $A \in \mathfrak{A}$. Then $\mathcal{S}$ is a bounded linear transformation of $\mathfrak{A}$ (with uniform norm) onto $\mathcal{H}$. Clearly $\mathcal{S}^{-1}(\mathcal{D})$ is a linear subspace of $\mathfrak{A}$. Moreover, $B \in \mathcal{S}^{-1}(\mathcal{D})$ and $C \in \mathfrak{A}$ imply $\mathcal{S}(C B) = C B x_0$ and $B x_0 \in \mathcal{D}$; this shows that $C B \in \mathcal{S}^{-1}(\mathcal{D})$. Hence $\mathcal{S}^{-1}(\mathcal{D})$ is a left ideal of $\mathfrak{A}$. If $\mathcal{S}^{-1}(\mathcal{D})$ were a proper ideal, then the closure of $\mathcal{S}^{-1}(\mathcal{D})$ would also be proper, (Proposition 0.12), and there would exist a non-empty open subset $\mathcal{U}$ of $\mathfrak{A}$ such that $\mathcal{U} \cap \mathcal{S}^{-1}(\mathcal{D}) = \emptyset$. By the open mapping theorem, $\mathcal{S}(\mathcal{U})$ would be open, and this would contradict the fact that $\mathcal{D}$ is dense. Therefore $\mathcal{S}^{-1}(\mathcal{D}) = \mathfrak{A}$, and $\mathcal{D} = \mathcal{S}(\mathcal{S}^{-1}(\mathcal{D})) = \mathcal{H}$. $\quad \square$

Corollary 9.10. *The only strictly cyclic transitive algebra is $\mathcal{B}(\mathcal{H})$.*

Proof. This follows immediately from Corollary 9.9. $\quad \square$

There is an approach to the finite-dimensional case of the reductive algebra problem that is very different from the above.

Theorem 9.11. *If $\mathfrak{A}$ is a reductive algebra, and if there exists a col-lection $\{\mathcal{M}_i\}_{i=1}^n \subset Lat\,\mathfrak{A}$ such that $\mathcal{H} = \sum_{i=1}^n \oplus \mathcal{M}_i$ and $\mathfrak{A}|\mathcal{M}_i = \mathcal{B}(\mathcal{M}_i)$ for all i, then $\mathfrak{A}$ is a von Neumann algebra.*

Proof. We proceed by induction on n; the case $n=1$ is trivial since $\mathcal{B}(\mathcal{H})$ is a von Neumann algebra.

Assume that the theorem is known for $n-1$, and consider the case n. We must show that $A \in \mathfrak{A}$ implies $A^* \in \mathfrak{A}$. It will be convenient to con-sider two cases separately.

Case (i): there exist distinct integers j and k between 1 and n and an algebra isomorphism ϕ of $\mathcal{B}(\mathcal{M}_j)$ onto $\mathcal{B}(\mathcal{M}_k)$ such that $A|\mathcal{M}_k = \phi(A|\mathcal{M}_j)$ for $A \in \mathfrak{A}$.

Since $\mathcal{B}(\mathcal{M}_j)$ and $\mathcal{B}(\mathcal{M}_k)$ are algebraically isomorphic, either $\mathcal{M}_j$ and $\mathcal{M}_k$ are both infinite-dimensional or both have the same finite dimension. Thus $\mathcal{M}_j$ and $\mathcal{M}_k$ can be identified with each other, and Theorem 7.9 implies that there is an invertible operator S mapping $\mathcal{M}_j$ onto $\mathcal{M}_k$ such that $\phi(B) = SBS^{-1}$ for $B \in \mathcal{B}(\mathcal{M}_j)$. We shall show that S can be taken to be unitary.

Let $\mathcal{M}$ be the subspace of $\mathcal{H}$ consisting of all vectors of the form $\sum_{i=1}^n \oplus x_i$ such that $x_i \in \mathcal{M}_i$ and $x_k = Sx_j$. The definition of ϕ implies that $\mathcal{M} \in Lat\,\mathfrak{A}$, and it follows that $\mathcal{M}^\perp \in Lat\,\mathfrak{A}$. A trivial computation gives

$$\mathcal{M}^\perp = \left\{ \sum_{i=1}^n \oplus x_i : x_i = 0 \text{ for } i \neq j,k \text{ and } x_j = -S^* x_k \right\}.$$

The fact that $A\mathcal{M}^\perp \subset \mathcal{M}^\perp$ for $A \in \mathfrak{A}$, together with the assumption that $\mathfrak{A}|\mathcal{M}_j = \mathcal{B}(\mathcal{M}_j)$, implies that

$$-S^* SBS^{-1} = -BS^* \quad \text{for all } B \in \mathcal{B}(\mathcal{M}_j).$$

Hence $S^* SB = BS^* S$ for $B \in \mathcal{B}(\mathcal{M}_j)$, and it follows that $S^* S = \lambda$ for some positive number λ. Let $U = \lambda^{-\frac{1}{2}} S$. Then clearly U is unitary and $\phi(B) = UBU^{-1}$ for $B \in \mathcal{B}(\mathcal{M}_j)$.

We next show that $\mathfrak{A}|\mathcal{M}_k^\perp$ is reductive. Lemma 9.1 implies that every invariant subspace of $\mathfrak{A}|\mathcal{M}_k^\perp$ is reducing; we must also prove that $\mathfrak{A}|\mathcal{M}_k^\perp$ is weakly closed. This follows easily, for if

$$\{A_{1,\alpha} \oplus \cdots \oplus A_{k-1,\alpha} \oplus A_{k+1,\alpha} \oplus \cdots \oplus A_{n,\alpha}\}$$

is a net in $\mathfrak{A}|\mathcal{M}_k^\perp$ which converges to

$$A_1 \oplus \cdots \oplus A_{k-1} \oplus A_{k+1} \oplus \cdots \oplus A_n$$

in the weak topology, then the net

$$_\circ\{A_{1,\alpha}\oplus\cdots\oplus A_{k-1,\alpha}\oplus UA_{j,\alpha}U^{-1}\oplus A_{k+1,\alpha}\oplus\cdots\oplus A_{n,\alpha}\}$$

is contained in $\mathfrak{A}$ and converges weakly to

$$A_1\oplus\cdots\oplus A_{k-1}\oplus UA_jU^{-1}\oplus A_{k+1}\oplus\cdots\oplus A_n.$$

Therefore $A_1\oplus\cdots\oplus A_{k-1}\oplus A_{k+1}\oplus\cdots\oplus A_n$ is in $\mathfrak{A}|\mathcal{M}_k^\perp$.

Since $\mathfrak{A}|\mathcal{M}_k^\perp$ is reductive, the inductive hypothesis implies that $\mathfrak{A}|\mathcal{M}_k^\perp$ is self-adjoint. Now if

$$A = A_1\oplus\cdots\oplus A_{k-1}\oplus A_k\oplus A_{k+1}\oplus\cdots\oplus A_n$$

is in $\mathfrak{A}$, then $A_k = UA_jU^{-1}$. Since $\mathfrak{A}|\mathcal{M}_k^\perp$ is self-adjoint, the operator $A_1^*\oplus\cdots\oplus A_{k-1}^*\oplus A_{k+1}^*\oplus\cdots\oplus A_n^*$ is in $\mathfrak{A}|\mathcal{M}_k^\perp$, and it follows that the operator

$$A_1^*\oplus\cdots\oplus A_{k-1}^*\oplus UA_j^*U^{-1}\oplus A_{k+1}^*\oplus\cdots\oplus A_n$$

is in $\mathfrak{A}$. Since U is unitary, this last operator is A^*.

Case (ii): there do not exist integers j and k as in case (i).

Consider a pair (j,k) of distinct integers between 1 and n. If $A|\mathcal{M}_j=0$ implies $A|\mathcal{M}_k=0$ for $A\in\mathfrak{A}$, then we can define a map from $\mathcal{B}(\mathcal{M}_j)$ onto $\mathcal{B}(\mathcal{M}_k)$ by setting $\phi(A|\mathcal{M}_j) = A|\mathcal{M}_k$ for $A\in\mathfrak{A}$; (recall that $\mathfrak{A}|\mathcal{M}_j = \mathcal{B}(\mathcal{M}_j)$). Such a ϕ is clearly an algebra homomorphism. If, in addition, $A|\mathcal{M}_k=0$ implies $A|\mathcal{M}_j=0$ for $A\in\mathfrak{A}$, then ϕ is an isomorphism. Thus the assumption that there does not exist a pair of integers as in case (i) implies that for each pair (j,k) of integers between 1 and n there is an $A\in\mathfrak{A}$ such that exactly one of the operators $A|\mathcal{M}_j$ and $A|\mathcal{M}_k$ is 0.

We now prove that there is an $A\in\mathfrak{A}$ such that $A|\mathcal{M}_i$ is different from 0 for exactly one integer i between 1 and n. Choose $A\in\mathfrak{A}$ different from 0 for which the number of integers i with $A|\mathcal{M}_i=0$ is maximal. By permuting the indices we can assume that

$$A = A_1\oplus\cdots\oplus A_m\oplus 0\oplus\cdots\oplus 0$$

with $A_i\neq 0$ for $1\leq i\leq m$. We must prove that $m=1$.

If $m>1$, then, by interchanging the indices 1 and m if necessary, we can assume that $\mathfrak{A}$ contains an operator C such that $C|\mathcal{M}_1\neq 0$ and $C|\mathcal{M}_m=0$. Let $\mathcal{I}$ denote the set of all $B\in\mathcal{B}(\mathcal{M}_1)$ such that $B=\hat{B}|\mathcal{M}_1$ for some $\hat{B}\in\mathfrak{A}$ with $\hat{B}|\mathcal{M}_m=0$. The fact that $\mathfrak{A}|\mathcal{M}_i=\mathcal{B}(\mathcal{M}_i)$ for all i implies that $\mathcal{I}$ is a two sided ideal in $\mathcal{B}(\mathcal{M}_1)$. Since $\mathcal{I}$ is not $\{0\}$, $\mathcal{I}$ contains all operators of finite rank, (by Theorem 7.8). Let $\mathcal{J}$ denote the set of all operators $B\in\mathcal{B}(\mathcal{M}_1)$ such that $B=\hat{B}|\mathcal{M}_1$ for some $\hat{B}\in\mathfrak{A}$ with $\hat{B}|\mathcal{M}_i=0$ for $i>m$. Then $\mathcal{J}$ is also a two-sided ideal in $\mathcal{B}(\mathcal{M}_i)$, and $\mathcal{J}$ is not $\{0\}$ since $A_1\in\mathcal{J}$. Thus $\mathcal{J}$ contains all operators of finite rank.

Let P be any projection of finite rank in $\mathscr{B}(\mathscr{M}_1)$ other than 0. Since $P \in \mathscr{I} \cap \mathscr{J}$, there exist operators of the form

$$P \oplus B_2 \oplus \cdots \oplus B_{m-1} \oplus 0 \oplus B_{m+1} \oplus B_{m+2} \oplus \cdots \oplus B_n$$

and

$$P \oplus C_2 \oplus \cdots \oplus C_{m-1} \oplus C_m \oplus 0 \oplus 0 \oplus \cdots \oplus 0$$

in $\mathfrak{A}$. Hence the product

$$P \oplus B_2 C_2 \oplus \cdots \oplus B_{m-1} C_{m-1} \oplus 0 \oplus 0 \oplus \cdots \oplus 0$$

of these two operators is in $\mathfrak{A}$. This contradicts the minimality of m, and we conclude that $m = 1$.

Since $m = 1$ the ideal $\mathscr{J}$ constructed above is the set of all operators B on $\mathscr{M}_1$ such that $\hat{B}|\mathscr{M}_1 = B$ and $\hat{B}|\mathscr{M}_i = 0$ when $i > 1$ for some $\hat{B} \in \mathfrak{A}$. Thus $\mathscr{J}$ is weakly closed, and, since every operator is in the weak closure of the set of operators of finite rank, it follows that $\mathscr{J} = \mathscr{B}(\mathscr{M}_1)$. Thus $\mathfrak{A}$ is the direct sum of $\mathfrak{A}|\mathscr{M}_1$ and $\mathfrak{A}|\mathscr{M}_1^\perp$. In particular $\mathfrak{A}|\mathscr{M}_1^\perp$ is weakly closed, and, by Lemma 9.1, $\mathfrak{A}|\mathscr{M}_1^\perp$ is reductive. The inductive hypothesis implies that $\mathfrak{A}|\mathscr{M}_1^\perp$ is self-adjoint, and therefore $\mathfrak{A} = \mathscr{B}(\mathscr{M}_1) \oplus (\mathfrak{A}|\mathscr{M}_1^\perp)$ is a von Neumann algebra. $\square$

This theorem applies, in particular, to the case where every $\mathscr{M}_i$ is finite-dimensional; (the proof can be somewhat simplified in this case, as there is no need to consider weak closures of subalgebras or ideals). This gives an alternate proof for the finite-dimensional case.

Corollary 9.12. *If $\mathfrak{A}$ is a reductive subalgebra of $\mathscr{B}(\mathscr{H})$ and $\mathscr{H}$ is finite-dimensional, then $\mathfrak{A}$ is a von Neumann algebra.*

Proof. If $\mathfrak{A}$ is transitive, then $\mathfrak{A} = \mathscr{B}(\mathscr{H})$ by Burnside's theorem, (Corollary 8.6). If $\mathfrak{A}$ has a non-trivial invariant subspace, then it has a minimal one, $\mathscr{M}_1$, since $\mathscr{H}$ is finite-dimensional. Lemma 9.1 implies that $\mathfrak{A}|\mathscr{M}_1^\perp$ is reductive. If $\mathfrak{A}|\mathscr{M}_1^\perp$ is not transitive, it has a minimal invariant subspace, $\mathscr{M}_2$. Proceeding in this manner we obtain a decomposition $\mathscr{H} = \sum_{i=1}^{n} \oplus \mathscr{M}_i$ such that $\mathfrak{A}|\mathscr{M}_i$ is transitive for all i. Burnside's theorem implies that $\mathfrak{A}|\mathscr{M}_i = \mathscr{B}(\mathscr{M}_i)$ for each i. Thus Theorem 9.11 applies. $\square$

We will generalize Theorem 8.21 to prove that a reductive algebra which contains an abelian von Neumann algebra of finite uniform multiplicity is self-adjoint.

Lemma 9.13. *If $A \in \mathscr{B}(\mathscr{H})$ and $\mathscr{M}$ is a subspace such that there is a net $\{P_\alpha\}$ of projections converging strongly to 1 with $P_\alpha \mathscr{M} \subset \mathscr{M}$ and $P_\alpha \mathscr{M} \in \operatorname{Lat} P_\alpha A P_\alpha$ for all α, then $\mathscr{M} \in \operatorname{Lat} A$.*

Proof. Let $x \in \mathcal{M}$ and $y \in \mathcal{M}^{\perp}$; we must prove that $(Ax, y) = 0$. Since $P_{\alpha}\mathcal{M} \subset \mathcal{M}$, $P_{\alpha}y$ is orthogonal to $P_{\alpha}\mathcal{M}$, and therefore

$$0 = (P_{\alpha}AP_{\alpha}x, P_{\alpha}y) = (AP_{\alpha}x, P_{\alpha}y) \quad \text{for all } \alpha.$$

Thus

$$
\begin{aligned}
|(Ax, y)| &= |(Ax, y) - (AP_{\alpha}x, P_{\alpha}y)| \\
&= |(Ax, y) - (Ax, P_{\alpha}y) + (Ax, P_{\alpha}y) - (AP_{\alpha}x, P_{\alpha}y)| \\
&\leq |(Ax, y - P_{\alpha}y)| + |(Ax - AP_{\alpha}x, P_{\alpha}y)| \\
&\leq \|Ax\| \, \|y - P_{\alpha}y\| + \|x - P_{\alpha}x\| \, \|A^*P_{\alpha}y\|.
\end{aligned}
$$

Hence $(Ax, y) = 0$. $\quad\square$

Lemma 9.14. *If $\mathfrak{A}$ is reductive, $T \in \mathfrak{A}'$, and T is similar to a normal operator, then $T \in (\mathfrak{A}^*)'$.*

Proof. Suppose that $T = S^{-1}NS$, where N is normal and $N = \int \lambda dE_{\lambda}$, (cf. Theorem 1.12). For $A \in \mathfrak{A}$, $AT = TA$ implies $(SAS^{-1})N = N(SAS^{-1})$. Hence, for each E_{λ}, $AS^{-1}E_{\lambda}S = S^{-1}E_{\lambda}SA$ for all $A \in \mathfrak{A}$ by Theorem 1.16. The range and the nullspace of the projection $S^{-1}E_{\lambda}S$ are therefore each invariant under $\mathfrak{A}$, thus also under $\mathfrak{A}^*$, and it follows that $A^*S^{-1}E_{\lambda}S = S^{-1}E_{\lambda}SA^*$ for $A \in \mathfrak{A}$, (cf. Theorem 0.2). Therefore, for $A \in \mathfrak{A}$, $SA^*S^{-1}N = NSA^*S^{-1}$ or $A^*T = TA^*$. $\quad\square$

Theorem 9.15. *If $\mathfrak{A}$ is reductive, $\mathfrak{R}$ is a m.a.s.a., k is a positive integer and $\mathfrak{A} \supset \mathfrak{R}^{(k)}$, then $\mathfrak{A}$ is self-adjoint.*

Proof. By Lemma 7.1, it suffices to show that $\mathfrak{A}^{(n)}$ is reductive for all n, or that $A \in \mathfrak{A}$ implies $Lat\,\mathfrak{A}^{(n)} \subset Lat(A^*)^{(n)}$ for all n. This is the case for $n = 1$ by hypothesis; assume that $Lat\,\mathfrak{A}^{(n-1)} \subset Lat(A^*)^{(n-1)}$ for all $A \in \mathfrak{A}$ and let $\mathcal{M} \in Lat\,\mathfrak{A}^{(n)}$.

Define

$$\mathcal{M}_1 = \{x_1 \oplus x_2 \oplus \cdots \oplus x_n \in \mathcal{M} : x_1 = 0\}.$$

Then $\mathcal{M}_1 \in Lat\,\mathfrak{A}^{(n)}$, and the induction hypothesis implies that $\mathcal{M}_1^{\perp} \in Lat(\mathfrak{A}^*)^{(n)}$. We now define $\mathcal{N} = \mathcal{M} \ominus \mathcal{M}_1$; then $\mathcal{N} \in Lat\,\mathfrak{A}^{(n)}$, and it suffices to show that $\mathcal{N} \in Lat(A^*)^{(n)}$ for all $A \in \mathfrak{A}$.

The special form of $\mathcal{N}$ implies that it is a graph invariant subspace of $\mathfrak{A}^{(n)}$:

$$\mathcal{N} = \{x \oplus T_1 x \oplus T_2 x \oplus \cdots \oplus T_{n-1} x : x \in \mathcal{D}\}.$$

Fix a strong basic neighbourhood $\mathscr{V}$ of 1:

$$\mathscr{V} = \{A \in \mathscr{B}(\mathscr{H}) : \|Ax_i - x_i\| < \varepsilon \text{ for } i = 1, \ldots, m\}.$$

Let $\mathscr{U} = \{A \in \mathscr{B}(\mathscr{H}) : \|Ax_i - x_i\| < \varepsilon/2 \text{ for } i = 1, \ldots, m\}$. Now $\mathcal{N} \in Lat\,\mathfrak{R}^{(nk)}$, and, by Lemma 8.20, there exists a projection $Q_0 \in \mathfrak{R}$ such that if $Q = (Q_0)^{(k)}$, then $Q \in \mathscr{U}$ and $\tilde{Q} Q^{(n)} \mathscr{N}$ is closed for every special projection $\tilde{Q}$ on $\mathscr{H}^{(n)}$. Fix such a Q. As in the proof of Theorem 8.21 it follows that $Q T_i Q \in (\mathfrak{R}^{(k)})'$ for $i = 1, \ldots, n-1$, and that $Q \mathscr{D} \in Lat \, \mathfrak{R}^{(k)}$, $Q \mathscr{D} \subset \mathscr{D}$. Moreover, $\mathscr{N} \in Lat \, \mathfrak{A}^{(n)}$ and $Q^{(n)} \in \mathfrak{A}^{(n)}$ implies $Q^{(n)} \mathscr{N} \subset \mathscr{N}$ and $Q^{(n)} \mathscr{N} \in Lat \, Q^{(n)} A^{(n)} Q^{(n)}$ for all $A \in \mathfrak{A}$.

Since $(Q T_i Q) \in (\mathfrak{R}^{(k)})'$, Theorem 7.23 implies that there is a projection $Q_1 \in \mathfrak{R}$ such that $Q_1^{(k)} \in \mathscr{U}$ and $Q_1^{(k)} Q T_i Q Q_1^{(k)} = N_i + S_i$, where N_i is a nilpotent operator, S_i is similar to a normal operator, and $S_i N_i = N_i S_i$. Let $P = Q_1^{(k)} Q$. Then $P \in \mathscr{V}$, $P^{(n)} \mathscr{N} \subset \mathscr{N}$, $P^{(n)} \mathscr{N} \in Lat \, P^{(n)} \mathfrak{A}^{(n)} P^{(n)}$, and

$$P^{(n)} \mathscr{N} = \{x \oplus P T_1 P x \oplus \cdots \oplus P T_{n-1} P x : x \in P \mathscr{D}\}.$$

Note also that $P \in \mathfrak{A}$ implies $(P \mathfrak{A} P)|P \mathscr{H}$ is a reductive algebra. For each $A \in \mathfrak{A}$, $P A P$ commutes with $P T_i P = N_i + S_i$, and, by Theorem 7.24, it follows that each $P A P$ commutes with each N_i and each S_i. Now Lemmas 9.3 and 9.14 imply that $P A^* P$ commutes with $P T_i P$ for each i; i.e., $P^{(n)} \mathscr{N} \in Lat \, P^{(n)} (\mathfrak{A}^*)^{(n)} P^{(n)}$. Since there is such a P in every strong neighbourhood $\mathscr{V}$ of 1 (and thus such a $P^{(n)}$ in every strong neighbourhood of the identity on $\mathscr{H}^{(n)}$), it follows from Lemma 9.13 that $\mathscr{N} \in Lat \, (\mathfrak{A}^*)^{(n)}$. Hence $\mathfrak{A}^{(n)}$ is reductive for all n, by induction, and $\mathfrak{A}$ is self-adjoint. $\quad\square$

Another sufficient condition that a reductive algebra be self-adjoint is given in Corollary 9.22.

9.2 Reflexive Operator Algebras

Definition. If $\mathscr{F}$ is any collection of subspaces of $\mathscr{H}$, then $Alg \, \mathscr{F}$ is the collection of operators $A \in \mathscr{B}(\mathscr{H})$ such that $\mathscr{F} \subset Lat \, A$.

Obviously for any collection $\mathscr{F}$ of subspaces, $Alg \, \mathscr{F}$ is a weakly closed subalgebra of $\mathscr{B}(\mathscr{H})$ which contains 1. If $\mathfrak{A}$ is any subset of $\mathscr{B}(\mathscr{H})$, then $\mathfrak{A} \subset Alg \, Lat \, \mathfrak{A}$.

Definition. The algebra $\mathfrak{A} \subset \mathscr{B}(\mathscr{H})$ is *reflexive* if $\mathfrak{A} = Alg \, Lat \, \mathfrak{A}$.

The reflexive subalgebras of $\mathscr{B}(\mathscr{H})$ are those that are determined by their invariant subspaces. It is trivial to see that $Alg \, \mathscr{F}$ is reflexive for each collection $\mathscr{F}$ of subspaces of $\mathscr{H}$.

The transitive and reductive algebra problems can be rephrased as problems about reflexive algebras. The transitive algebra problem is: if $\mathfrak{A}$ is a transitive algebra must $\mathfrak{A}$ be reflexive? The reductive algebra problem is equivalent to the question: if $\mathfrak{A}$ is a reductive algebra must $\mathfrak{A}$ be reflexive?

This section is devoted to establishing a number of sufficient conditions that a subalgebra of $\mathscr{B}(\mathscr{H})$ be reflexive.

Theorem 9.16. *If $\mathfrak{A}$ is a reflexive algebra, then the algebras $\mathfrak{A}^* = \{A^*: A \in \mathfrak{A}\}$ and $\{S^{-1}AS: A \in \mathfrak{A}\}$, (for any invertible operator S), are also reflexive. Moreover, the direct sum of any number of reflexive algebras is reflexive.*

Proof. Each of these assertions has a trivial verification.   □

Theorem 9.17. *Every von Neumann algebra is reflexive.*

Proof. This is an easy consequence of the double commutant theorem. For suppose that $\mathfrak{A}$ is a von Neumann algebra. We must show that $B \in \mathscr{B}(\mathscr{H})$ and $Lat\,\mathfrak{A} \subset Lat\,B$ implies $B \in \mathfrak{A}$. Fix such a B. Then every reducing subspace of $\mathfrak{A}$ reduces B, hence every projection in $\mathfrak{A}'$ commutes with B. Thus, by Theorem 7.3, $B \in \mathfrak{A}''$, and the double commutant theorem (Theorem 7.5) implies that $B \in \mathfrak{A}$.   □

Note that Theorem 9.17 is really equivalent to the double commutant theorem; (in fact, our proof of the double commutant theorem proceeded by first establishing Theorem 9.17). Thus the sufficient conditions for reflexivity presented below can be regarded as analogues of the double commutant theorem in various contexts.

Theorem 9.18. *If $\mathfrak{A}$ is a weakly closed algebra containing 1 and there exists a unitary operator U taking $\mathscr{H}$ onto $\mathscr{H}^{(2)}$ such that $A^{(2)} = UAU^{-1}$ for all $A \in \mathfrak{A}$, then $\mathfrak{A}$ is reflexive.*

Proof. Suppose that $B \in \mathscr{B}(\mathscr{H})$ and $Lat\,\mathfrak{A} \subset Lat\,B$. By Theorem 7.1 it suffices to show that $Lat\,\mathfrak{A}^{(n)} \subset Lat\,B^{(n)}$ for all n. Fix $n > 1$. Since there exists a U as in the hypothesis, there also exists a unitary operator V such that $A^{(3)} = VAV^{-1}$ for $A \in \mathfrak{A}$; $(A \oplus A \oplus A = UAU^{-1} \oplus A$, which is unitarily equivalent to $A \oplus A)$. In the same way it can be seen that there is a unitary operator V such that $A^{(n)} = VAV^{-1}$ for $A \in \mathfrak{A}$.

To prove that $Lat\,\mathfrak{A}^{(n)} \subset Lat\,B^{(n)}$, let $\mathscr{M} \in Lat\,\mathfrak{A}^{(n)}$, and let V be the unitary operator implementing the unitary equivalence of $\mathfrak{A}^{(n)}$ and $\mathfrak{A}: A^{(n)} = VAV^{-1}$ for $A \in \mathfrak{A}$. Then $V^{-1}\mathscr{M} \in Lat\,\mathfrak{A}$, and therefore $V^{-1}\mathscr{M} \in Lat\,B$. It follows that $\mathscr{M} \in Lat\,VBV^{-1}$. To complete the proof we need only show that $VBV^{-1} = B^{(n)}$. Since V is unitary, the fact that $VAV^{-1} = A^{(n)}$ implies $VA^*V^{-1} = (A^*)^{(n)}$, and we conclude that $VCV^{-1} = C^{(n)}$ for every operator C in the von Neumann algebra generated by $\mathfrak{A}$. The fact that $Lat\,\mathfrak{A} \subset Lat\,B$ implies, in particular, that every reducing subspace of $\mathfrak{A}$ reduces B, and it follows from the double commutant theorem that B is in this von Neumann algebra. Hence $VBV^{-1} = B^{(n)}$.   □

Definition. The operator A is an *inflation* if there exists an operator C such that A is unitarily equivalent to $\sum_{i=1}^{\infty} \oplus C_i$ with $C_i = C$ for all i.

Corollary 9.19. *If A is an inflation, then the weakly closed algebra generated by $\{A,1\}$ is reflexive.*

Proof. Clearly there is a unitary operator U such that $A^{(2)} = UAU^{-1}$. It follows that $B^{(2)} = UBU^{-1}$ for all B in the weakly closed algebra generated by $\{A,1\}$, and Theorem 9.18 gives the result. $\square$

We shall prove that weakly closed algebras of normal operators are reflexive. We require the fact that every such algebra is commutative.

Lemma 9.20. *If $\mathfrak{A}$ is a linear manifold in $\mathscr{B}(\mathscr{H})$ consisting of normal operators, then $AB = BA$ whenever A and B are in $\mathfrak{A}$.*

Proof. Given A and B in $\mathfrak{A}$,

$$2(B^*A - AB^*) = (A+B)^*(A+B) - (A+B)(A+B)^*$$
$$+ i[(A+iB)^*(A+iB) - (A+iB)(A+iB)^*],$$

(as multiplying shows). The fact that every member of $\mathfrak{A}$ is normal implies that the right side is 0. Hence $B^*A = AB^*$, and Fuglede's theorem (Corollary 1.18) implies $AB = BA$. $\square$

Theorem 9.21. *If $\mathfrak{A}$ is a weakly closed algebra of normal operators which contains 1, then $\mathfrak{A}$ is reflexive.*

Proof. The proof is similar to the proof of Theorem 9.18. Suppose that $Lat\,\mathfrak{A} \subset Lat\,B$. To prove that $Lat\,\mathfrak{A}^{(n)} \subset Lat\,B^{(n)}$ it suffices to show that every cyclic invariant subspace of $\mathfrak{A}^{(n)}$, (i.e., subspace of the form $\overline{\{A^{(n)}x : A \in \mathfrak{A}\}}$ for some $x \in \mathscr{H}^{(n)}$), is in $Lat\,B^{(n)}$. Let $\mathscr{M} = \overline{\{A^{(n)}x : A \in \mathfrak{A}\}}$. Lemma 9.20 implies $\mathfrak{A}$ is abelian, and therefore Fuglede's theorem, (Corollary 1.18), implies that the von Neumann algebra $\mathfrak{B}$ generated by $\mathfrak{A}$ is abelian. Now, by Theorem 7.12, there is a Hermitian operator H which generates $\mathfrak{B}$ as a von Neumann algebra.

Let E_λ denote the spectral measure of H; (cf. Section 1.4). Then Theorem 1.14 implies that the spectral measure of $H^{(n)}$ is $E_\lambda^{(n)}$, where $E^{(n)}(S) = (E(S))^{(n)}$ for each Borel set S. There is some $z \in \mathscr{H}$ such that $E(S) = 0$ if and only if $(E(S)z,z) = 0$. (One way to see this is by representing H as M_ϕ on $\mathscr{L}^2(X,\mu)$, using Theorem 1.6, and taking z to be the function identically equal to 1 on X).

Define the measures v_1 and v_2 by setting $v_1(S) = (E^{(n)}(S)x,x)$ and $v_2(S) = (E(S)z,z)$ for each Borel set S. Then $v_2(S) = 0$ implies that $E(S) = 0$ and, hence, that $v_1(S) = 0$. Thus Theorem 1.15 shows that there is some $y \in \mathscr{H}$ such that $(E^{(n)}(S)x,x) = (E(S)y,y)$ for every Borel set S. Now for each pair (j,k) of natural numbers.

$$((H^{(n)})^j x, (H^{(n)})^k x) = ((H^{(n)})^{j+k}x, x) = \int \lambda^{j+k} d(E_\lambda^{(n)}x, x)$$
$$= \int \lambda^{j+k} d(E_\lambda y, y) = (H^{j+k}y, y) = (H^j y, H^k y).$$

Thus if $\mathcal{N}$ is the smallest reducing subspace of H containing y and $\mathcal{N}_n$ is the smallest reducing subspace of $H^{(n)}$ containing x, then $H|\mathcal{N}$ is unitarily equivalent to $H^{(n)}|\mathcal{N}_n$. Let U be a unitary operator taking $\mathcal{N}$ onto $\mathcal{N}_n$ such that $U(H|\mathcal{N})U^{-1} = H^{(n)}|\mathcal{N}_n$.

Note that $\mathcal{N} \in Lat\,\mathfrak{B}$ implies that $\mathcal{N}$ reduces $\mathfrak{A}$, and similarly $\mathcal{N}_n$ reduces $\mathfrak{A}^{(n)}$. Also $x \in \mathcal{N}_n$ implies $\mathcal{M} \subset \mathcal{N}_n$. Since $U(H|\mathcal{N})U^{-1} = H^{(n)}|\mathcal{N}_n$, it follows that $U(C|\mathcal{N})U^{-1} = C^{(n)}|\mathcal{N}_n$ for all $C \in \mathfrak{B}$. The double commutant theorem (Theorem 7.5) implies $B \in \mathfrak{B}$ so that $U(B|\mathcal{N})U^{-1} = B^{(n)}|\mathcal{N}_n$.

We now easily verify the fact that $\mathcal{M} \in Lat\,B^{(n)}$. For $\mathcal{M} \in Lat\,\mathfrak{A}^{(n)}$ implies $\mathcal{M} \in Lat(\mathfrak{A}^{(n)}|\mathcal{N}_n)$, and thus $U^{-1}\mathcal{M} \in Lat(\mathfrak{A}|\mathcal{N})$. Therefore $U^{-1}\mathcal{M} \in Lat\,\mathfrak{A}$, and $U^{-1}\mathcal{M} \in Lat\,B$. Hence $\mathcal{M} \in Lat(U(B|\mathcal{N})U^{-1}) = Lat(B^{(n)}|\mathcal{N})$, and $\mathcal{M} \in Lat\,B^{(n)}$. $\square$

Corollary 9.22. *A reductive algebra of normal operators is an abelian von Neumann algebra.*

Proof. This follows immediately from Theorem 9.21 and Lemma 9.20. $\square$

There is a result about certain algebras of non-normal operators which is very similar to the above theorem.

Theorem 9.23. *If $\mathfrak{A}$ is a weakly closed algebra of analytic Toeplitz operators containing 1, then $\mathfrak{A}$ is reflexive.*

Proof. The proof of this theorem is very much like the proof of Theorem 9.21, with the role of the Hermitian operator H played by the unilateral shift. Let S denote the unilateral shift of multiplicity 1. It follows immediately from Theorem 3.3 that the algebra $\mathfrak{B}$ of all analytic Toeplitz operators is the weakly closed algebra generated by $\{S,1\}$.

Suppose that $Lat\,\mathfrak{A} \subset Lat\,B$ for some B. Then $Lat\,\mathfrak{B} \subset Lat\,\mathfrak{A}$ implies $Lat\,S \subset Lat\,B$. By Proposition 3.9, $BS = SB$, and Theorem 3.4 implies $B \in \mathfrak{B}$.

Let n be a positive integer, and $\mathcal{M}$ be the cyclic invariant subspace of $\mathfrak{A}^{(n)}$ generated by x. The unilateral shift $S^{(n)}$ of multiplicity n is contained in $\mathfrak{B}^{(n)}$. Let $\mathcal{N}_n$ be the invariant subspace of $S^{(n)}$ generated by x. Note that $\mathcal{M} \subset \mathcal{N}_n$. Now Theorem 3.33 implies that $S^{(n)}|\mathcal{N}_n = USU^{-1}$ for some unitary operator U taking $\mathcal{H}^2$ onto $\mathcal{N}_n$. Then $UCU^{-1} = C^{(n)}|\mathcal{N}_n$ for all $C \in \mathfrak{B}$, and it follows that $U^{-1}\mathcal{M} \in Lat\,\mathfrak{A}$. Thus $U^{-1}\mathcal{M} \in Lat\,B$, and $\mathcal{M} \in Lat\,UBU^{-1}$. But $UBU^{-1} = B^{(n)}|\mathcal{N}_n$ and, hence, $\mathcal{M} \in Lat\,B^{(n)}$. $\square$

The next sufficient condition for reflexivity is another generalization of Corollary 8.10.

Theorem 9.24. *If $\mathfrak{A}$ is a weakly closed algebra containing a m.a.s.a., and if $Lat\,\mathfrak{A}$ is totally ordered, then $\mathfrak{A}$ is reflexive.*

Proof. Let $\mathfrak{R}$ be the m.a.s.a. contained in $\mathfrak{A}$. We divide the proof into several steps.

Assertion (i): if $\mathscr{M}$ is an invariant linear manifold of $\mathfrak{A}^{(2)}$ such that $0 \oplus x \in \mathscr{M}$ implies $x = 0$, then there is a complex number λ such that $\mathscr{M} = \{x \oplus \lambda x : x \in \mathscr{D}\}$ for some invariant linear manifold $\mathscr{D}$ of $\mathfrak{A}$.

To prove this first observe that $\mathscr{M} = \{x \oplus Tx : x \in \mathscr{D}\}$ for some linear transformation T and domain $\mathscr{D}$. We must show that T is a multiple of the identity on $\mathscr{D}$. Now $\mathscr{D}$ is invariant under $\mathfrak{A}$, and so is the closure of $\mathscr{D}$. Let P denote the projection of $\mathscr{H}$ onto the closure of $\mathscr{D}$. The transformation T commutes with $\mathfrak{A}$, and, since $P \in \mathfrak{R}$, it thus commutes with P. Hence the range of T is a subset of $P\mathscr{H}$, and we can regard T as a densely defined operator on $P\mathscr{H}$. Then T commutes with PRP for every $R \in \mathfrak{R}$. It follows from Theorem 7.16 that, for each strong neighbourhood $\mathscr{U}$ of the identity on $P\mathscr{H}$ there is a projection $Q \in \mathscr{U} \cap \mathfrak{R}$ such that $Q\tilde{T}Q \in \mathfrak{R}$, where $\tilde{T}$ is the closure of T. If every such $Q\tilde{T}Q$ is a multiple of Q, then T is a multiple of the identity on $\mathscr{D}$.

If $Q\tilde{T}Q$ were not a multiple of Q there would exist a Borel set S such that $E(S)$ is a non-trivial spectral projection of $Q\tilde{T}|(Q\mathscr{H})$; (cf. the proof of Corollary 1.17). By Theorem 7.16 we can assume that QAQ commutes with $Q\tilde{T}Q$ for every $A \in \mathfrak{A}$. Hence Fuglede's Theorem (Theorem 1.16) implies that $QAQE(S) = E(S)QAQ$ for all $A \in \mathfrak{A}$. If $x \in E(S)\mathscr{H}$ and $y \in (1 - E(S))Q\mathscr{H}$ are non-zero vectors and $A \in \mathfrak{A}$, then

$$(Ax, y) = (AE(S)x, Qy) = (QAQE(S)x, y)$$
$$= (E(S)QAQx, y) = 0.$$

Similarly, $(Ay, x) = 0$. Therefore if $\mathscr{M}$ and $\mathscr{N}$ are the closures of $\{Ax : A \in \mathfrak{A}\}$ and $\{Ay : A \in \mathfrak{A}\}$ respectively, $x \in \mathscr{M} \setminus \mathscr{N}$ and $y \in \mathscr{N} \setminus \mathscr{M}$. Thus $\mathscr{M}$ and $\mathscr{N}$ are not comparable, which contradicts the hypothesis that $Lat\,\mathfrak{A}$ is totally ordered. This proves assertion (i).

Assertion (ii): if $\mathscr{M}$ is an invariant linear manifold of $\mathfrak{A}^{(n)}$ (for any $n \geq 2$), and if $0 \oplus 0 \oplus \cdots \oplus 0 \oplus x \in \mathscr{M}$ implies $x = 0$, then there exist complex numbers $\lambda_1, \ldots, \lambda_{n-1}$ and an invariant linear manifold $\mathscr{D}$ of $\mathfrak{A}^{(n-1)}$ such that

$$\mathscr{M} = \left\{x_1 \oplus \cdots \oplus x_{n-1} \oplus \sum_{i=1}^{n-1} \lambda_i x_i : x_1 \oplus \cdots \oplus x_{n-1} \in \mathscr{D}\right\}.$$

The case $n = 2$ is assertion (i). Suppose that the result is known for $n - 1$, and let $\mathscr{M}$ be such an invariant linear manifold of $\mathfrak{A}^{(n)}$. Then

$$\mathscr{M}_1 = \{x_2 \oplus \cdots \oplus x_n : 0 \oplus x_2 \oplus \cdots \oplus x_n \in \mathscr{M}\}$$

is an invariant linear manifold of $\mathfrak{A}^{(n-1)}$, and meets the last coordinate space only in $\{0\}$. By the inductive hypothesis $\mathscr{M}_1$ consists entirely of vectors of the form $x_2 \oplus \cdots \oplus x_{n-1} \oplus \sum\limits_{i=2}^{n-1} \lambda_i x_i$ for some complex numbers $\lambda_2, \ldots, \lambda_{n-1}$. (Of course, $\mathcal{M}_1$ could consist of 0 alone; in that case choose any scalars $\lambda_2, \ldots, \lambda_{n-1}$.)

Now, since $\mathcal{M}$ does not meet the last coordinate space, it has the form

$$\mathcal{M} = \{y \oplus Ty : y \in \mathcal{D}\},$$

where $\mathcal{D}$ is the linear manifold in $\mathcal{H}^{(n-1)}$ defined by

$$\mathcal{D} = \{x_1 \oplus \cdots \oplus x_{n-1} : x_1 \oplus \cdots \oplus x_{n-1} \oplus x \in \mathcal{M} \text{ for some } x\},$$

and T is some linear transformation from $\mathcal{D}$ into the last coordinate space. Since $\mathcal{M}$ is invariant under $\mathfrak{A}^{(n)}$, $\mathcal{D}$ is invariant under $\mathfrak{A}^{(n-1)}$ and $T A^{(n-1)} = A T$ for each $A \in \mathfrak{A}$.

Define the linear transformation T_1 taking $\mathcal{D}$ into the last coordinate space by setting

$$T_1(x_1 \oplus \cdots \oplus x_{n-1}) = T(x_1 \oplus \cdots \oplus x_{n-1}) - \sum\limits_{i=2}^{n-1} \lambda_i x_i,$$

where the λ_i are as previously determined. Then $T_1 A^{(n-1)} = A T_1$ for $A \in \mathfrak{A}$, and

$$\mathcal{M} = \left\{ x_1 \oplus \cdots \oplus x_{n-1} \oplus \left(T_1(x \oplus \cdots \oplus x_{n-1}) + \sum\limits_{i=2}^{n-1} \lambda_i x_i \right) : (x_1 \oplus \cdots \oplus x_{n-1}) \in \mathcal{D} \right\}.$$

Now $x_1 = 0$ implies $T_1(x_1 \oplus \cdots \oplus x_{n-1}) = 0$. Hence $T_1(x_1 \oplus \cdots \oplus x_{n-1})$ is determined by x_1; let $T_1(x_1 \oplus \cdots \oplus x_{n-1}) = S x_1$. Then S is a linear transformation with domain

$$\mathcal{D}_1 = \{x_1 : x_1 \oplus x_2 \oplus \cdots \oplus x_{n-1} \in \mathcal{D} \text{ for some } x_2 \oplus \cdots \oplus x_{n-1}\}.$$

Since $T_1 A^{(n-1)} = A T_1$ it follows that S commutes with all the operators in $\mathfrak{A}$. The case $n=2$ (assertion (i) above) implies that $S x = \lambda_1 x$ for some complex number λ_1.

Hence

$$\mathcal{M} = \left\{ x_1 \oplus \cdots \oplus x_{n-1} \oplus \left(\sum\limits_{i=1}^{n-1} \lambda_i x_i \right) : x_1 \oplus \cdots \oplus x_{n-1} \in \mathcal{D} \right\},$$

and assertion (ii) is established.

Assertion (iii): if $Lat\,\mathfrak{A} \subset Lat\,B$, then $Lat\,\mathfrak{A}^{(n)} \subset Lat\,B^{(n)}$ for all positive integers n.

Assertion (iii) implies the theorem, by Theorem 7.1. We prove assertion (iii) by induction. Suppose that, for every algebra $\mathfrak{A}$ satisfying the hypotheses of the theorem, $Lat\,\mathfrak{A} \subset Lat\,B$ implies $Lat\,\mathfrak{A}^{(n-1)} \subset Lat\,B^{(n-1)}$. We must prove that $Lat\,\mathfrak{A} \subset Lat\,B$ implies $Lat\,\mathfrak{A}^{(n)}$

$\subset Lat\,B^{(n)}$. Suppose that $Lat\,\mathfrak{A} \subset Lat\,B$ and $\mathcal{M} \in Lat\,\mathfrak{A}^{(n)}$. We consider two cases.

Case (a): there is some coordinate space such that the intersection of $\mathcal{M}$ and the coordinate space consists of the zero vector alone. By rearranging if necessary we can assume that this coordinate space is the nth one; i.e., that $0 \oplus \cdots \oplus 0 \oplus x \in \mathcal{M}$ implies $x = 0$. By assertion (ii), then, there are complex numbers $\lambda_1, \ldots, \lambda_{n-1}$ such that

$$\mathcal{M} = \left\{ x_1 \oplus x_2 \oplus \cdots \oplus x_{n-1} \oplus \left(\sum_{i=1}^{n-1} \lambda_i x_i \right) : (x_1 \oplus x_2 \oplus \cdots \oplus x_{n-1}) \in \mathcal{D} \right\}$$

for some linear manifold $\mathcal{D} \subset \mathcal{H}^{(n-1)}$. Since $\mathcal{M}$ is closed it follows that $\mathcal{D}$ is closed, and therefore $\mathcal{D} \in Lat\,\mathfrak{A}^{(n-1)}$. By the inductive hypothesis $\mathcal{D} \in Lat\,B^{(n-1)}$. The relation between $\mathcal{M}$ and $\mathcal{D}$ obviously implies $\mathcal{M} \in Lat\,B^{(n)}$.

Case (b): $\mathcal{M}$ intersects every coordinate space in a subspace other than $\{0\}$. In this case, for each i, $1 \le i \le n$, those members of $\mathcal{M}$ all of whose components except the ith are 0 form a subspace

$$\{0\} \oplus \cdots \oplus \{0\} \oplus \mathcal{M}_i \oplus \{0\} \oplus \cdots \oplus \{0\}.$$

Then $\mathcal{M}_i \in Lat\,\mathfrak{A}$ for each i. Since $Lat\,\mathfrak{A}$ is totally ordered, one of the $\{\mathcal{M}_i\}$ is the smallest; suppose that $\mathcal{M}_{i_0} \subset \mathcal{M}_i$ for all i. We can now reduce this case to case (a) by "dividing out" $\mathcal{M}_{i_0}$. That is, let P denote the projection of $\mathcal{H}$ onto $\mathcal{M}_{i_0}^\perp$. Let $\hat{\mathfrak{A}} = \{PA|(P\mathcal{H}) : A \in \mathfrak{A}\}$, $\hat{\mathfrak{R}} = \{PR|(P\mathcal{H}) : R \in \mathfrak{R}\}$, and $\hat{B} = PB|(P\mathcal{H})$. Then $\hat{\mathfrak{A}}$ is a weakly closed subalgebra of $\mathcal{B}(P\mathcal{H})$, (since $P \in \mathfrak{A}$), and $\hat{\mathfrak{R}}$ is a m.a.s.a. contained in $\hat{\mathfrak{A}}$. Clearly $Lat\,\hat{\mathfrak{A}}$ is totally ordered and $Lat\,\hat{\mathfrak{A}} \subset Lat\,\hat{B}$. Let $\hat{\mathcal{M}} = \mathcal{M} \ominus \mathcal{M}_{i_0}^{(n)}$. Then $\hat{\mathcal{M}} \in Lat(\hat{\mathfrak{A}})^{(n)}$, and $\hat{\mathcal{M}}$ meets the i_0^{th} coordinate space only in the zero vector. By case (a) it follows that $\hat{\mathcal{M}} \in Lat\,\hat{B}^{(n)}$. Since $\mathcal{M}_{i_0}^{(n)}$ is obviously in $Lat\,B^{(n)}$, it follows that $\mathcal{M} \in Lat\,B^{(n)}$. This completes the proof of case (b), and we conclude that $Lat\,\mathfrak{A}^{(n)} \subset Lat\,B^{(n)}$ for all n. Thus $B \in \mathfrak{A}$. $\quad\square$

Corollary 9.25. *If A is a unicellular operator, then there exists a Hermitian operator H such that the weakly closed algebra generated by A and H is $Alg\,Lat\,A$.*

Proof. Since $Lat\,A$ is totally ordered, the projections onto members of $Lat\,A$ commute with each other. Let $\mathfrak{R}$ be any m.a.s.a. containing all the projections onto invariant subspaces of A. By Theorem 7.12, $\mathfrak{R}$ has a Hermitian generator H. Let $\mathfrak{A}$ denote the weakly closed algebra generated by $\{A, H\}$. Then $\mathfrak{A}$ contains $\mathfrak{R}$, and $Lat\,\mathfrak{A} = Lat\,A$. Theorem 9.24 implies that $\mathfrak{A}$ is reflexive; hence

$$\mathfrak{A} = Alg\,Lat\,\mathfrak{A} = Alg\,Lat\,A. \quad\square$$

The following example is a particular instance of Corollary 9.25.

Example 9.26. *For* $\alpha \in [0,1]$ *let*

$$\mathcal{M}_\alpha = \{f \in \mathcal{L}^2(0,1) : f = 0 \text{ a.e. on } [0,\alpha]\} .$$

Let V and M denote the Volterra operator and the operator consisting of multiplication by the independent variable on $\mathcal{L}^2(0,1)$ *respectively. Then the weakly closed algebra generated by* $\{V,M\}$ *is* $Alg\{\mathcal{M}_\alpha\}$.

Proof. As shown in Example 4.11, the weakly closed algebra generated by M is the m.a.s.a. $\mathcal{L}^\infty$. Obviously $\{\mathcal{M}_\alpha\} \subset Lat\, M$, and it is easily seen (even without Theorem 4.14) that $Lat\{V,M\}$ is $\{\mathcal{M}_\alpha\}$. Thus the result follows from Theorem 9.24 in the same way that Corollary 9.25 does. $\square$

The following examples show that none of the other partial solutions of the transitive algebra problem presented in Chapter 8 generalize as Corollary 8.10 did to give a sufficient condition for reflexivity.

Example 9.27. *Let* $\{e_n\}_{n=0}^\infty$ *be an orthonormal basis for* $\mathcal{H}$, *and let*

$$\mathfrak{A} = \{A \in \mathcal{B}(\mathcal{H}) : (Ae_i, e_i) = (Ae_j, e_j) \text{ for all } i,j$$
$$\text{and } (Ae_i, e_j) = 0 \text{ whenever } i > j\} .$$

Then $\mathfrak{A}$ *is not reflexive.*

Proof. It is easily verified that the only non-trivial invariant subspaces of $\mathfrak{A}$ are subspaces of the form $\bigvee_{i=N}^\infty \{e_i\}$ for non-negative integers N. Hence $Alg\, Lat\, \mathfrak{A} = \{A \in \mathcal{B}(\mathcal{H}) : (Ae_i, e_j) = 0 \text{ whenever } i > j\}$, and $\mathfrak{A}$ is not reflexive. $\square$

Note that the algebra $\mathfrak{A}$ of Example 9.27 is weakly closed, contains the identity, and has a discrete totally ordered invariant subspace lattice. Moreover, $\mathfrak{A}$ contains the unilateral shift of multiplicity 1 and many operators of finite rank. Also $\mathfrak{A}^*$ is not reflexive, (since $\mathfrak{A}$ is not), although it contains all Donoghue operators.

The next example is similar to the previous one.

Example 9.28. *Let* $\mathfrak{A}$ *denote the algebra of all operators on* $\mathcal{H}^{(2)}$ *of the form* $\begin{pmatrix} A & B \\ 0 & A \end{pmatrix}$ *with* $A, B \in \mathcal{B}(\mathcal{H})$. *Then* $\mathfrak{A}$ *is not reflexive.*

Proof. It is easily seen that the only non-trivial invariant subspace of $\mathfrak{A}$ is $\mathcal{H} \oplus \{0\}$. Hence

$$Alg\, Lat\, \mathfrak{A} = \left\{ \begin{pmatrix} A & B \\ 0 & C \end{pmatrix} : A, B, C \in \mathcal{B}(\mathcal{H}) \right\} . \quad \square$$

Note that the algebra of Example 9.28 contains $\mathfrak{R}^{(2)}$ for every m.a.s.a. $\mathfrak{R}$, many operators of finite rank, all unilateral shifts of even multiplicity, etc.

9.3 Triangular Operator Algebras

The notion of a "triangular operator algebra" is a generalization to infinite-dimensional spaces of subalgebras of the algebra of triangular matrices which contain the diagonal matrices.

Definition. The subalgebra $\mathfrak{T}$ of $\mathcal{B}(\mathcal{H})$ is *triangular* if $\mathfrak{T} \cap \mathfrak{T}^*$ is a m.a.s.a. If $\mathfrak{T}$ is triangular, then $\mathfrak{T} \cap \mathfrak{T}^*$ is the *diagonal* of $\mathfrak{T}$.

Note that triangular operator algebras are not necessarily closed in any topology. There are some examples of triangular operator algebras on infinite-dimensional spaces which are completely analogous to the finite-dimensional examples.

Examples 9.29. *Let $\{e_n\}_{n=0}^{\infty}$ be an orthonormal basis for $\mathcal{H}$. If $\mathfrak{T}$ is any subalgebra of $\mathcal{B}(\mathcal{H})$ which contains all diagonal operators relative to $\{e_n\}$, (i.e., all operators A with $(Ae_n, e_m) = 0$ for $m \neq n$), and which is contained in the upper triangular operators relative to $\{e_n\}$, (i.e., those operators A such that $(Ae_n, e_m) = 0$ for $m > n$), then $\mathfrak{T}$ is a triangular algebra.*

Proof. The collection of all operators which are diagonal relative to $\{e_n\}$ is a m.a.s.a., and clearly $\mathfrak{T} \cap \mathfrak{T}^*$ is equal to this collection of operators. $\square$

The following lemma is useful in producing other examples of triangular operator algebras.

Lemma 9.30. *If $\mathfrak{T}$ is a subalgebra of $\mathcal{B}(\mathcal{H})$, $\mathfrak{R}$ is a m.a.s.a., and $\mathfrak{R} \subset \mathfrak{T}$, then $\mathfrak{T}$ is triangular if and only if every Hermitian operator in $\mathfrak{T}$ is contained in $\mathfrak{R}$.*

Proof. If $\mathfrak{T}$ is triangular then every Hermitian operator in $\mathfrak{T}$ is in $\mathfrak{T} \cap \mathfrak{T}^*$. Since $\mathfrak{R}$ is a m.a.s.a. and $\mathfrak{R} \subset \mathfrak{T} \cap \mathfrak{T}^*$, $\mathfrak{R} = \mathfrak{T} \cap \mathfrak{T}^*$.

Conversely, suppose that each Hermitian operator in $\mathfrak{T}$ is in $\mathfrak{R}$. Then if A and A^* are in $\mathfrak{T}$, so are the Hermitian operators $A + A^*$ and $i(A - A^*)$. Hence these operators, and thus also A and A^*, are in $\mathfrak{R}$. Thus $\mathfrak{T} \cap \mathfrak{T}^* = \mathfrak{R}$. $\square$

Which reflexive algebras are triangular?

Theorem 9.31. *Let $\mathcal{F}$ be a collection of subspaces and let $\mathfrak{A}$ denote the von Neumann algebra generated by the set of projections onto the elements of $\mathcal{F}$. Then $\mathrm{Alg}\,\mathcal{F}$ is triangular if and only if $\mathfrak{A}$ is a m.a.s.a.*

Proof. Suppose that $\mathfrak{A}$ is a m.a.s.a. If H is Hermitian and is in $\mathrm{Alg}\,\mathcal{F}$, then $H \in \mathfrak{A}'$ and therefore $H \in \mathfrak{A}$. Lemma 9.30 implies that $\mathrm{Alg}\,\mathcal{F}$ is triangular.

If $\mathrm{Alg}\,\mathcal{F}$ is triangular, then $(\mathrm{Alg}\,\mathcal{F}) \cap (\mathrm{Alg}\,\mathcal{F})^*$ is a m.a.s.a. $\mathfrak{R}$. Note that $\mathfrak{R}$ is the set of operators which are reduced by all the subspaces in $\mathscr{F}$; i.e., $\mathfrak{R}=\mathfrak{A}'$. Hence $\mathfrak{R}'=\mathfrak{A}''$, and, since $\mathfrak{R}'=\mathfrak{R}$ and $\mathfrak{A}''=\mathfrak{A}$, it follows that $\mathfrak{R}=\mathfrak{A}$. ☐

Definition. The triangular algebra $\mathfrak{T}$ is *maximal* if it is not properly contained in any other triangular algebra.

The only maximal triangular algebras on finite-dimensional spaces are the algebras of all the upper triangular matrices relative to some orthonormal basis; (Proposition 9.12).

Theorem 9.32. *Every triangular algebra is contained in a maximal triangular algebra.*

Proof. If $\mathfrak{T}$ is a triangular algebra with diagonal $\mathfrak{R}$, and if $\{\mathfrak{T}_\alpha\}$ is a chain of triangular algebras each of which contains $\mathfrak{T}$, then the diagonal of $\mathfrak{T}_\alpha$ is $\mathfrak{R}$ for all α. If $A\in(\bigcup\mathfrak{T}_\alpha)\cap(\bigcup\mathfrak{T}_\alpha^*)$ then there is some α_0 such that $A\in\mathfrak{T}_{\alpha_0}\cap\mathfrak{T}_{\alpha_0}^*$, and this shows that $A\in\mathfrak{R}$. Zorn's lemma now gives the result. ☐

"Upper triangular blocks" can be added to triangular algebras without destroying triangularity.

Lemma 9.33. *If $\mathfrak{T}$ is a triangular algebra, P is a projection onto an invariant subspace of $\mathfrak{T}$, and B is an operator such that $B=PB(1-P)$, then the algebra generated by $\mathfrak{T}$ and B is triangular.*

Proof. Note that $T\in\mathfrak{T}$ implies $BTB=PB(1-P)TPB(1-P)=0$, since $(1-P)TP=0$. Moreover $B^2=0$. Thus the algebra generated by $\mathfrak{T}$ and B consists of all operators of the form

$$S = S_1 + S_2 B S_3 + \cdots + S_{2k} B S_{2k+1}$$

with $S_i\in\mathfrak{T}$ for all i. By Lemma 9.30 it suffices to show that each Hermitian operator of this form is in $\mathfrak{T}\cap\mathfrak{T}^*$.

Let S be as above with $S=S^*$. Since the range of P is in $(Lat\,\mathfrak{T})\cap(Lat\,B)$ and S is Hermitian, $PS=SP$. Also

$$PS_{2i}BS_{2i+1}(1-P) = PS_{2i}PB(1-P)S_{2i+1}(1-P)$$
$$= S_{2i}PB(1-P)S_{2i+1} = S_{2i}BS_{2i+1}$$

for each i. Therefore

$$0 = PS(1-P) = PS_1(1-P) + PS_2 BS_3(1-P) + \cdots + PS_{2k}BS_{2k+1}(1-P)$$
$$= PS_1(1-P) + S_2 BS_3 + \cdots + S_{2k}BS_{2k+1}$$
$$= PS_1(1-P) + S - S_1.$$

Thus $S=S_1-PS_1(1-P)$. Since $P\in\mathfrak{T}\cap\mathfrak{T}^*$, $S\in\mathfrak{T}$, and it follows that $S\in\mathfrak{T}\cap\mathfrak{T}^*$. ☐

Theorem 9.34. *If $\mathfrak{T}$ is a maximal triangular algebra, then $Lat\,\mathfrak{T}$ is totally ordered.*

Proof. Suppose that $Lat\,\mathfrak{T}$ contains the non-comparable subspaces $\mathscr{M}$ and $\mathscr{N}$. Let P and Q be the projections onto $\mathscr{M}$ and $\mathscr{N}$ respectively. Then P and Q are in $\mathfrak{T} \cap \mathfrak{T}^*$, and PQ is the projection with range $\mathscr{M} \cap \mathscr{N}$. Since $\mathscr{M}$ and $\mathscr{N}$ are non-comparable, $P - PQ \neq 0$ and $Q - PQ \neq 0$. Thus there exists a non-zero operator B such that $B = (P - QP)B(Q - QP)$. Then $B = P[(1-Q)BQ](1-P)$, and Lemma 9.33 and the maximality of $\mathfrak{T}$ implies that $B \in \mathfrak{T}$. But $B^* = Q[(1-P)B^*P](1-Q)$, and the same reasoning shows that $B^* \in \mathfrak{T}$. Therefore $B \in \mathfrak{T} \cap \mathfrak{T}^*$, B commutes with P and Q, and it follows that $B = 0$, contrary to the assumption. $\quad\square$

The triangular algebras that turn out to be most tractable are those that have large invariant subspace lattices in the following sense.

Definition. The triangular algebra $\mathfrak{T}$ is *hyperreducible* if the weakly closed algebra generated by the set of projections onto invariant subspaces of $\mathfrak{T}$ is a m.a.s.a.

Note that the weakly closed algebra generated by the set of projections onto invariant subspaces of $\mathfrak{T}$ is contained in $\mathfrak{T} \cap \mathfrak{T}^*$; hence $\mathfrak{T}$ is hyperreducible if and only if this weakly closed algebra is $\mathfrak{T} \cap \mathfrak{T}^*$. Hyperreducible maximal triangular algebras can be completely described.

Theorem 9.35. *Hyperreducible maximal triangular algebras are reflexive.*

Proof. Let $\mathfrak{T}$ be a hyperreducible maximal triangular algebra. Then $Alg\,Lat\,\mathfrak{T}$ is a hyperreducible triangular algebra, by Theorem 9.31. Since $\mathfrak{T}$ is maximal it follows that $\mathfrak{T} = Alg\,Lat\,\mathfrak{T}$. $\quad\square$

Thus the hyperreducible maximal triangular algebras are those of the form $Alg\,\mathscr{F}$, where $\mathscr{F}$ is a chain of subspaces such that the projections onto the members of $\mathscr{F}$ generate a m.a.s.a. In particular, hyperreducible maximal triangular algebras are weakly closed. The converse of this gives a sufficient condition for hyperreducibility.

Theorem 9.36. *Weakly closed maximal triangular algebras are hyperreducible.*

Proof. If $\mathfrak{T}$ is a weakly closed maximal triangular algebra, then $Lat\,\mathfrak{T}$ is totally ordered, by Theorem 9.34. Since $\mathfrak{T}$ contains the m.a.s.a. $\mathfrak{T} \cap \mathfrak{T}^*$, Theorem 9.24 implies $\mathfrak{T}$ is reflexive. The result then follows from Theorem 9.31. $\quad\square$

There are a number of other results about triangular algebras—see Section 9.5.

9.4 Additional Propositions

Proposition 9.1. An operator A is an inflation if and only if there exists a unilateral shift S (of some multiplicity) such that A commutes with both S and S^*.

Proposition 9.2. If A has reducing subspaces $\mathcal{M}$ and $\mathcal{N}$ such that $\mathcal{M} \perp \mathcal{N}$ and A is unitarily equivalent to each of $A|\mathcal{M}$ and $A|\mathcal{N}$, then A is unitarily equivalent to an inflation.

Proposition 9.3. If $\mathfrak{A}$ is reflexive so is $\mathfrak{A}^{(n)}$ for each positive integer n.

Proposition 9.4. If $\mathcal{H}$ is finite-dimensional and $\mathfrak{A}$ is a subalgebra of $\mathcal{B}(\mathcal{H})$ which contains a m.a.s.a., then $\mathfrak{A}$ is reflexive.

Proposition 9.5. If $\mathfrak{A}$ is a reductive algebra containing a compact operator whose nullspace is finite-dimensional, then there exists a collection $\{\mathcal{M}_i\}_{i=1}^{\infty}$ of pairwise orthogonal reducing subspaces for $\mathfrak{A}$ such that $\mathcal{H} = \sum \oplus \mathcal{M}_i$ and $\mathfrak{A}|\mathcal{M}_i$ is transitive for all i; (cf. Theorem 5.9).

Proposition 9.6. If $\mathfrak{A}$ is a weakly closed algebra which contains all the diagonal operators relative to a given orthonormal basis, then $\mathfrak{A}$ is reflexive.

Proposition 9.7. If $\eta(\sigma(A)) \cap \eta(\sigma(B)) = \emptyset$, then the weakly closed algebras generated by $\{A, 1\}$ and $\{B, 1\}$ are each reflexive if and only if the weakly closed algebra generated by $\{A \oplus B, 1\}$ is reflexive.

Proposition 9.8. Let Q_i be a cyclic nilpotent operator on a space of finite dimension n_i for $i = 1, \ldots, k$. Suppose that $n_1 \geqq n_2 \geqq \cdots$, and let $Q = \sum_{i=1}^{k} \oplus Q_i$. Then the algebra generated by $\{Q, 1\}$ is reflexive if and only if $n_1 = n_2$ or $n_1 = n_2 + 1$; (we are disregarding the trivial case where $k = 1$).

Propositions 9.7 and 9.8 and the Jordan canonical form theorem give a complete description of the operators on finite-dimensional spaces which generate reflexive algebras.

Proposition 9.9. Every hyperreducible triangular algebra is contained in a hyperreducible maximal triangular algebra.

Proposition 9.10. Every triangular algebra whose diagonal is totally atomic, (i.e., there is an orthonormal basis consisting of eigenvectors of all the operators in the diagonal), is hyperreducible.

Proposition 9.11. If X is the unit circle, μ is Lebesgue measure on X, α is an irrational number and $(Uf)(t) = f(e^{2\pi i}t)$, then the algebra generated by $\mathscr{L}^\infty \cup \{U\}$ is triangular and has no non-trivial invariant subspaces.

Proposition 9.12. If $\mathfrak{A}$ is a maximal triangular algebra on a finite-dimensional space, then there is an orthonormal basis such that $\mathfrak{A}$ is the algebra of all operators with upper triangular matrices relative to this basis.

Proposition 9.13. A reductive algebra with the property that the span of the ranges of the finite-rank operators in it is the entire space is necessarily self-adjoint.

Proposition 9.14. A reductive algebra containing an injective compact operator is self-adjoint; (cf. Proposition 9.5, Proposition 9.13, and Lemma 8.22).

9.5 Notes and Remarks

The reductive algebra problem was first raised in Radjavi-Rosenthal [7], (where reductive algebras were called "Hermitian algebras"). Lemmas 9.1 and 9.4 are from Radjavi-Rosenthal [7], while Lemmas 9.2 and 9.3 are from Feintuch-Rosenthal [1]. A (much earlier, different) proof of Theorem 9.5 was found by Cater [1]. Theorem 9.7 and Corollaries 9.8 and 9.9 are due to Feintuch-Rosenthal [1]. The relevance of considering strictly cyclic algebras was pointed out in earlier work of Lambert [2] who obtained Corollary 9.10. Corollary 9.10 has been generalized to algebras of "finite strict multiplicity" by Herrero [1], and it is observed in Feintuch-Rosenthal [1] that Corollary 9.9 generalizes to this case also.

Theorem 9.11 and the proof of the finite-dimensional case given in Corollary 9.12 are from Radjavi-Rosenthal [7], as is Lemma 9.13. Lemma 9.14 was suggested by the work of Hoover [3]. The case $k=1$ of Theorem 9.15 is due to Radjavi-Rosenthal [7] and, independently, to Sul'man [1], the case where $\mathfrak{A}$ is transitive and k is arbitrary (i.e., Theorem 8.21) is due to Douglas-Pearcy [3], and the general case is due to Hoover [3]. The proof in the text was motivated by Hoover [3], although it is quite different in many respects.

Hoover [2] has shown that a reductive algebra with the property that the von Neumann algebra generated by the projections onto its invariant subspaces is either properly infinite or has a cyclic vector must be self-adjoint. Nordgren-Rosenthal [1] generalizes Theorem 8.18 to obtain the result that reductive algebras containing unilateral shifts of finite multiplicity are self-adjoint; they also prove Proposition 9.13

and show that a reductive algebra generated by operators of finite rank is self-adjoint. This result is generalized to reductive algebras generated by compact operators in Radjavi Rosenthal [9].

Define a *reductive operator* as an operator all of whose invariant subspaces are reducing; (equivalently, as an operator with the property that the weakly closed algebra that it generates is a reductive algebra). Then the *reductive operator problem* is the question: is every reductive operator normal? By Corollary 9.22 this is equivalent to the question: is every singly-generated reductive algebra self-adjoint? Theorem 5.9 states that every reductive polynomially compact operator is normal, and it follows immediately from Lemma 9.14 that a reductive operator which is similar to a normal operator is itself normal. Dyer and Porcelli [1] have obtained the remarkable result that the invariant subspace and reductive operator problems are equivalent; (cf. Dyer-Pedersen-Porcelli [1]). Also Nordgren, Radjavi and Rosenthal [2] have strengthened Corollaries 6.15 and 6.17 to the result that reductive operators T with the property that $T - T^*$ or $1 - T^*T$ are in $\mathscr{C}_p$ are necessarily normal. These results have been generalized in Jafarian [1] and Radjabalipour [1].

The first result on reflexive algebras was Theorem 9.21, which is due to Sarason [4]. This paper stimulated all the subsequent work on reflexive algebras. The word "reflexive" was suggested by Halmos, and first appeared in Radjavi-Rosenthal [1]. Theorem 9.18, which is implicit in Sarason [4], was pointed out in Radjavi-Rosenthal [2]. Lemma 9.20, which allows the removal of the hypothesis of commutativity in Sarason's formulation of Theorem 9.21, is from Radjavi-Rosenthal [2]. Corollary 9.22 is from Sarason [4]; the case where the algebra is singly generated was independently discovered by Goodman [1]. A generalization of Theorem 9.21 appears in Loginov-Sul'man [1]. Theorem 9.23 is also from Sarason [4]. Theorem 9.24, generalizing Arveson's Theorem (Corollary 8.10), is due to Radjavi-Rosenthal ([1], [2]), as is Corollary 9.25 and Example 9.26. Conway [1] has obtained variants of Theorem 9.24 motivated by the Wedderburn theorem. Examples 9.27 and 9.28 are from Rosenthal [6]; similar examples are in Halmos [12] and Conway [1]. Deddens [1] has proven that the weakly closed algebra generated by an isometry and the identity is reflexive. Longstaff [1] has used the ideas of Corollary 9.25 and the main result of Erdos [4] to prove that every reflexive algebra with totally-ordered invariant subspace lattice is generated by two operators. Liberzon-Sul'man [1] contains an investigation of reflexivity for "symmetric" algebras on Pontrjagin spaces.

Triangular operator algebras were introduced by Kadison-Singer [1], which contains Lemma 9.30, Theorem 9.32, Lemma 9.33, Theorem

9.34, Theorem 9.35, Proposition 9.10, Proposition 9.11, and many other fundamental results. Theorem 9.31 is from Ringrose [3]. Theorem 9.36, which was conjectured in the unpublished sequel to Kadison-Singer [1], is due to Rosenthal [5]; it was independently found by Tsuji [1]. Arveson [4] has recently generalized this result by proving that every weakly closed triangular algebra is hyperreducible. Arveson [4] also includes a sufficient condition that an algebra containing a m.a.s.a. be reflexive which generalizes Theorem 9.24, the case $k=1$ of Theorem 9.15, and Proposition 9.6. In addition, Arveson [4] gives an example of a weakly closed algebra which contains a m.a.s.a. but is not reflexive.

Proposition 9.1 is from Halmos [8], while Proposition 9.2 is from Radjavi-Rosenthal [2]. Propositions 9.4 and 9.6 are due to Davis-Radjavi-Rosenthal [1], and Proposition 9.8 is due to Deddens-Fillmore [1]. Proposition 9.9 was independently observed by Arveson and Rosenthal. Proposition 9.14 is due to Rosenthal [8]; it has been generalized by Azoff [1].

Other results on triangular algebras are in Kadison-Singer [1], Erdos [1], Ringrose [4], and Hopenwasser ([1], [2]). Ringrose ([2], [3]) has made an interesting study of reflexive algebras with totally-ordered invariant subspace lattices, which he calls "nest algebras". In particular, he has obtained a characterization of the radical of nest algebras that can be very useful, (cf. Erdos-Longstaff [1]). Other related results can be found in Erdos ([2], [3], [4]) and Lance [1]. Arveson ([2], [5], [6]) has obtained interesting results about other classes of non-self-adjoint operator algebras. A generalization of Theorem 9.11 is contained in Radjavi [2].

Chapter 10. Some Unsolved Problems

The fundamental problems that motivated most of the work in this book remain unsolved. We discuss certain special cases of these problems that might be more tractable than the general cases, and we also consider some related questions. Our list of unsolved problems is representative rather than exhaustive; a number of other problems will no doubt occur to the reader.

10.1 Normal Operators

We have obtained some information about the invariant subspaces of normal operators, but there are still a number of unsolved problems. A very general problem is: what lattices are attained by normal operators? Any such lattice must be self-dual (Theorem 4.9), and it is not hard to see that such a lattice (on a separable space) must have a countable order-dense subset. What other properties must a lattice attainable by a normal operator satisfy?

A more specific question: if A is normal and $\mathcal{M}$ is an atom in *Lat A*, must $\mathcal{M}$ be one-dimensional? Halmos [7] initiated the study of *subnormal* operators; i.e., operators that are restrictions of normal operators to invariant subspaces, and he asked this question in the form: does every subnormal operator (on a space of dimension greater than 1) have a non-trivial invariant subspace? Bram [1] has shown that whenever T is a subnormal operator with a cyclic vector, there is a finite Borel measure μ on $\sigma(T)$ such that T is unitarily equivalent to M_z restricted to the span in $\mathcal{L}^2(\sigma(T), \mu)$ of the polynomials in z. This shows that the question of the existence of invariant (or hyperinvariant) subspaces of subnormal operators is equivalent to an interesting and difficult problem in classical analysis. This approach to the problem has been used by Wermer [3], Yoshino [1], and Brennan [1] to obtain certain special cases, but the general problem is still unsolved.

Every subnormal operator is also hyponormal, (Halmos [7]), and therefore a more general question is whether or not hyponormal operators have invariant subspaces. Putnam [2] has shown that every hypo-normal operator whose spectrum has (two-dimensional Lebesgue) measure 0 is normal.

10.2 Attainable Lattices

In considering the question of which abstract lattices are attainable it seems reasonable to start with totally-ordered lattices. It is easily seen, (as R.G. Douglas has observed), that every totally-ordered lattice of subspaces (of a separable space) has a countable order-dense subset. Hence it follows from a basic result in lattice theory (cf. Birkhoff [1]) that every attainable chain is isomorphic to a closed subset of the closed unit interval with its standard ordering. Is there an example of a closed subset of $[0,1]$ which is *not* attainable?

The only totally-ordered lattices which are known to be attainable are $\omega+n$, $[0,1]+n$, $\omega+\omega+1$ and their duals. Halmos has asked whether the lattice $1+{}^*\omega+\omega+1$ is attainable. In particular, he has asked whether the bilateral weighted shift with weights $\{2^{-|n|}\}_{n=-\infty}^{\infty}$ has any invariant subspaces other than the obvious ones. C. Pearcy [2] has raised the specific question of whether or not the lattice $[-1,0]+1$ $+[2,3]$ is attainable. Some operator of the form $\begin{pmatrix} V & C \\ 0 & A \end{pmatrix}$, where V is the Volterra operator and A is a Donoghue operator, may have lattice $[0,1]+\omega+1$. A related question from Rosenthal [7] is still unanswered: if $\mathscr{L}$ is an attainable totally-ordered lattice must $\mathscr{L}+1$ be attainable?

It was observed in Rosenthal [7] that certain lattices such as those in the accompanying figure, (where going up indicates increasing in the lattice), are not attainable if every operator has an invariant subspace.

For if A is an operator such that $Lat\,A$ is isomorphic to either of these lattices, then $\mathcal{M}$ must correspond to a finite-dimensional subspace, (if some quotient operator is not to be an operator on a space of dimension greater than 1 which has no non-trivial invariant subspace). But then the initial segment $[0,\mathcal{M}]$ of $Lat\,A$ would be the invariant-subspace lattice of an operator on a finite-dimensional space, which would imply (Theorem 4.6) that $[0,\mathcal{M}]$ is self-dual, although it obviously is not. Can it be shown (independently of the invariant subspace problem) that either of these lattices is not attainable?

10.3 Existence of Invariant Subspaces

The known results on existence of invariant subspaces suggest certain obvious generalizations. Can Corollary 6.15 be improved to show that every operator with compact imaginary part has a non-trivial invariant subspace? Can existence of invariant subspaces be established for operators T of the form $T = N + K$ where N is normal and K has finite rank, or, more generally, $K \in \mathscr{C}_p$? Or, more generally still, for operators T such that $T^*T - TT^*$ has finite rank or is in $\mathscr{C}_p$? Can the operators which commute with compact operators be characterized (cf. Corollary 8.24)? What sufficient conditions can be found that an operator be quasi-similar to an operator which commutes with a compact operator (cf. Corollary 8.24 and Theorem 6.19)? Does the relation $ST = NS$, where N is normal and S is injective and has dense range, imply that T has invariant subspaces (cf. Theorem 6.19)? If A and B are quasi-similar and A has an invariant subspace must B have an invariant subspace (cf. Theorem 6.19)? These questions have all been considered by a number of authors.

Say that the operator A *has property* $\mathscr{S}$ if $A|\mathcal{M}$ has a non-trivial invariant subspace whenever $\mathcal{M} \in Lat\,A$ and the dimension of $\mathcal{M}$ is greater than 1. A question raised in Rosenthal [7] is: if A has property $\mathscr{S}$ must $A \oplus A$ have property $\mathscr{S}$? An affirmative answer would imply that $A^{(n)}$ has property $\mathscr{S}$ whenever A does, and this would include Corollary 6.18 as well as a number of other results.

Halmos [9] has asked whether every quasitriangular operator has an invariant subspace; the result of Apostol-Foiaş-Voiculescu ([1], [2]) discussed in Section 5.6 shows that this is equivalent to the invariant subspace problem.

There have been a number of attempts to find an operator without non-trivial invariant subspaces. An interesting kind of operator was considered by E. Bishop: a *Bishop operator* is an operator on $\mathscr{L}^2(0,1)$ defined as follows. Let α be a fixed irrational number and define

$$(B_\alpha f)(x) = x f(x + \alpha),$$

where $x + \alpha$ is computed modulo $[0, 1]$. A. M. Davie has shown (in work not yet published) that a result similar to Theorem 6.4 can be used to show that B_α has a non-trivial invariant subspace whenever α is in a certain set of measure 1 which contains all algebraic numbers. Does every B_α have a non-trivial invariant subspace?

10.4 Reducing Subspaces and von Neumann Algebras

Finding reducing subspaces is generally easier than finding invariant subspaces of a given operator. Nonetheless there are some difficult problems associated with reducing subspaces.

The reducing subspaces of an operator A are the subspaces whose projections are in the commutant of the von Neumann algebra generated by A. Thus the lattice of reducing subspaces is the "projection lattice" of a von Neumann algebra. Is every projection lattice of a von Neumann algebra the lattice of reducing subspaces of some operator? This is equivalent (by the double-commutant theorem) to the well-known unsolved problem: is every von Neumann algebra (on a separable space) generated by a single operator? There are a number of known partial results in addition to Theorem 7.12—see Pearcy ([1], [3]), Suzuki-Saito [1], Topping [1], Saito ([2], [3]), Behncke [1], Wogen [1] and Douglas-Pearcy [5]. The general problem, however, is still unsolved.

The irreducible operators are dense in $\mathscr{B}(\mathscr{H})$, (Theorem 7.10). Halmos [15] has asked whether the reducible operators, (i.e., the operators with non-trivial reducing subspaces), are also dense. As Halmos has pointed out, the answer is obviously negative on finite-dimensional spaces, and affirmative on non-separable spaces (where every operator is reducible). What is the answer for operators on infinite-dimensional separable spaces?

It is sometimes difficult to determine the reducing subspaces of given operators. Nordgren [4] has studied reducing subspaces of analytic Toeplitz operators, and has shown that the analytic Toeplitz operator T_ϕ induced by the function $\phi \in \mathscr{H}^\infty$ has a non-trivial reducing subspace if $\phi = \psi \circ h$, where $\psi \in \mathscr{H}^\infty$ and h is a non-constant inner function other than a single Blaschke factor. Nordgren [4] asked whether the converse holds; i.e., if ϕ is not of this form, must T_ϕ be irreducible? Although some additional information has been obtained by Ball [1], the question remains unresolved.

There are a number of other operators whose reducing subspaces have not been characterized, and there are undoubtedly many solvable problems in this area.

If $\mathfrak{A}$ is a von Neumann algebra and $A \in \mathfrak{A}$, say that *A has an invariant subspace in* $\mathfrak{A}$ if there is a Hermitian projection $P \in \mathfrak{A}$ such that $P \neq 0,1$ and $AP = PAP$. A question raised independently by W.B. Arveson and D. Topping is: if $\mathfrak{A}$ is a von Neumann algebra and $A \in \mathfrak{A}$, must A have an invariant subspace in $\mathfrak{A}$? If $\mathfrak{A} = \mathscr{B}(\mathscr{H})$ this is the invariant subspace problem. The question immediately reduces to the case where $\mathfrak{A}$ is a factor, (i.e., $\mathfrak{A} \cap \mathfrak{A}'$ consists of multiples of the identity), for otherwise every operator in $\mathfrak{A}$ has reducing subspaces in $\mathfrak{A}$. Perhaps the problem is more tractable for factors of type II or type III than for those of type I (i.e., $\mathscr{B}(\mathscr{H})$).

10.5 Transitive and Reductive Algebras

The proof of each of the special cases of the transitive algebra problem presented in Chapter 8 involves using properties of given operators in the algebra to reduce the problem to an application of Arveson's lemma. Other cases where it is reasonable to expect to be able to apply this technique include those where the algebra contains the following kinds of operators: injective weighted shifts, (the result of Shields-Wallen [1] that the commutant is the weakly closed algebra generated by the operator in this case might be relevant; the result holds (by Corollary 9.10) for strictly cyclic shifts (see Lambert [1])); or normal operators of finite uniform multiplicity, (this question arose in conversation with Eric Nordgren). It does not even seem easy to prove that $\mathscr{B}(\mathscr{H})$ is the only transitive algebra containing the bilateral shift.

The transitive algebra problem could also be considered for algebras generated by special kinds of operators. If $\mathfrak{A}$ is a transitive algebra generated by Hermitian operators, then $\mathfrak{A}$ is a von Neumann algebra and hence (Theorem 8.7) must equal $\mathscr{B}(\mathscr{H})$. What is the situation if $\mathfrak{A}$ is generated by normal operators?

Arveson [1] has asked whether $Lat\,\mathfrak{A}^{(2)} = Lat(\mathscr{B}(\mathscr{H}))^{(2)}$ implies $\mathfrak{A} = \mathscr{B}(\mathscr{H})$. Equivalently, if the only closed densely-defined operators which commute with a transitive algebra $\mathfrak{A}$ are the multiples of the identity, does it follow that the only graph transformations for $\mathfrak{A}$ are the multiples of the identity?

Any partial solution of the transitive algebra problem produces a question about reductive algebras. All the presently known results in the transitive case have been generalized to the reductive case; if any of the above problems about transitive algebras are solved, then the corresponding results should be investigated in the reductive case.

The algebraic analogue of the reductive algebra problem is still unsolved. Namely, if $\mathfrak{A}$ is a weakly closed algebra with the property that every invariant linear manifold of $\mathfrak{A}$ is invariant under $\mathfrak{A}^*$, does it follow that $\mathfrak{A} = \mathfrak{A}^*$? The answer is affirmative for the algebra generated by a single operator A: in fact, Fillmore [2] has proven that A^* is a polynomial in A if every invariant linear manifold of A is invariant under A^*.

Von Neumann algebras have the property that $\mathfrak{A}|\mathscr{M}$ is weakly closed whenever $\mathscr{M}$ reduces $\mathfrak{A}$ (cf. Dixmier [1]). Do reductive algebras have this property? An affirmative answer would have several applications (cf. Radjavi-Rosenthal [7]).

If A is quasi-similar to a normal operator, and if the algebra generated by A is reductive, is A necessarily normal?

10.6 Reflexive Algebras

Any general characterization of reflexive algebras seems to be very difficult. Is (the algebra generated by) $A \oplus A$ reflexive for every $A \in \mathscr{B}(\mathscr{H})$? A stronger statement, conjectured independently by D. Sarason and P. Rosenthal, would be: $Lat\,A \subset Lat\,B$ and $AB = BA$ implies that B is in the weakly closed algebra generated by A and 1, (cf. Proposition 4.1). J. Erdos has asked whether operators with the property that their invariant subspaces are all spanned by eigenvectors are necessarily reflexive; (certain special cases of this are contained in Gillespie-West [1]). It would suffice to show, as Erdos has pointed out, that the hypothesis implies that every invariant subspace of $A^{(2)}$ is spanned by eigenvectors.

B. Moore, III and E. Nordgren have, in unpublished work, investigated reflexivity for compressions of the unilateral shift. Let T be the compression of the unilateral shift to $(\phi \mathscr{H}^2)^\perp$ for a given inner function ϕ. Moore and Nordgren have shown that T is reflexive if ϕ is a Blaschke product with zeroes of multiplicity 1, and that T is not reflexive if ϕ has a multiple zero or if the measure associated with the singular part of ϕ has an atom. They have asked: if ϕ is singular and the measure associated with ϕ is totally non-atomic, is T necessarily reflexive?

There must be many other reasonable sufficient conditions for reflexivity.

Another area that requires more study is commutants of weakly closed algebras. Which operators have the property that the algebras they generate are equal to their double commutants? Reductive normal operators (by Corollary 9.22 and Theorem 7.5) and algebraic operators (Turner [1]) have this property.

10.7 Triangular Algebras

One of the questions of Kadison-Singer [1] is: is every operator in some hyperreducible maximal triangular algebra? (Equivalently, (by Proposition 9.9), is every operator in some hyperreducible triangular algebra?) That is, given A, does there exist a chain in $Lat\,A$ such that the projections onto the members of the chain generate a m.a.s.a.? J. Erdos has observed that an affirmative answer would imply that every singly-generated reductive algebra is self-adjoint. For if $\mathfrak{A}$ is generated by A, if $\{P_\alpha\}$ is a collection of projections onto invariant subspaces of $\mathfrak{A}$, and if $\{P_\alpha\}$ generates a m.a.s.a., then $A \in \{P_\alpha\}'$ and A is in the m.a.s.a. On the other hand, even though the latter assertion is true for compact A (cf. Theorem 5.9), it is still not known whether every compact operator is in a hyperreducible triangular algebra; (see Erdos [1] for a partial result).

J.R. Ringrose has raised a very interesting question related to the above. Suppose that $\{\mathcal{M}_\alpha\}$ is a chain of subspaces such that the collection of Hermitian projections onto the $\{\mathcal{M}_\alpha\}$ generates a m.a.s.a. and that S is an invertible operator. Must the collection of Hermitian projections onto the $\{S\mathcal{M}_\alpha\}$ generate a m.a.s.a.? The answer is easily seen to be yes in the case that the $\{\mathcal{M}_\alpha\}$ generate a totally-atomic m.a.s.a. The answer is not known in the case where the $\{\mathcal{M}_\alpha\}$ are the invariant subspaces of the Volterra operator; (the work of Erdos [4] suggests that a solution in this case might lead to the general result). It would be very interesting if there were a counterexample in this case. For if S is an invertible operator and $\{\mathcal{M}_\alpha\}$ is the set of invariant subspaces of the Volterra operator V, then $\{S\mathcal{M}_\alpha\} = Lat(SVS^{-1})$. If $\{S\mathcal{M}_\alpha\}$ did not generate a m.a.s.a., then SVS^{-1} would be an operator that is not contained in any hyperreducible triangular algebra.

Another question from Kadison-Singer [1] that remains unanswered is: does every operator lie in some triangular algebra? The triangular algebra could have no invariant subspaces, (cf. Proposition 9.11), so this would not even imply existence of invariant subspaces. The answer is affirmative in the case of compact operators, (Erdos [1]).

J. Erdos has asked two interesting questions about triangular algebras. Is every maximal triangular algebra uniformly closed? If P is a projection in $\mathfrak{T} \cap \mathfrak{T}^*$ and T is a maximal triangular algebra, is $P\mathfrak{T}|(P\mathcal{H})$ a maximal triangular algebra? (Note that $P\mathfrak{T}|(P\mathcal{H})$ is obviously triangular. Also, as the unpublished sequel to Kadison-Singer [1] points out, the result is easily established for the case $P = P_1 - P_2$ with $\{P_1\mathcal{H}, P_2\mathcal{H}\} \subset Lat\,\mathfrak{T}$.)

References

Ahern, P. R., Clark, D. N.: [1] On functions orthogonal to invariant subspaces. Acta Math. **124**, 191—204 (1970).

[2] On star-invariant subspaces. Bull. Amer. Math. Soc. **76**, 629—632 (1970).

Andô, T.: [1] On hyponormal operators. Proc. Amer. Math. Soc. **14**, 290—291 (1963).

[2] Note on invariant subspaces of a compact normal operator. Arch. Math. **14**, 337—340 (1963).

Apostol, C.: [1] On the growth of resolvent, perturbation and invariant subspaces. Rev. Roumaine Math. Pures Appl. **16**, 161—172 (1971).

[2] A theorem on invariant subspaces. Bull. Acad. Polon. Sci. **16**, 181—185 (1968).

Apostol, C., Foiaş, C., Voiculescu, D.: [1] Some results on non-quasitriangular operators IV. Rev. Roumaine Math. Pures Appl. **18**, 487—514 (1973).

[2] Structure spectrale des operateurs non-quasitriangulaires, Comptes Rendus Acad. Sci. Paris (Séries A) **276**, 49—52 (1973).

Apostol, T. M.: [1] Mathematical Analysis. Reading, Mass.: Addison-Wesley 1957.

Aronszajn, N., Smith, K. T.: [1] Invariant subspaces of completely continuous operators. Ann. of Math. **60**, 345—350 (1954).

Arveson, W. B.: [1] A density theorem for operator algebras. Duke Math. J. **34**, 635—647 (1967).

[2] Analyticity in operator algebras. Amer. J. Math. **89**, 578—642 (1967).

[3] Lattices of invariant subspaces. Bull. Amer. Math. Soc. **78**, 515—519 (1972).

[4] Operator algebras and invariant subspaces. To appear.

[5] Subalgebras of C^*-algebras I. Acta Math. **123**, 141—224 (1969).

[6] Subalgebras of C^*-algebras II. Acta Math. **128**, 271—308 (1972).

Arveson, W. B., Feldman, J.: [1] A note on invariant subspaces. Michigan Math. J. **15**, 60—64 (1968).

Azoff, E.: [1] Compact operators in reductive algebras. To appear.

Ball, J. A.: [1] Hardy space expectation operators and reducing subspaces. To appear.

Banach, S.: [1] Théorie des opérations linéaires. Warsaw: Monografje Matematyczne 1932.

Barnes, B. A.: [1] Density theorems for algebras of operators and annihilator Banach algebras. Michigan Math. J. **19**, 149—155 (1972).

Bartle, R. G.: [1] Spectral localization of operators in Banach space. Math. Ann. **153**, 261—269 (1964).

Behncke, H.: [1] Topics in C^*- and von Neumann algebras. Lectures on Operator Algebras. Lecture Notes in Mathematics 247. Berlin-Heidelberg-New York: Springer 1972.

Berberian, S. K.: [1] Introduction to Hilbert Space. New York: Oxford University Press 1961.
[2] Notes on Spectral Theory. Princeton: Van Nostrand 1966.
[3] Note on a theorem of Fuglede and Putnam. Proc. Amer. Math. Soc. **10**, 175—182 (1959).
[4] A note on hyponormal operators. Pacific J. Math. **12**, 1171—1175 (1962).

Bernstein, A. R.: [1] Invariant subspaces of polynomially compact operators. Pacific J. Math. **21**, 445—465 (1967).
[2] Invariant subspaces for certain commuting operators on Hilbert space. Ann. of Math. **95**, 253—260 (1972).

Bernstein, A. R., Robinson, A.: [1] Solution of an invariant subspace problem of K. T. Smith and P. R. Halmos. Pacific J. Math. **16**, 421—431 (1966).

Beurling, A.: [1] On two problems concerning linear transformations in Hilbert space. Acta Math. **81**, 239—255 (1949).

Birkhoff, G.: [1] Lattice Theory. Amer. Math. Soc. Colloq. Pub., Vol. 25. Second Edition, New York 1948.

Bram, J.: [1] Subnormal operators. Duke Math. J. **22**, 75—94 (1955).

de Branges, L.: [1] Hilbert Spaces of Entire Functions. Englewood Cliffs: Prentice-Hall 1968.
[2] Some Hilbert spaces of analytic functions II. J. Math. Anal. Appl. **11**, 44—72 (1965).

de Branges, L., Rovnyak, J.: [1] Square Summable Power Series. New York: Holt, Rinehart and Winston 1966.
[2] Canonical models in quantum scattering theory. In: Perturbation Theory and its Applications in Quantum Mechanics. New York: Wiley 1966, 295—391.

Brennan, J. E.: [1] Invariant subspaces and rational approximation. J. Functional Analysis **7**, 285—310 (1971).

Brickman, L., Fillmore, P. A.: [1] The invariant subspace lattice of a linear transformation. Can. J. Math. **19**, 810—822 (1967).

Brodskiĭ, M. S.: [1] Triangular and Jordan Representations of Linear Operators. Transl. of Mathematical Monographs, Vol. 32. Providence: Amer. Math. Soc. 1971.
[2] On a problem of I. M. Gelfand. Uspehi Mat. Nauk. **12**, 129—132 (1957).

Brodskiĭ, M. S., Livsic, M. S.: [1] Spectral analysis of non-self-adjoint operators and intermediate systems. Amer. Math. Soc. Transl. (2) **13**, 265—346 (1960).

Brown, A.: [1] The unitary equivalence of binormal operators. Amer. J. Math. **76**, 414—439 (1954).

Brown, A., Halmos, P. R.: [1] Algebraic properties of Toeplitz operators. J. Reine Angew. Math. **231**, 89—102 (1963).

Caradus, S. R.: [1] Universal operators and invariant subspaces. Proc. Amer. Math. Soc. **23**, 526—527 (1969).

Cater, F. S.: [1] Lectures on Real and Complex Vector Spaces. Philadelphia: W. B. Saunders 1966.

Colojoară, I., Foiaș, C.: [1] Theory of Generalized Spectral Operators. New York: Gordon and Breach 1968.

Conway, J. B.: [1] On algebras of operators with totally ordered lattices of invariant subspaces. Proc. Amer. Math. Soc. **28**, 163—168 (1971).

Crimmins, T., Rosenthal, P.: [1] On the decomposition of invariant subspaces. Bull. Amer. Math. Soc. **73**, 97—99 (1967).

Davis, Ch.: [1] Generators of the ring of bounded operators. Proc. Amer. Math. Soc. **6**, 970—972 (1955).

Davis, Ch., Radjavi, H., Rosenthal, P.: [1] On operator algebras and invariant subspaces. Canad. J. Math. **21**, 1178—1181 (1969).

Deckard, D., Douglas, R. G., Pearcy, C.: [1] On invariant subspaces of quasitriangular operators. Amer. J. Math. **91**, 637—647 (1969).

Deckard, D., Pearcy, C.: [1] Another class of invertible operators without square roots. Proc. Amer. Math. Soc. **14**, 445—449 (1963).

[2] On rootless operators and operators without logarithms. Acta Sci. Math. (Szeged) **28**, 1—7 (1967).

[3] On matrices over the ring of continuous complex-valued functions on a Stonian space. Proc. Amer. Math. Soc. **14**, 322—328 (1963).

Deddens, J. A.: [1] Every isometry is reflexive. Proc. Amer. Math. Soc. **28**, 509—512 (1971).

[2] A necessary condition for quasitriangularity. Proc. Amer. Math. Soc. **32**, 630—631 (1972).

Deddens, J., Fillmore, P.: [1] Reflexive linear transformations. J. Lin. Alg. To appear.

Dieudonné, J.: [1] Foundations of Modern Analysis. New York: Academic Press 1960.

Dixmier, J.: [1] Les algèbres d'opérateurs dans l'espace Hilbertien. Second Edition. Paris: Gauthier-Villars 1969.

[2] Les C^*-algebres et leurs représentations. Second Edition. Paris: Gauthier-Villars 1969.

[3] Les opérateurs permutables á l'opérateur integral, Fas. 2. Portugal. Math. **8**, 73—84 (1949).

[4] Étude sur les variétés et les opérateurs de Julia. Bull. Soc. Math. France **77**, 11—101 (1949).

Donoghue, W. F.: [1] The lattice of invariant subspaces of a completely continuous quasi-nilpotent transformation. Pacific J. Math. **7**, 1031—1035 (1957).

Douglas, R. G.: [1] Structure theory for operators, I. J. Reine Angew. Math. **232**, 180—193 (1968).

[2] On the hyperinvariant subspaces for isometries. Math. Z. **107**, 297—300 (1968).

[3] On majorization, factorization and range inclusion of operators in Hilbert space. Proc. Amer. Math. Soc. **17**, 413—416 (1966).

Douglas, R. G., Pearcy, C.: [1] On a topology for invariant subspaces. J. Functional Analysis **2**, 323—341 (1968).

[2] A note on quasitriangular operators. Duke Math. J. **37**, 177—188 (1970).

[3] Hyperinvariant subspaces and transitive algebras. Michigan Math. J. **19**, 1—12 (1972).

[4] A characterization of thin operators. Acta Sci. Math. (Szeged) **29**, 295—297 (1968).

[5] Von Neumann algebras with a single generator. Michigan Math. J. **16**, 21—26 (1969).

Douglas, R. G., Pearcy, C., Salinas, N.: [1] Hyperinvariant subspaces via topological properties of lattices. Michigan Math. J. To appear.

Douglas, R. G., Shapiro, H. S., Shields, A. L.: [1] Cyclic vectors and invariant subspaces of the backwards shift. Ann. Inst. Fourier (Grenoble) **20**, 37—76 (1970).

Dowson, H. R.: [1] On an unstarred operator algebra. J. London Math. Soc. (2) **5**, 489—492 (1972).

Dunford, N.: [1] Spectral operators. Pacific J. Math. **4**, 321—354 (1954).

[2] Spectral theory. Bull. Amer. Math. Soc. **49**, 637—651 (1943).

[3] Spectral theory I. Convergence to projections. Trans. Amer. Math. Soc. **54**, 185—217 (1943).

[4] Spectral theory II. Resolutions of the identity. Pacific J. Math. **2**, 559—614 (1952).

Dunford, N., Schwartz, J. T.: [1] Linear Operators. Part I: General Theory. New York: Interscience 1957.

[2] Linear Operators. Part II: Spectral Theory. New York: Interscience 1963.

[3] Linear Operators. Part III: Spectral Operators. New York: Interscience 1971.

Duren, P. L.: [1] Theory of H^p Spaces. New York: Academic Press 1970.

Dyer, J., Porcelli, P.: [1] Concerning the invariant subspace problem. Notices Amer. Math. Soc. **17**, 788 (1970).

Dyer, J., Pedersen, E., Porcelli, P.: [1] An equivalent formulation of the invariant subspace conjecture. Bull. Amer. Math. Soc. **78**, 1020—1023 (1972).

Eidelheit, M.: [1] On isomorphisms of rings of linear operators. Studia Math. **9**, 97—105 (1940).

Embry, M. R.: [1] An invariant subspace theorem. Proc. Amer. Math. Soc. **32**, 331—332 (1972).

Erdos, J. A.: [1] Some results on triangular operator algebras. Amer. J. Math. **98**, 85—92 (1967).

[2] Operators of finite rank in nest algebras. J. London Math. Soc. **43**, 391—397 (1968).

[3] An abstract characterization of nest algebras. Quart. J. Math. Oxford Ser. (2) **22**, 47—63 (1971).

[4] Unitary invariants for nests. Pacific J. Math. **23**, 229—256 (1967).

Erdos, J. A., Longstaff, W. E.: [1] The convergence of triangular integrals of operators on Hilbert space. Indiana University Math. J. **22**, 929—938 (1973).

Feintuch, A., Rosenthal, P.: [1] Remarks on reductive operator algebras. Israel J. Math. To appear.

Fillmore, P. A.: [1] Notes on Operator Theory. New York: Van Nostrand Reinhold 1970.

[2] On invariant linear manifolds. Proc. Amer. Math. Soc. To appear.

Fillmore, P., Topping, D.: [1] Sums of irreducible operators. Proc. Amer. Math. Soc. **20**, 131—133 (1969).

Fillmore, P., Williams, J. P.: [1] On operator ranges. Advances in Math. **7**, 254—281 (1971).

Flaschka, H.: [1] Invariant subspaces of abstract multiplication operators. Indiana University Math. J. **21**, 413—418 (1971).

Foguel, S. R.: [1] Normal operators of finite multiplicity. Comm. Pure Appl. Math. **11**, 297—313 (1958).

Foiaş, C.: [1] A remark on the universal model for contractions of G. C. Rota. Acad. R. P. Romîne Baza Cerc. Şti. Timişoara Stud. Cerc. Şti. Tehn. **13**, 349—352 (1963).

[2] Invariant para-closed subspaces. Indiana University Math. J. **20**, 897—900 (1971).

[3] Invariant para-closed subspaces. Indiana University Math. J. **21**, 887—906 (1972).

Foiaş, C. Williams, J. P.: [1] Some remarks on the Volterra operator. Proc. Amer. Math. Soc. **31**, 177—184 (1972).

Fuglede, B.: [1] A commutativity theorem for normal operators. Proc. Nat. Acad. Sci. U.S.A. **36**, 35—40 (1950).

Gardner, L. T.: [1] On isomorphisms of C^*-algebras. Amer. J. Math. **87**, 384—396 (1965).

Gehér, L.: [1] Cyclic vectors of a cyclic operator span the space. Proc. Amer. Math. Soc. **33**, 109—110 (1972).

Gelfand, I. M.: [1] Normierte Ringe. Mat. Sb. **9**, 3—24 (1941).
[2] Zur Theorie der Charaktere der Abelschen topologischen Gruppen. Rec. Math., N.S. **9**, 49—50 (1941).

Gellar, R.: [1] Two sublattices of weighted shift invariant subspaces. To appear.
[2] Cyclic vectors and parts of the spectrum of a weighted shift. Trans. Amer. Math. Soc. **146**, 69—85 (1969).

Gerlach, E.: [1] Generalized invariant subspaces for linear operators. Studia Math. **42**, 87—90 (1972).

Gilfeather, F.: [1] Operator-valued roots of abelian analytic functions. To appear.

Gillespie, T. A.: [1] An invariant subspace theorem of J. Feldman. Pacific J. Math. **26**, 67—72 (1968).

Gillespie, T. A., West, T. T.: [1] Operators generating weakly compact groups. Indiana University Math. J. **21**, 671—688 (1972).

Godement, R.: [1] Théorème Taubériene et théorie spectrale. Ann. Sci. École Norm. Sup. (3) **64**, 119—138 (1947).

Gohberg, I. C., Kreĭn, M. G.: [1] Introduction to the Theory of Linear Non-selfadjoint Operators. Transl. of Mathematical Monographs, Vol. 18. Providence: Amer. Math. Soc. 1969.
[2] Theory and Applications of Volterra operators in Hilbert Space. Transl. of Mathematical Monographs, Vol. 24. Providence: Amer. Math. Soc. 1970.

Goldberg, S.: [1] Unbounded Linear Operators. New York: McGraw-Hill 1966.

Goluzin, G. M.: [1] Geometric Theory of Functions of a Complex Variable. Providence: Amer. Math. Soc. 1969.

Goodman, R.: [1] Invariant subspaces for normal operators. J. Math. Mech. **15**, 123—128 (1966).

Grabiner, S.: [1] Weighted shifts and Banach algebras of power series. Amer. J. Math. To appear.

Halmos, P. R.: [1] Measure Theory. Princeton, New Jersey: D. Van Nostrand Co. 1950.
[2] Introduction to Hilbert Space. Second Edition. New York: Chelsea 1957.
[3] A Hilbert Space Problem Book. Princeton, New Jersey: D. Van Nostrand Co. 1967.
[4] Finite-dimensional Vector Spaces. Princeton, New Jersey: D. Van Nostrand Co. 1958.
[5] What does the spectral theorem say? Amer. Math. Monthly **70**, 241—247 (1963).
[6] Commutativity and spectral properties of normal operators. Acta Sci. Math. (Szeged) **12**, 153—156 (1950).
[7] Normal dilations and extensions of operators. Summa Brasil. Math. **2**, 125—134 (1950).
[8] Shifts on Hilbert spaces. J. Reine Angew. Math. **208**, 102—112 (1961).
[9] Quasitriangular operators. Acta Sci. Math. (Szeged) **29**, 283—293 (1968).
[10] Reflexive lattices of subspaces. J. London Math. Soc. (2) **4**, 257—263 (1971).
[11] Invariant subspaces of polynomially compact operators. Pacific J. Math. **16**, 433—437 (1966).
[12] Ten problems in Hilbert space. Bull. Amer. Math. Soc. **76**, 887—933 (1970).

[13] Invariant subspaces, Abstract Spaces and Approximation. Proc. M.R.I. Oberwolfach, pp. 26—30. Basel: Birkhäuser 1968.

[14] Capacity in Banach algebras. Indiana University Math. J. **20**, 855—863 (1971).

[15] Irreducible operators. Michigan Math. J. **15**, 215—223 (1968).

Halmos, P. R., Kakutani, S.: [1] Products of symmetries. Bull. Amer. Math. Soc. **64**, 77—78 (1958).

Halmos, P. R., Lumer, G.: [1] Square roots of operators II. Proc. Amer. Math. Soc. **5**, 589—595 (1954).

Halmos, P. R., Lumer, G., Schäffer, J. J.: [1] Square roots of operators. Proc. Amer. Math. Soc. **4**, 142—149 (1953).

Halmos, P. R., McLaughlin, J. E.: [1] Partial isometries. Pacific J. Math. **13**, 585—596 (1962).

Harrison, K. J.: [1] A new attainable lattice. Proc. Amer. Math. Soc. To appear.
[2] Transitive atomic lattices of subspaces. Indiana University Math. J. **21**, 621—642 (1972).

Harrison, K. J., Radjavi, H., Rosenthal, P.: [1] A transitive medial subspace lattice. Proc. Amer. Math. Soc. **28**, 119—121 (1971).

Helson, H.: [1] Lectures on Invariant Subspaces. New York: Academic Press 1964.

Helson, H., Lowdenslager, D.: [1] Invariant subspaces. Proc. Internat. Symp. Linear Spaces, Jerusalem. New York: Pergamon 1961.

Helton, J. W.: [1] Unitary operators on a space with an indefinite inner product. J. Functional Analysis **6**, 412—440 (1970).

Herrero, D.: [1] Transitive operator algebras containing a subalgebra of finite strict multiplicity. Rev. Un. Mat. Argentina **26**, 77—83 (1972).
[2] A pathological lattice of invariant subspaces. J. Functional Analysis **11**, 131—137 (1972).

Herrero, D., Salinas, N.: [1] Analytically invariant and bi-invariant subspaces. Trans. Amer. Math. Soc. **173**, 117—136 (1972).

Hille, E.: [1] On the theory of characters of groups and semigroups in normed vector rings. Proc. Nat. Acad. Sci. U.S.A. **30**, 58—60 (1944).

Hille, E., Phillips, R. S.: [1] Functional Analysis and Semi-groups. Amer. Math. Soc. Colloq. Pub., Vol. 31. Providence 1957.

Hoffman, K.: [1] Banach Spaces of Analytic Functions. Englewood Cliffs: Prentice-Hall 1962.

Hoffman, K., Kunze, R.: [1] Linear Algebra. Englewood Cliffs: Prentice-Hall 1961.

Hoover, T. B.: [1] Hyperinvariant subspaces for n-normal operators. Acta Sci. Math. (Szeged) **32**, 109—119 (1971).
[2] Operator algebras with reducing invariant subspaces. Pacific J. Math. **44**, 173—180 (1973).
[3] Operator algebras with complemented invariant subspace lattices. Indiana University Math. J. **22**, 1029—1035 (1973).

Hopenwasser, A.: [1] Completely isometric maps and triangular operator algebras. Proc. London Math. Soc. (3) **25**, 96—114 (1972).
[2] Isometries on irreducible triangular operator algebras. Math. Scand. **30**, 136—140 (1972).

Hsu, N. H.: [1] Invariant subspaces of polynomially compact operators in Banach spaces. Yokohama Math. J. **15**, 11—15 (1967).

Iohvidov, I. S., Krein, M. G.: [1] Spectral theory of operators in spaces with an indefinite metric. Amer. Math. Soc. Transl. (2) **13**, 105—175 (1960).

Ismagilov, R. S.: [1] Rings of operators in a space with an indefinite metric. Soviet Math. Dokl. 7, 1460—1462 (1966).

Jacobson, N.: [1] Structure theory of simple rings without finiteness assumptions. Trans. Amer. Math. Soc. 57, 228—245 (1945).
[2] Structure of Rings. Amer. Math. Soc. Colloq. Pub., Vol. 37. Providence 1956.
[3] Lectures in Abstract Algebra II: Linear Algebra. Princeton: D. Van Nostrand 1953.

Jafarian, A. A.: [1] Spectral decompositions of operators on Banach spaces. Dissertation. University of Toronto 1973.

Kadison, R. V.: [1] On the orthogonalization of operator representations. Amer. J. Math. 78, 600—621 (1955).
[2] Irreducible operator algebras. Proc. Nat. Acad. Sci. U.S.A. 43, 273—276 (1957),

Kadison, R. V., Singer, I.: [1] Triangular operator algebras. Amer. J. Math. 82, 227—259 (1960).

Kalisch, G. K.: [1] On similarity, reducing manifolds, and unitary equivalence of certain Volterra operators. Ann. of Math. 66, 481—494 (1957).
[2] A functional analytic proof of Titchmarsh's convolution theorem. J. Math. Anal. Appl. 5, 176—183 (1962).

Kaplansky, I.: [1] Infinite Abelian Groups. Ann Arbor: University of Michigan Press 1954.
[2] Rings of Operators. New York: W. A. Benjamin, Inc. 1968.

Kitano, K.: [1] Invariant subspaces of some non-self-adjoint operators. Tôhoku Math. J. 20, 313—322 (1968).
[2] A note on invariant subspaces. Tôhoku Math. J. 21, 144—151 (1969).

Kocan, D.: [1] A characterization of some spectral manifolds for a class of operators. Illinois J. Math. 16, 359—369 (1972).

Korenblgum, B. I.: [1] Invariant subspaces of the shift operator in certain weighted Hilbert spaces of sequences. Soviet Math. Dokl. 13, 272—275 (1972).

Krein, M. G.: [1] A new application of the fixed point principle in the theory of operators on a space with an indefinite metric. Soviet Math. Dokl. 5, 224—228 (1964).

Lambert, A.: [1] Strictly cyclic weighted shifts. Proc. Amer. Math. Soc. 29, 331—336 (1971).
[2] Strictly cyclic operator algebras. Dissertation. University of Michigan 1970.
[3] Strictly cyclic operator algebras. Pacific J. Math. 39, 717—727 (1971).

Lance, E. C.: [1] Some properties of nest algebras. Proc. London Math. Soc. (3) 19, 45—68 (1969).

Lax, P.: [1] Translation invariant spaces. Acta Math. 101, 163—178 (1959).

Leaf, G. K.: [1] A spectral theory for a class of linear operators. Pacific J. Math. 13, 141—155 (1963).

Liberzon, V. I., Sul'man, V. S.: [1] Operator irreducible symmetric algebras of operators in the Π^1 Pontrjagin Space. Math. USSR-Izv. 5, 1168—1179 (1971).

Livsic, M. S.: [1] On the spectral resolution of linear nonselfadjoint operators. Amer. Math. Soc. Transl. (2) 5, 67—114 (1957).

Ljubič, Ju. I., Macaev, V. I.: [1] On operators with a separable spectrum. Amer. Math. Soc. Transl. (2) 47, 89—129 (1965).

Loginov, A. I., Sul'man, V. S.: [1] On Sarason's theorem and the Radjavi-Rosenthal hypothesis. Soviet Math. Dokl. 13, 928—930 (1972).

Lomonosov, V. J.: [1] Invariant subspaces for operators commuting with compact operators. Functional Anal. and Appl. 7 (1973). To appear.

Longstaff, W. E.: [1] Some results on nest algebras. Dissertation. University of Toronto 1972.

Lorch, E. R.: [1] The spectrum of a linear transformation. Trans. Amer. Math. Soc. **52**, 238—248 (1942).

[2] The integral representation of weakly almost-periodic transformations in reflexive vector spaces. Trans. Amer. Math. Soc. **49**, 18—40 (1941).

Lumer, G., Rosenblum, M.: [1] Linear operator equations. Proc. Amer. Math. Soc. **10**, 32—41 (1959).

Macaev, V. I.: [1] On a class of completely continuous operators. Soviet Math. Dokl. **2**, 972—975 (1961).

Maeda, F.-Y.: [1] Generalized spectral operators on locally convex spaces. Pacific J. Math. **13**, 177—192 (1963).

[2] Generalized scalar operators whose spectra are contained in a Jordan curve. Illinois J. Math. **10**, 431—459 (1966).

Masuda, K.: [1] On the existence of invariant subspaces in spaces with indefinite metric. Proc. Amer. Math. Soc. **32**, 440—444 (1972).

Meyer-Nieberg, P.: [1] Invariante Unterräume von polynomkompakten Operatoren. Arch. Math. (Basel) **19**, 180—182 (1968).

[2] Quasitriangulierbare Operatoren und invariante Untervektorräume stetiger linearer Operatoren. Arch. Math. (Basel) **22**, 186—199 (1971).

Naimark, M. A.: [1] Normed Rings. The Netherlands: Noordhoff 1959.

von Neumann, J.: [1] Zur Algebra der Functionaloperationen und Theorie der Normalen Operatoren. Math. Ann. **102**, 370—427 (1930).

[2] Approximative properties of matrices of high finite order. Portugal. Math. **3**, 1—62 (1942).

[3] Allgemeine Eigenwerttheorie Hermitescher Funktionaloperatoren. Math. Ann. **102**, 49—131 (1929).

Nikolskii, N. K.: [1] Invariant subspaces of certain completely continuous operators. Vestnik Leningrad Univ. (Math. 1) **7**, 68—77 (1965) (Russian).

[2] Invariant subspaces of weighted shift operators. Math. USSR-Sb. **3**, 159—176 (1967).

Nordgren, E. A.: [1] Closed operators commuting with a weighted shift. Proc. Amer. Math. Soc. **24**, 424—428 (1970).

[2] Invariant subspaces of a direct sum of weighted shifts. Pacific J. Math. **27**, 587—598 (1968).

[3] Transitive operator algebras. J. Math. Anal. Appl. **32**, 639—643 (1970).

[4] Reducing subspaces of analytic Toeplitz operators. Duke Math. J. **34**, 175—182 (1967).

Nordgren, E., Radjavi, H., Rosenthal, P.: [1] On density of transitive algebras. Acta Sci. Math. (Szeged) **30**, 175—179 (1969).

[2] On operators with reducing invariant subspaces. Amer. J. Math. To appear.

Nordgren, E., Rosenthal, P.: [1] Algebras containing unilateral shifts or finite-rank operators. Duke Math. J. To appear.

Olsen, C. L.: [1] A structure theorem for polynomially compact operators. Amer. J. Math. **93**, 686—698 (1971).

[2] Thin operators in a von Neumann algebra. Acta Sci. Math. (Szeged). To appear.

[3] A characterization of thin operators in a von Neumann algebra. Proc. Amer. Math. Soc. To appear.

Pearcy, C.: [1] W^*-algebras with a single generator. Proc. Amer. Math. Soc. **13**, 831—832 (1962).

[2] Some unsolved problems in operator theory. manuscript. June 1972.

[3] On certain von Neumann algebras which are generated by partial isometries. Proc. Amer. Math. Soc. **15**, 393—395 (1964).

Pearcy, C., Salinas, N.: [1] An invariant subspace theorem. Michigan Math. J. **20**, 21—31 (1973).

Phillips, R. S.: [1] The extension of dual subspaces invariant under an algebra. Proc. of the International Symposium on Linear Spaces, pp. 366—398. Israel 1960.

Plafker, S.: [1] Spectral representations for a general class of operators on a locally convex space. Illinois J. Math. **13**, 573—582 (1969).

Pontrjagin, L. S.: [1] Hermitian operators in spaces with indefinite metric. Izv. Akad. Nauk SSSR Ser. Math. **8**, 243—280 (1944) (Russian).

Porcelli, P.: [1] Linear Spaces of Analytic Functions. Chicago: Rand McNally 1966.

Potapov, V. P.: [1] The multiplicative structure of J-contractive matrix functions. Amer. Math. Soc. Transl. **15**, 131—243 (1960).

Privalov, I. I.: [1] Randeigenschaften analytischer Funktionen. Berlin: Deutscher Verlag 1956.

Putnam, C. R.: [1] On normal operators in Hilbert space. Amer. J. Math. **73**, 357—362 (1951).

[2] An inequality for the area of hyponormal spectra. Math. Z. **116**, 323—330 (1970).

Radjabalipour, M.: [1] Growth conditions, spectral operators, and reductive operators. To appear.

Radjavi, H.: [1] Every operator is the sum of two irreducible ones. Proc. Amer. Math. Soc. **21**, 251—252 (1969).

[2] Isomorphisms of transitive operator algebras. To appear.

Radjavi, H., Rosenthal, P.: [1] Invariant subspaces and weakly closed algebras. Bull. Amer. Math. Soc. **74**, 1013—1014 (1968).

[2] On invariant subspaces and reflexive algebras. Amer. J. Math. **91**, 683—692 (1969).

[3] The set of irreducible operators is dense. Proc. Amer. Math. Soc. **21**, 256 (1969).

[4] Matrices for operators and generators of $\mathscr{B}(\mathscr{H})$. J. London Math. Soc. (2) **2**, 557—560 (1970).

[5] Hyperinvariant subspaces for spectral and n-normal operators. Acta Sci. Math. (Szeged) **32**, 121—126 (1971).

[6] On reflexive algebras of operators. Indiana University Math. J. **20**, 935—937 (1971).

[7] A sufficient condition that an operator algebra be self-adjoint. Canad. J. Math. **23**, 588—597 (1971).

[8] Invariant subspaces for products of Hermitian operators. Proc. Amer. Math. Soc. To appear.

[9] On transitive and reductive operator algebras. To appear.

Ricardo, H.J.: [1] Invariant subspaces, an expository dissertation. Yeshiva University 1972.

Rickart, C. E.: [1] Banach Algebras. Princeton, New Jersey: D. Van Nostrand Co. 1960.

[2] The uniqueness of norm problem in Banach algebras. Ann. of Math. **51**, 615—628 (1950).

Riesz, F.: [1] Les systèmes d'équations linéaires à une infinité d'inconnues. Paris: Gauthier-Villars 1913.

Riesz, F., Riesz, M.: [1] Über die Randwerte einer analytischen Funktion. Quartrième Congrès des Math. Scand., pp. 27—44 Stockholm: 1916.

Riesz, F., Sz.-Nagy, B.: [1] Functional Analysis. New York: Ungar 1955.

Ringrose, J. R.: [1] Super-diagonal forms for compact linear operators. Proc. London Math. Soc. (3) **12**, 367—384 (1962).

[2] On some algebras of operators. Proc. London Math. Soc. (3) **15**, 61—83 (1965).

[3] On some algebras of operators II. Proc. London Math. Soc. (3) **16**, 385—402 (1966).

[4] Algebraic isomorphisms between ordered bases. Amer. J. Math. **83**, 463—478 (1961).

[5] Compact Non-Self-Adjoint Operators. New York: Van Nostrand Reinhold Co. 1971.

Robinson, A.: [1] Non-Standard Analysis. Amsterdam: North Holland 1966.

Rosenblum, M.: [1] On the operator equation $BX - XY = Q$. Duke Math. J. **23**, 263—269 (1956).

[2] On a theorem of Fuglede and Putnam. J. London Math. Soc. **33**, 376—377 (1958).

Rosenthal, P.: [1] A note on unicellular operators. Proc. Amer. Math. Soc. **19**, 505—506 (1968).

[2] Completely reducible operators. Proc. Amer. Math. Soc. **19**, 826—830 (1968).

[3] Examples of invariant subspace lattices. Duke Math. J. **37**, 103—112 (1970).

[4] Remarks on invariant subspace lattices. Canad. Math. Bull. **12**, 639—643 (1969).

[5] Weakly closed maximal triangular algebras are hyperreducible. Proc. Amer. Math. Soc. **24**, 220 (1970).

[6] Problems on invariant subspaces and operator algebras. Proc. of 1970, Tihany Colloquium on Hilbert Space Operators and Operator Algebras. Tihany, Hungary: Colloquia Mathematica Societatis János Bolyoi 5 1972.

[7] On Lattices of Invariant Subspaces. Dissertation. University of Michigan 1967.

[8] On reductive algebras containing compact operators. To appear.

Rota, G.-C.: [1] On models for linear operators. Comm. Pure Appl. Math. **13**, 469—472 (1960).

Rovnyak, J.: [1] Ideals of square-summable power-series II. Proc. Amer. Math. Soc. **16**, 209—212 (1965).

Rudin, W.: [1] Real and Complex Analysis. New York: McGraw-Hill 1966.

Saitô, T.: [1] Some remarks on Andô's theorems. Tôhoku Math. J. (2) **18**, 404—409 (1966).

[2] On generators of von Neumann algebras. Michigan Math. J. **15**, 373—376 (1968).

[3] Generators of von Neumann algebras. Lectures on Operator Algebras. Lecture Notes in Mathematics 247. Berlin-Heidelberg-New York: Springer 1972.

Sakai, S.: [1] C^*-Algebras and W^*-Algebras. Ergebnisse der Math. 60. Berlin-Heidelberg-New York: Springer 1971.

Sarason, D.: [1] The $\mathcal{H}^p$ Spaces of an Annulus. Mem. Amer. Math. Soc. **56** (1965).

[2] A remark on the Volterra operator. J. Math. Anal. Appl. **12**, 244—246 (1965).

[3] Generalized interpolation in $\mathcal{H}^\infty$. Trans. Amer. Math. Soc. **127**, 179—203 (1967).

[4] Invariant subspaces and unstarred operator algebras. Pacific J. Math. **17**, 511—517 (1966).

[5] Invariant subspaces. In: Studies in Operator Theory. A.M.S. Surveys. To appear.

Schäffer, J. J.: [1] More about invertible operators without roots. Proc. Amer. Math. Soc. **16**, 213—219 (1965).

Schue, J.R.: [1] The structure of hyperreducible triangular algebras. Proc. Amer. Math. Soc. **15**, 766—772 (1964).

Schwartz, J. T.: [1] *W**-Algebras. New York: Gordon and Breach 1967.

[2] Subdiagonalization of operators in Hilbert space with compact imaginary part. Comm. Pure Appl. Math. **15**, 159—172 (1962).

Scroggs, J. E.: [1] Invariant subspaces of a normal operator. Duke Math. J. **26**, 95—111 (1959).

Segal, I. E.: [1] Decompositions of Operator Algebras, I, II. Mem. Amer. Math. Soc. **9**, (1951).

Sherman, M.J.: [1] Invariant subspaces containing all analytic directions. J. Functional Analysis **3**, 164—172 (1969).

[2] Invariant subspaces containing all constant directions. J. Functional Analysis **8**, 82—85 (1971).

Shields, A. L.: [1] A note on invariant subspaces. Michigan Math. J. **17**, 231—233 (1970).

Shields, A. L., Wallen, L. J.: [1] The commutants of certain Hilbert space operators. Indiana University Math. J. **20**, 777—788 (1971).

Stampfli, J. G.: [1] Hyponormal operators. Pacific J. Math. **12**, 1453—1458 (1962).

[2] On hyponormal and Toeplitz operators. Math. Ann. **183**, 328—336 (1969).

[3] A local spectral theory for operators IV. Invariant subspaces. Indiana University Math. J. **22**, 159—167 (1972).

Sul'man, V.S.: [1] On reflexive operator algebras. Math. USSR-Sb. **16**, 181—189 (1972).

Suzuki, N.: [1] Reduction theory of operators on Hilbert space. Indiana University Math. J. **20**, 953—958 (1971).

Suzuki, N., Saito, T.: [1] On the operators which generate continuous von Neumann algebras. Tôhoku Math. J. **15**, 277—280 (1963).

Sz.-Nagy, B.: [1] Completely continuous operators with uniformly bounded iterates. Publ. Math. Debrecen **4**, 82—92 (1959).

[2] On uniformly bounded linear transformations in Hilbert space. Acta. Sci. Math. (Szeged) **11**, 152—157 (1947).

[3] Spektraldarstellung linear transformationen des Hilbertschen Raumes. Ergebnisse der Math. 5. Berlin: Springer 1942.

Sz.-Nagy, B., Foiaş, C.: [1] Harmonic Analysis of Operators on Hilbert Space. Budapest: Akadémiai Kiadó 1970.

[2] Sur les contractions de l'espace de Hilbert V. Translations bilatérales. Acta Sci. Math. (Szeged) **23**, 106—129 (1962).

[3] Sur les contractions de l'espace de Hilbert IX. Factorisations de la fonction caracteristique. Sous-espaces invariants. Acta Sci. Math. (Szeged) **25**, 283—316 (1964) and **26**, 193—196 (1965).

Szücs, J., Máté, E., Foiaş, C.: [1] Non-existence of unicellular operators on certain non-separable Banach spaces. Acta Sci. Math. (Szeged) **31**, 297—300 (1970).

Taylor, A. E.: [1] Introduction to Functional Analysis. New York: John Wiley and Sons 1958.

[2] Analysis in complex Banach spaces. Bull. Amer. Math. Soc. **49**, 652—669 (1943).

[3] Spectral theory of closed distributive operators. Acta Math. **84**, 189—224 (1950).

Titchmarsh, E. C.: [1] The Theory of Functions. Second Edition. London: Oxford University Press 1939.

[2] The zeros of certain integral functions. Proc. London Math. Soc. (2) **25**, 283—302 (1926).

Topping, D.: [1] Lectures on von Neumann Algebras. London: Van Nostrand Reinhold 1971.

[2] On linear combinations of special operators. J. Algebra **10**, 516—521 (1968).

Tsuji, K.: [1] On nest algebras of operators. Proc. Japan Acad. **46**, 337—340 (1970).

Turner, T. R.: [1] Double commutants of algebraic operators. Proc. Amer. Math. Soc. **33**, 415—419 (1972).

Wermer, J.: [1] On invariant subspaces of normal operators. Proc. Amer. Math. Soc. **3**, 270—277 (1952).

[2] The existence of invariant subspaces. Duke Math. J. **19**, 615—622 (1952).

[3] Report on subnormal operators. In: Report of an International Conference on Operator Theory and Group Representations. New York 1955.

Weyl, H.: [1] Über beschränkte quadratische Formen, deren Differenz vollstetig ist. Rend. Circ. Mat. Palermo **27**, 373—392 (1909).

Wilder, R. L.: [1] Topology of Manifolds. Amer. Math. Soc. Colloq. Pub., Vol. 32. Providence 1949.

Wogen, W.: [1] On generators for von Neumann algebras. Bull. Amer. Math. Soc. **75**, 146—152 (1969).

Wold, H.: [1] A Study in the Analysis of Stationary Time Series. Stockholm: Almqvist and Wiksell 1938.

Wolf, F.: [1] Operators in Banach space which admit a generalized spectral decomposition. Nederl. Akad. Wetensch. Indag. Math. **19**, 302—311 (1957).

Yood, B.: [1] Additive groups and linear manifolds of transformations between Banach spaces. Amer. J. Math. **71**, 663—677 (1949).

Yoshino, T.: [1] Subnormal operator with a cyclic vector. Tôhoku Math. J. **21**, 47—55 (1969).

List of Symbols

Author Index

Subject Index

Ergebnisse der Mathematik und ihrer Grenzgebiete

Prices are subject to change without notice